Springer Series on Polymer and Composite Materials

Series editor

Susheel Kalia, Dehradun, India

More information about this series at http://www.springer.com/series/13173

Susheel Kalia
Editor

Polymeric Hydrogels as Smart Biomaterials

Springer

Editor
Susheel Kalia
Department of Chemistry
ACC Wing, Indian Military Academy
Dehradun, Uttarakhand
India

ISSN 2364-1878 ISSN 2364-1886 (electronic)
Springer Series on Polymer and Composite Materials
ISBN 978-3-319-25320-6 ISBN 978-3-319-25322-0 (eBook)
DOI 10.1007/978-3-319-25322-0

Library of Congress Control Number: 2015953788

Springer Cham Heidelberg New York Dordrecht London

Printed on acid-free paper

Springer International Publishing AG Switzerland is part of Springer Science+Business Media (www.springer.com)

Preface

Hydrogels are considered as the most promising types of polymers being used for mankind. Hydrogels are three-dimensional, hydrophilic, polymeric networks that can absorb, swell, and retain large quantities of water or aqueous fluids. Natural polysaccharides are biodegradable, nontoxic, low cost, and renewable and can potentially be used as key ingredients for the production of biomaterials for mankind. Metal nanoparticles, nanofibrils, or nanowhiskers embedded polymeric hydrogels are a new class of materials and have attracted great attention due to their unique properties and applications in various fields including pharmaceutics and biomedicine. Conducting polymer hydrogels are materials with added advantages such as electrical conductivity. In recent years, considerable attention has been paid to the modification of crosslinked hydrogels with conducting polymers since this offers a facile methodology to combine the superior properties of conducting polymers with the highly crosslinked hydrogels. Conducting polymeric hydrogels can be used in many applications like electro-sensors, capacitors to electromechanical actuators and artificial muscles. Antimicrobial polymeric hydrogels loaded with antibiotics, antimicrobial polymers, or metal nanoparticles are very useful for mankind. Polysaccharide-based hydrogels have found extensive application in various fields, including agriculture, wastewater treatment, electronics, pharmaceutical, and biomedical applications.

The various chapters in this volume have been contributed by prominent researchers from industry, academia, and research laboratories across the world. The chapter "Polymeric Hydrogels: A Review of Recent Developments" of this book deals with the recent research done in the area of hydrogels, modified hydrogels, hydrogel composites, and nanocomposites. This chapter also provides comprehensive details of research studies of natural and synthetic hydrogels. Fabrication of electroconductive hydrogels from conducting polymers either as a single component or as an additive to conventional hydrogel networks are reviewed in the chapter "Conductive Polymer Hydrogels". Applications of conducting polymer hydrogels are also reported in this chapter. The chapter "Polysaccharide-Based Hydrogels as Biomaterials" covers the recent developments and advances in hydrogels derived from natural polysaccharides as biomaterials. Chitosan, seaweed, hyaluronic acid, and dextran polysaccharide-based hydrogels are discussed in this chapter. Applications of polysaccharide-based biomaterials as scaffold, cell encapsulation, and wound dressing are also reported here. The chapter "Protein-Based Hydrogels" reports on protein-based hydrogels. Emerging applications and technologies for protein-based hydrogels are also briefly mentioned here. The role of sterculia gum as a promising biodegradable material in the development of various biomedical applications including drug delivery applications, wound dressing applications, etc., are discussed in the chapter "Sterculia Gum-Based Hydrogels for Drug Delivery Applications."

The chapter "Antimicrobial Polymeric Hydrogels" deals with natural and synthetic antimicrobial hydrogels loaded with antibiotics, antimicrobial polymers or peptides, and metal nanoparticles. This chapter summarizes the significant and recent progress in the manufacture and application of antimicrobial hydrogels. The chapter "Biopolymer-Based Hydrogels for Decontamination for Organic Waste" highlights the role of biopolymer-based hydrogels for decontamination of organic waste. Synthesis of chitosan and starch-based hydrogels via graft copolymerization is reported in the chapter "Chitosan and Starch-Based Hydrogels Via Graft Copolymerization." This chapter also includes the concept and methods of graft copolymerization. Applications of these hydrogels as adsorbent, ion exchangers, superabsorbent polymers, and in the pharmaceutical and medical fields are also reviewed in this chapter.

This book covers scientific, technological, and practical concepts concerning the research, development, and realization of polymeric hydrogels as smart biomaterials. This book will be a very useful tool for scientists, academicians, research scholars, biomaterial engineers, and for pharmaceutical industries. This book can also be supportive for undergraduate and postgraduate students at various pharmaceutical institutes and researchers from R&D polymer laboratories working in this area.

The editor would like to express gratitude to all prominent contributors to this book, who have provided excellent contributions. The editor thanks his research team, which helped him in the editorial works. Finally, he gratefully acknowledges the permissions to reproduce copyright materials from a number of sources.

July 2015 Susheel Kalia

Contents

About the Editor

Susheel Kalia is Associate Professor and Head of the Department of Chemistry at Army Cadet College Wing of the Indian Military Academy, Dehradun, India. He was visiting researcher in the Department of Civil, Chemical, Environmental and Materials Engineering at University of Bologna, Italy in 2013. Kalia has around 65 research papers in international journals along with 80 publications in national and international conferences and many book chapters. Kalia's other editorial activities include work as a reviewer and memberships of editorial boards for various international journals. He is an Editor of 'Springer Series on Polymer and Composite Materials'. Additionally, he has edited a number of books with renowned publishers such as Springer, Wiley and Scrivener Publication, etc. He has been invited to many international and national conferences to deliver his work. He is also a member of a number of professional organizations, including the Asian Polymer Association, Indian Cryogenics Council, the Society for Polymer Science, Indian Society of Analytical Scientists, Him Science congress Association, and the International Association of Advanced Materials. Presently, Kalia's research is in the field of biocomposites, nanocomposites, conducting polymers, cellulose nanofibers, inorganic nanoparticles, hybrid materials, hydrogels, and cryogenics.

Polymeric Hydrogels: A Review of Recent Developments

Shivani Bhardwaj Mishra and Ajay Kumar Mishra

Abstract Hydrogels are special types of polymers that have enormous capacity to absorb large volumes of water. Hydrogels are natural as well as man-made. To suit to a type of an application, hydrogel can be modified to tailor made properties that can be exploited for natural, applied, and medical sciences. This chapter deals with the recent research done in the area of hydrogels, modified hydrogels, hydrogel composites, and nanocomposites. General trends of the thrust areas where hydrogels have prime role of importance were biomedical and health care. However, the other areas such as environmental aspects for the utility of hydrogels have also been an area of interest among researchers across the globe.

Keywords Hydrogels · Composites · Nanocomposites · Applications

1 Introduction

Hydrogel is a water loving gel which is formally known as hydrogel which has an ability to change its chemical structure that induces the volume change as per the physical conditions such as pH, temperature, salt concentration, electric field, and solvent quantity thus making these materials as stimuli-responsive smart polymers [1]. The most important property is the swelling behaviour of the cross-linked three-dimensional polymer networks. It has been well reported that the cross linking and the charge densities of these polymer networks directly affects the swelling and elastic behaviour [2]. Hydrogel exhibit an important nonideal-feature of spatial gel inhomogeneity referring to the inhomogeneous cross-linked density distribution that decreases the optical clarity, strength, ionization degree, electrostatic repulsion, and mobile counter ions of hydrogels [3–5].

S.B. Mishra (✉) · A.K. Mishra
Nanotechnology and Water Sustainability Research Unit, College of Engineering, Science and Technology, University of South Africa, Florida Science Campus, Johannesburg, South Africa
e-mail: bshivani73@gmail.com; bhards@unisa.ac.za

S. Kalia (ed.), *Polymeric Hydrogels as Smart Biomaterials*,
Springer Series on Polymer and Composite Materials,
DOI 10.1007/978-3-319-25322-0_1

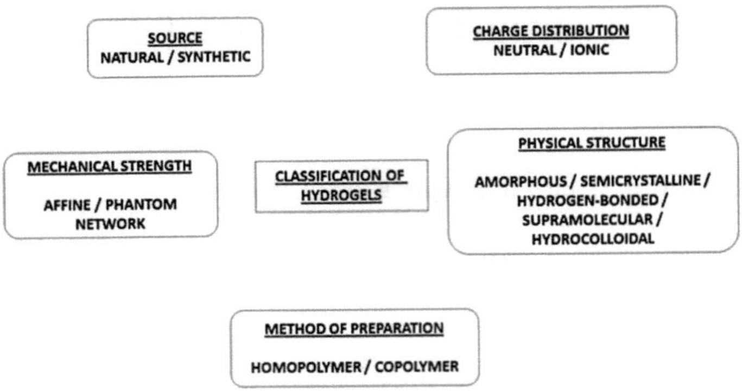

Scheme 1 Classification of hydrogels

The hydrogels are of various types which are classified into charged distribution, method of preparation, mechanical strength and physical structure [6]. The basic classifications of hydrogels are shown in Scheme 1. These polymers are biodegradable and biocompatible in nature. Based on the properties, these hydrogels find its utility well in the field of commercial healthcare and biomedical science such as implants, inserts, osmotic devices, implants and tablets [7], tissue expander [8], and contact lenses [9]. Besides this, these polymers have found an extensive use in the field of environmental science whereby researchers have used it for the decontamination of different type of pollutants [10].

It has been reported that hydrogel properties are influenced by the liquid composition and hydrogel structure [11]. Table 1 briefly describes about the hydrogel properties.

The hydrogels can be synthesized via cross linking of polymers by heat/chemical reactions/photopolymerization/radiations (X-rays/gamma rays/electron beams). This chapter describes about the recent work done in the field of hydrogels, modified or functionalized hydrogels, as a composite and nanocomposite.

Table 1 Effect of hydrogel structure and liquid composition on hydrogel properties

Swelling capacity ↑	Swelling rate ↑	Wet strength ↑
Hydrogel structure: very hydrophilic polymers, ionic polymers containing monovalent ions, lower crosslink density, hydrophilic crosslinkers	*Hydrogel structure*: more hydrophilic, higher crosslink density, more porosity, open pores, interconnected pores	*Hydrogel structure*: high crosslink density (there is an optimum crosslink density at which, the mode of hydrogel failure changes from ductile to brittle), low porosity, more hydrophobicity
Liquid composition: more solvents, less salts, low ionic strength, less numbers of di and trivalent cations	*Liquid composition*: presence of permeation enhancers for more hydrophobic hydrogel, more solvents	*Liquid composition*: more nonsolvents, more salts

2 Recent Progress in Research and Development for Different Types of Hydrogels

2.1 Native/Pristine Hydrogels

2.1.1 Natural Hydrogels

The native or pristine hydrogels are the hydrogels that have been used in its natural state to study its properties and applications. Most of the natural hydrogels fall into this category such as natural gums, proteins, peptides, cellulosic materials, collagen, hyaluronic acid, and polysaccharides, viz., schelaro glucan, alginates, cellulose, xyloglucan, etc.

Spinal cord injury is defect in the spinal cord that arise temporarily or permanently leading to different types of diseased from pain to paralysis. Natural hydrogel play an important role for improving tissue regeneration and central nervous system repair [12–14] as it holds the characteristics mechanical strength comparative to that of tissue, porosity that allows cell infiltration and transplantation, biocompatibility toward cell attachment and tissue growth and targeted in situ drug delivery [15, 16]. These hydrogels can be tailored to obtain channels that can target the nerve guidance and sustained drug release. Table 2 provides brief information about naturally derived hydrogels and its use in spinal cord injuries [17]. The ultimate objective of using these natural hydrogel is to substitute the diseased section of spinal cord with structural matrix.

In another work, pH of gelation influenced the pH-dependent charge and iso-electric point of the thermally generated hydrogels of whey proteins [18]. Loss of charged groups during gelation probably will reduce swelling ration and increase the acidity of alkaline hydrogels. Silk elastin like protein polymer analogue SELP 47-K and SELP 415-K were used to study real time imaging to gene expression of adenovirus embedded in this biopolymer hydrogel. It was reported that polymer structure and concentration affected the release of adenovirus. The hydrogel concentration was found to be directly proportional to the stiffness and inversely proportional to inter fibril distance affecting the release [19].

Silk hydrogels were prepared with variety of silk concentrations and were analyzed for water retention, mechanical strength and cytotoxic behaviour toward human mesenchymal stem cells [20]. The network structure of these hydrogels so formed was β-sheet structure. It was reported that the bond water promoted cell adhesion protiens in the cellular matrix whereas the bulk water disrupts the same.

Polysaccharide-based hydrogels have been studied for soil conditioning for the controlled release of nutrients [21, 22].

Selective electrostatic complexation of chitosan with positively charged methylene blue and negatively charged drugs allura red and methyl orange were carried out. It was observed that incorporation of these charged dyes were primarily based on charge of the cross linker used in these hydrogels [23, 24].

Table 2 Natural hydrogels investigated for spinal cord injury

Materials	Description	Acronym	Application in SCI
Agarose	Polysaccharide		Cell growth matrix[74]
			Encapsulation and delivery of neutrophobic factors[14]
			Controlled chondroitinase delivery[73]
			Support for nanoparticles delivery[14,74]
			Brain-derived neurotrophic factor (BDNF)
			Controlled delivery[75,75]
			Linear guidance (free dried)[70,71]
			Cell encapsulation for growth matrix[37]
	Co-methylcellulose	Agarose/MC	Nerve guidance[77]
Alginate	Polysaccharide		Anisotropic scaffold for axonal re-growth[72] Neutral cell growth matrix[66,69,87] Embryonic stem cell growth matrix[59]
Cellulose	Polysaccharide		Mesenchymal stem cell growth matrix[63]
Chitosan	Polysaccharide		Scaffold for cell adhesion and growth with polylysine[79] Scaffold for neurite regrowth with hyaluronic acid[80]
Collagen	Polypeptide		Polymeric channels[81] Filament bridges as growth substances[82] Cell growth matrix[83,84]
Fibrin	Linked proteins		Neural stem cell growth matrix[68]
Gelatin	Hydrolyzed collagen		Mesenchymal stem cell growth[64]
Gellan gum	Polysaccharide		Tubular, porous scaffold for axonal regrowth[85]
Hyaluronic acid	Polysaccharide	HA	Controlled delivery of neurotrophic factors[62]
			Scaffold for neurite regrowth[65,80]
			Controlled peptide delivery[26,86,87]
	Co-polylysine		Nogo 66 receptor antibody delivery system[88,89]
	Co-methylcellulose	HAMC	Intrathecal drug and growth factor delivery[90–95]
	Co-collagen		Neural stem cell career for cell therapies[61] Cell growth matrix[96]
Matrigel	Lamimin, collagen IV, heparin		Scaffold supporting cell adhesion and growth[38] Neural stem cell carrier for cell therapies[69]

(continued)

Table 2 (continued)

Materials	Description	Acronym	Application in SCI
Seleroglucan	Polysaccharide		Controlled drug delivery[97]
Xyloglucan	Polysaccharide		Scaffold supporting cell adhesion and growth[50,67]

2.1.2 Synthetic Pristine Hydrogels

The term native or pristine can also be applied to synthetic or man-made hydrogels like polyhydroxyethyl methacrylate, polyvinyl alcohol, polyacrylic acid, poly vinyl pyrrolidone etc. In this section authors would describe the recent research done in the field of native or pristine hydrogels. Synthetic pristine hydrogels are man-made hydrogels that have found much application in the field of science and technology. Most of the hydrogels are biocompatible and have been investigated for medical and health care products. Some of the synthetic hydrogels are.

Polyvinyl alcohol [PvOH] being nontoxic, noncancerous, and biocompatible synthetic hydrogels which has find its utility for developing contact lenses, the lining of artificial heart, soft tissue replacement, articular cartilage, skin, and pancreas [25]. Controlled release of many drugs such as ergotamine tartrate were carried our using polyvinyl alcohol [26].

Another very interesting synthetic polymer that has found a great place in the field of biomedical science is polyHydroxyethyl methacrylate [pHEMA]. This hydrogel is extensively used for contact lenses, protein/drug delivery devices, matrices for immobilization and separation of cells and molecules and scaffolds [27].

In another study, polyacrylic acid [PAA] was used to synthesize using electron beam radiation process for fabricating bioadhesive specimen that can be attached to mucosal surface for transmucosal drug delivery [28].

Table 3 giving brief information of some other application of synthetic pristine hydrogels.

2.2 *Modified Hydrogels*

Modified hydrogels or functionalized hydrogels are those hydrogels whose basic chemical structure at the surface is modified by copolymerization, grafting or merely blending so that additional functionalities can be added to provide tailor made properties in the hydrogel for any particular application.

Enormous research has been done and is being carried out to explore the newer hydrogels with improved properties. This section will focus on these modified or functionalized hydrogels and their various applications.

Table 3 Synthetic native/pristine hydrogels and their applications

Synthetic pristine hydrogel	Chemical structure	Application	References
Poly (hydroxyethyl methacrylate) [pHEMA]	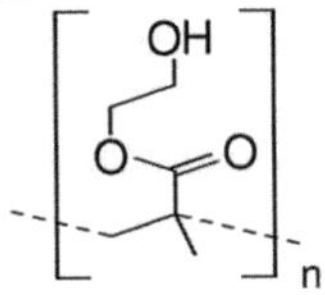	Spinal Cord regeneration, scaffolds for cell adhesion, and artificial cartilage production	[29–32]
Poly (ethylene glycol) [PEG]	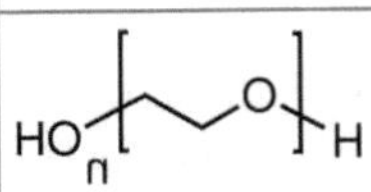	Drug, proteins, and biomolecules carrier	[33–36]
Poly (vinyl alcohol) [PVA]	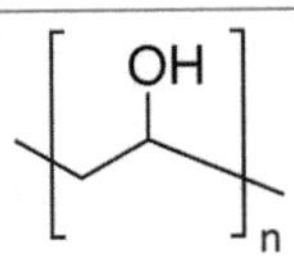	Injectable implants, endoprostheses, soft tissue, soft tissue filler in platic, reconstructive, and aesthetic surgery	[37, 38]
Poly (vinylpyrrolidone) [PVP]	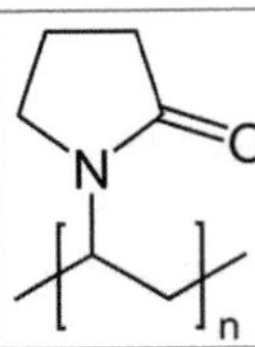	Wound healing applications	[39]
Polyimide	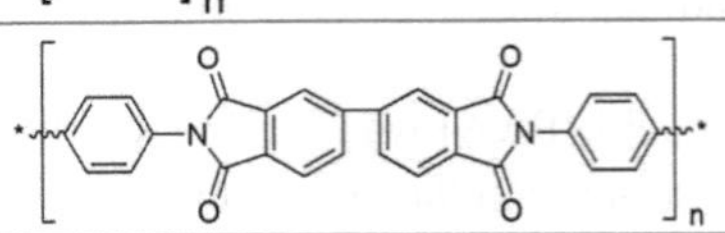	Plastic and reconstructive surgery	[40]

In a recent research reported, collagen hydrogel was modified with aldehyde modified dextran to increase the mechanical strength of the collagen hydrogel [41]. The modified hydrogel was found to be thermally stable with maximum compressive strength of 32.5 ± 0.6 kPa. The graphical abstract shown in Fig. 1 provides the details of the research reported.

Injectable chitosan hydrogel was incorporated with type II collagen and chondroitin sulphate to be used for cartilage tissue engineering [42]. It was observed that the modified chitosan hydrogel positively influenced the cell proliferation and promoted tissue regeneration.

Cucurbit [6] uril-conjugated hyaluronic acid (CB [6] -HA), diaminohexane conjugated HA (DAH-HA), and tags-CB [6] hydrogels were used for in situ supramolecular assembly and modular modifications under the skin of mice to study its application for cellular engineering. Figure 2 illustrates the various host guest interaction for these modularly modified hydrogels.

Fig. 1 Collagen (Col)/modified dextran (DAD) hydrogel for improved mechanical properties. Reprinted with permission from [41]. Copyright © 2014, American Chemical Society

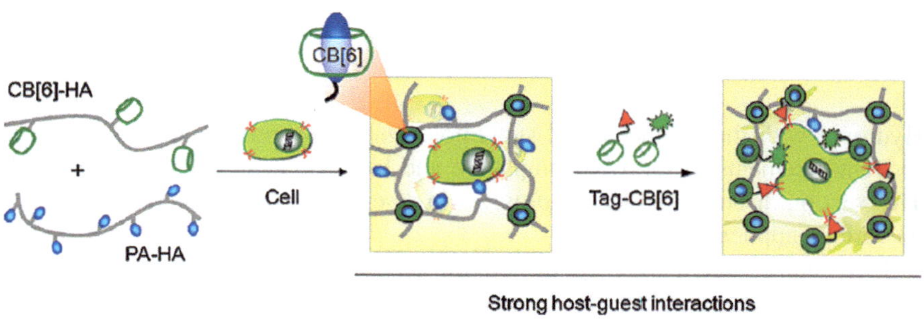

Fig. 2 Host-gust interaction of Cucurbit [6] uril-conjugated hyaluronic acid (CB [6]-HA), diaminohexaneconjugated HA (DAH-HA), and tags-CB [6]-based hydrogels. Reprinted with permission from [43]. Copyright © 2012, American Chemical Society

Cellulose-g-poly (acrylic acid) was synthesized using electron beam radiation for loading bovine serum albumin to study the drug release from these modified hydrogels to be used as oral delivery devices [44].

Hybrid hydrogels were synthesized by making use of molecular hydrogels and calcium alginate were prepared to study the preservation capacity and stability of thermosensitive enzyme. The molecular hydrogels provided the ultrastability to the hybrid hydrogel [45].

Salecan/polyacrylamide semi interpenetrating network hydrogel via radical copolymerisation/cryopolymerisation at subzero temperature showed homogeneity in pore distribution with a pore size in the range of 5–50 μm. The hydrogel was able to attain equilibrium in water in 260 s in 4 min and were found nontoxic toward CO7 cells [46].

A novel hybrid hydrogel was synthesized via photopolymerization using thermoresponsive poly(*N*-isopropylacrylamide) and redox responsive poly(ferrocenylsilane) (PFS) polymers which was later used to develop redox induced hydrogel silver composites. These composite showed an efficient antimicrobial behaviour with biocompatible behaviour [47].

In another study bacterial cellulose was used to functionalise poly (vinyl alcohol) to study the mechanical properties. It was found that excessive bacterial cellulose reduced the compressive modulus. On the contrary freezable bound water significantly improved this property as well as flexibility and stiffness [48].

Salecan, a type of water soluble glucan was reinforced into poly (vinylalcohol) matrix using freeze thaw method with homogenous porosity. Increase in salecan content improved the swelling property but negatively influenced the compressive modulus [49].

Poly(xylitol-co-maleate-co-PEG) was in situ synthesized using free radical polymerisation to be used as IHS career. The hydrogel showed significant elastic properties and cytocompatibility with controlled degradation, promoted in vitro cell adhesion and growth [50].

2.3 Hydrogel Composites

Hydrogel with three-dimensional supramolecular arrangement allows may option for reinforcements to form biocompatible and biodegradable composite materials. This section discusses about the progress in research focusing composite area and its various applications.

Free radical polymerization of sodium acrylate, *N*-isopropylacrylamide using *N*,*N*0-methylenebisacrylamide as crosslinking agent was done incorporating the light-absorbing carbon particles to be used as draw agent that could facilitate the forward osmosis process of desalination [51]. It was reported that hydrogel composite had shown a flux of 1.32 LMH in the first 0.5 h forward osmosis process. It was reported that hydrogel composite had shown a flux of 1.32 LMH in the first 0.5 h forward osmosis process.

Polyuronic acid, alginate and pectate were used as matrices that formed a coordinate covalent bond with iron (III) ions to produce visible light responsive hydrogel composites beads. The photoreactivity of the three different biopolymers were found to be in order mannuronic rich alginate > gluronic-rich alginate > pectate. The composite beads were prepared to study as a drug releasing vehicles for various organic/biological molecules such congo red dye, folic acid and chloroamphenicol [52]. Graphical presentation of the study shown in Fig. 3.

Dual approach starting from self-assembly process at room temperature and followed by oxidation polymerisation was used to prepared polypyrrole/grapheme oxide composite with an application of high energy storage device [53]. This composite was reported to show high specific capacitance of 473 F/g at 1 A/g of which 82 % was sustained even after 5000 cycles of charge–discharge.

Zeolite-metal organic composite framework were prepared using biodegradable monomers such 2-hydroxyethyl methacrylate (HEMA), 2,3-dihydroxypropyl methacrylate (DHPMA), *N*-vinyl-2-pyrolidinone (VP) and ethylene glycol dimethacrylate (EGDMA) using ultraviolet polymerization. These networks reinforced arrhythmic drug [54].

Electrophoresis technique was employed for in situ formation of polyacrylamide/hydroxyapatite composite hydrogel that showed a high extensibility

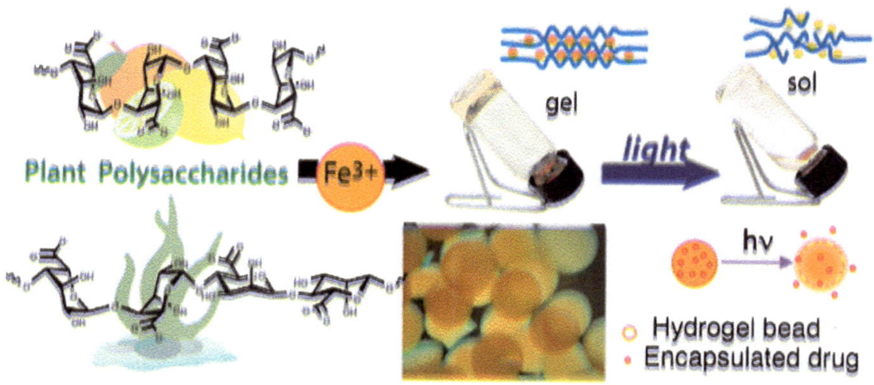

Fig. 3 Schematic representation of the plant polysaccharide and Fe (III)-based hydrogel composite. Reprinted with permission from [52]. Copyright © 2015. American Chemical Society

>2000 % with a tensile strength in the range of 0.1–1.0 MPa. The authors reported the cell adhesion studies of the composite hydrogel and reported that the electrophoresis approach provided the bioactive sites for osteoblasts [55].

Alginate/acrylamide gel precursor was used as matrix for reinforcing UV curable adhesives (Emax 904 Gel-SC) via single step process of three-dimensional printing in conjunction with digital modelling. The mechanical properties of composite hydrogel followed the "rule of mixtures" and a prototype meniscus cartilage was produced using this material to show it potential utility in the field of bioengineering [56].

Montmorillonite/xylose composite hydrogel was prepared by non-covalent strategy where grafting of methyl guanidine hydrochloride onto the xylose template. The guanidine ions were entangled to the exfoliated organic clay platelets. This clay/hydrogel composite had a significant swelling capacity with self-healing ability. For this, the composite hydrogel was divided into two equal halves and the ruptured surfaces were pressed together again that became single piece within few minutes. The authors reported the fact of to the mechanism whereby self-healing behaviour was considered to be due to non-covalent bonds of hydrogen bonds and electrostatic adsorption. The inorganic filler provided the thermal stability to the hydrogel composite [57].

In another fascinating research the hydrogel composites have been employed for the sensor application. In a recent study, chlorpyrifos was detected by acetylcholinesterase which was immobilized onto the carbon electrode prepared using ferrocene phenylalanine hydrogel composite. The chlorprifos was detected at a limit of 3 nM by the enzyme acetylcholinesterase upon the enzymatic conversion from acetylthiocholine to thiocholine. The electrochemical biosensor showed with respect to short response time, stability, detection limit and regeneration [58]. The authors reported the work to be the finest until the research was published.

In another investigation, calcium deficient hydroxyapatite was incorporated into poly (hydroxyl ethyl methacrylate) to produce thermoset hydrogel. The researcher reported that the final morphology of these thermoset hydrogels was dependent on the concentration of the monomer and solvent used [59]. The final product was recommended to be similar to be bone like material that might mimic the native vertebrae/annulus interface with an application in the field of a noninvasive IVD repair strategies.

Kinetic studies of ampicillin delivering targeting Staphylococcus aureus test strain, was the ultimate application for the nickel ferrite/polyacrylamide hydrogel prepared via polymerization crosslinking simultaneous method. Nickel ferrite particles were developed using sol gel technique to be used reinforcement. The drug release occurred for longer duration and showed to negative microbial results for the test strain [60].

Sodium alginate and graphite-based hydrogel composites were developed via homogeneous dispersion of graphite into sodium alginate matrix. The composite was found to have reticulate and layered structure with the swelling capacity dependent on graphite content [61].

In a recent review, the authors reported the functional hydrogel composite membranes [62]. They discussed about the recent work done in this field where these membrane had an application of membrane for protein separation, virus capture and antibody purification.

2.4 Hydrogel Nanocomposites

The nano size material in zero to three-dimensional type when reinforced into the hydrogel matrix gives rise to hydrogel nanocomposite. The nanoscale size of the reinforcement provides fascinating unique tailor made properties that is exploited by many application areas. This section deals with the work done in the area of hydrogel-based nanocomposites and its application areas.

Synthetic silicate platelets in a gelatin matrix were prepared as hydrogel nanocomposite to be used as a handy source of injectable haemostatic agent keeping in view for external injuries during out of hospital emergency conditions. The results obtained from this study suggested that these nanocomposite were to be promising in vitro and in vivo coagulants and the clay platelets remain exfoliated in the final injectable products [63]. Figure 4 provides a brief methodology and results obtained for this study.

Three-dimensional graphene/vanadium dioxide nanobelts composite hydrogels were prepared via hydrothermal approach for a potential use of energy storage device exhibited a capacitance of 239 F/g at 10 A g/1 with a capacitance retention of 56 % relative to 1 A/g1. The galvanostatic charge–discharge cycles at 10 A/g was carried out for a 5000-cycle test for the graphene/VO_2 composite hydrogels electrode was able to retain 92 % of the initial value [64].

Bioinert, non-fouling and biocompatible poly(ethylene glycol) diacrylate (PEGDA)/Laponite nanocomposite (NC) hydrogels were developed to support two and three-dimensional cell culture. This was carried out by simultaneous procedures

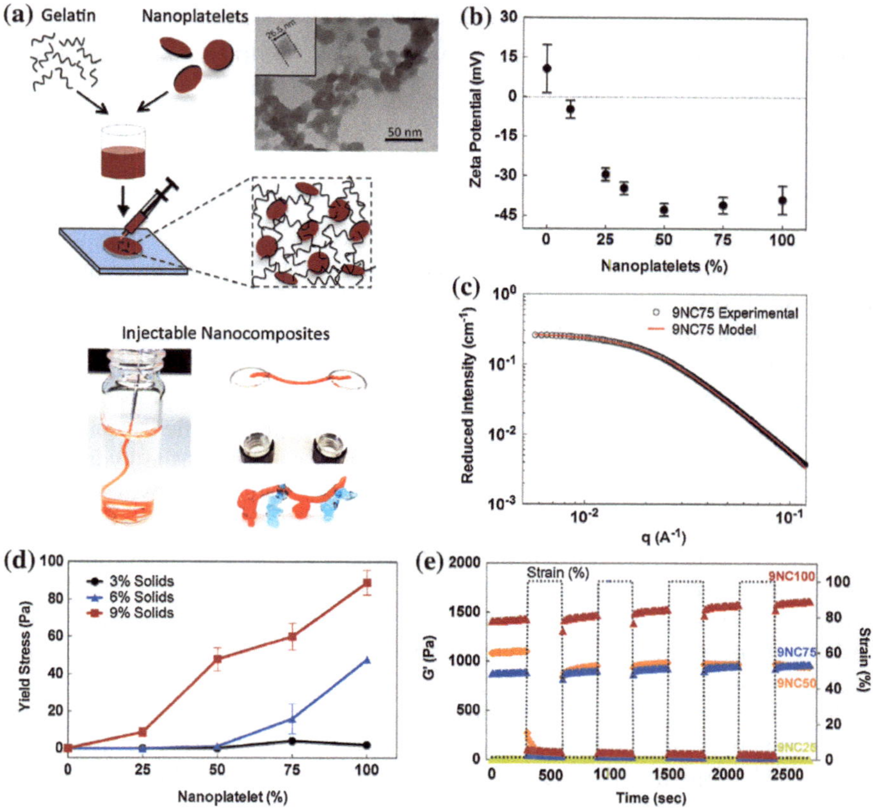

Fig. 4 **a** Methodology for the preparation of silicate gelatin hydrogel nanocomposite and TEM image of the product, **b** zeta potential measurements, **c** SAXS graphs for silica platelets, **d** yield stress for the hydrogel nanoplatelets, **e** results for the moduli measurements. Reprinted with permission from [63]. Copyright © 2015, American Chemical Society

of crosslinking of PEGDA oligomers along with the interaction with clay particles. The filler in this nanocomposite significantly improved the mechanical properties and provide much needed cell adhesion ability of the polymer. The nanocomposite hydrogel was for the first time applied to support cell encapsulated in 3D conditions [65].

Graphene oxide/poly (acrylic acid) hydrogel nanocomposite was synthesized for its utility as an adsorbent for decontamination of methylene blue, Cu^{2+} and Pb^{2+} from synthetic water. The nanocomposite hydrogels was found to well-defined 3D porous network that had a decent adsorption efficiency of as high as 1600 mg/g [66].

In a study, poly (acrylamide) hydrogel template was functionalised by DNA modified by acrydite. This functionalised hydrogel was used as a matrix for incorporation of gold nanoparticles to give rise to monolithic polymer/gold hydrogel nanocomposite for the colorimetric detection of DNA. The hydrogel

Table 4 Adsorption capacity of Rapheme Oxide-Chitosan-Manganese nanocomposite

Sample	Mn(II) pH = 4	Mn(II) pH = 6	Mn(II) pH = 8	Pb(II) pH = 6	Ni(II) pH = 6
Gr-Cs_5	0.58	0.8	1.38	1.14	0.58
Gr-Cs_{10}	0.64	0.84	1.46	1.22	0.62
Gr-Cs_{20}	0.68	0.88	1.58	1.4	0.84

showed a high capacity of gold nanoparticle adsorption and was able to detect ~0.1 nM target DNA [67]. The gold nanoparticles that catalyzed the reduction of silver ions had an addition benefit of detecting the target DNA up to 1 pM. The major advantage of the use of this hydrogel was the regeneration ability by simple heat treatment.

Graphene oxide–chitosan-manganese nanocomposites were prepared with a hierarchical porous structure of surface area 240 m^2/g and specific capacity of 190 F/g and electrosorptive capacity of 12.7 mg/g. This nanocomposite hydrogel was applied to uptake Ni(II), Cu(II) and Pd(II) ions. Table 4 provides different adsorption capacities by hydrogel nanocomposites [68].

Cellulose graphene nanocomposite was developed from the ionic liquids that were produced from mixture of wood pulp and graphite oxide respectively. The green synthesis methodology for developing these nanocomposites that were found to have improved mechanical strength with 0.5 wt% graphene oxides was up to 19.4 MPa, and thermal stability [69].

Montmorillonite was used as filler for starch grafted poly (methacrylic acid) were composed to study swelling behaviour, muco-adhesive behaviour and mechanical properties [70]. The increase in the content of clay increases the muco-adehsive properties and the grafting content decreases the hydrophilicity of this nanocomposite.

In a unique study, parchment samples contaminated by *Penicillium chrysogenum* and *Cladosporium cladosporioides* were cleaned and disinfected using gellan gum/titanium dioxide nanocomposite. This hydrogel nanocomposite was prepared by simple mixing of titanium dioxide dispersion in water in hydrogel solution. The authors claimed it to be an easy, low cost method for protecting, conserving and usability of archived documents in various libraries and elsewhere [71].

Polyethylene glycol methyl ether methacrylate was incorporated with silver nanoparticles to study the antimicrobial behaviour. It was concluded that the adsorption of the silver nanoparticles on the biomineralised composite was only responsible for the antimicrobial activity and has no negative cytotoxic effect against osteoblast MC 3T3 [72].

Magnetitie nanoparticles were synthesised via co-precipitation method and were functionalised with natural and biocompatible chondroitin-sulphate-A to be used as reinforcement for hyaluronate hydrogel. The functionalization of nanoparticle allowed it to get uniformly dispersed in hydrogel as shown in Fig. 5 revealing a comparison between pure and modified nanoparticles dispersion. The nanocomposite hydrogel was recommended for biomedical applications such as intra-articular injections in the treatment of osteoarthritis.

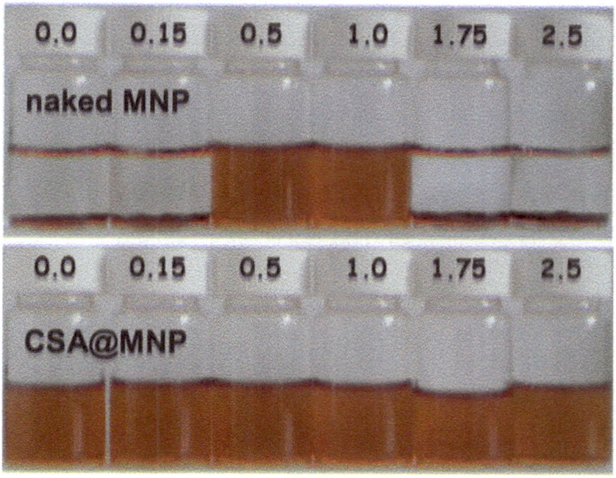

Fig. 5 Figure showing the colloidal stability of pure and modified magnetite nanoparticle in hyaluronate hydrogel. Reprinted with permission from [73]. Copyright © 2015, Elsevier

The charge neutralization was observed at 0.05 mmol/gHyA-addition, indictaing the surface charge heterogeneity that might occurred because of the patch-wise adsorption of HyA marking it to be the isoelectric point (IEP) and the size of HyA@MNP aggregates reached maximum (2300 nm) [73].

3 Summary

Hydrogels, the hydrophilic polymers that has extraordinary swelling capacity by the water to produce gels that have find many applications in the science and technology. Both natural and man-made or synthetic gels have unique properties that have been improvised by functionalization and reinforcing to produce its variety of forms as modified hydrogels, composites and nanocomposites.

The biodegradability, non-toxicity and biocompatibility have allowed its unlimited potential for its application in the field of diagnostics and health care. The only major limitation of these hydrogels is mechanical properties which resists its use for many other commercial applications.

This chapter provided comprehensive details of research studies for different types of hydrogel under main categories of pristine hydrogels, modified hydrogels, composite hydrogels and nanocomposite hydrogels. In general the utility of hydrogel in any form provided plenty of opportunities for further research studies.

4 Future Scope

Based on the research work done in the field of hydrogels, there lies tremendous potential for future scope. This may include the newer methodology for preparation of hydrogels with tailor made functional groups added onto polymer matrices designed for its novel applications. Biomimetic hydrogels with reduced toxicity should be explored for drug delivery devices, implants and tissue and molecular engineering and gene therapy. New methods promising green and eco-friendly and solvent less techniques of grafting and crosslinking of hydrogels to suit to objective of the study is another important area that needs much investigation.

Acknowledgements The authors thank national research foundation and University of South Africa for the financial support and other facilities.

References

1. Tanaka T (1978) Collapse of gels and the critical end point. Phys Rev Lett 40:820–823
2. Shibayama M, Tanaka T (1993) Phase transition and related phenomena of polymer gels. Adv Polym Sci 109:1–62
3. Lindemann B, Schroder UP, Oppermann W (1997) Influence of crosslinker reactivity on the formation of inhomogeneities in hydrogels. Macromolecules 30:4073–4077
4. Kizilay MY, Okay O (2004) Effect of swelling on spatial inhomogeneity in poly(acrylamide) gels formed at various monomer concentrations. Polymer 45:2567–2576
5. Gundogan N, Okay O, Oppermann W (2004) Swelling, elasticity and spatial inhomogeneity of poly(N, N-dimethylacrylamide) hydrogels formed at various polymer concentrations. Macromol Chem Phys 205:814–823
6. Peppas NA, Bures P, Leobandung W, Ichikawa H (2000) Hydrogels in pharmaceutical formulations. Euro J Pharma Biopharma 50:27–46
7. http://www.rxlist.com
8. Wiese KG, Heinemann DEH, Ostermeirer D, Peters JH (2001) Biomaterial properties and biocompatibility in cell culture of a novel self-inflating hydrogel tissue expander. J Biomed Mater Res 54:179–188
9. List of contact lenses allowed to be sold in the United States. Food and Drug Administration Website. http://www.fda.gov/cdrh/contactlenses/lenslist.html
10. Chen Q, Zhu L, Zhao C, Zheng J (2012) Hydrogels for removal of heavy metals from aqueous solution. J Environ Anal Toxicol, 2012 [Open access]
11. Omidian H, Park K (2012) Hydrogels. In: Siepmann J, Siegel R, Rathbone M (eds) Fundamentals and applications of controlled release drug delivery. Spinger, New York, pp 75–106
12. Hejcl A, Lesny P, Pradny M, Michalek J, Jendelova P, Stulik J, Sykova E (2008) Biocompatible hydrogels in spinal cord injury repair. Physiol Res 57:S121–S132
13. Zhong YH, Bellamkonda RV (2008) Biomaterials for the central nervous system. J R Soc Interface 5:957–975
14. Mano JF, Silva GA, Azevedo HS, Malafaya PB, Sousa RA, Silva S, Boesel LF, Oliveira JM, Santos TC, Marques AP, Neves NM, Reis RL (2007) Natural origin biodegradable systems in tissue engineering and regenerative medicine: present status and some moving trends. J R Soc Interface 4:999–1030
15. Chung HJ, Park TG (2007) Surface engineered and drug releasing pre-fabricated scaffolds for tissue engineering. Adv Drug Delivery Rev 59:249–262
16. Willerth SM, Sakiyama-Elbert SE (2007) Approaches to neural tissue engineering using scaffolds for drug delivery. Adv Drug Delivery Rev 59:325–338
17. Perale G, Rossi F, Sundstrom E, Bacchiega S, Masi M, Forloni G, Veglianese P (2011) Hydrogels in spinal cord injury repair strategies. ACS Chem Neurosci 2:336–345

18. Betz M, Hormansperger J, Fuchs T, Kulozik U (2012) Swelling behaviour, charge and mesh size of thermal protein hydrogels as influenced by pH during gelation. Soft Matter 8:2477
19. von Wald Cresce A, Dandu R, Burger A, Cappello J, Ghandehari H (2008) Characterization and real-time imaging of gene expression of adenovirus embedded silk-elastinlike protein polymer hydrogels. Mol Pharmaceu 5:891–897
20. Numata K, Katashima T, Sakai T (2011) State of water, molecular structure, and cytotoxicity of silk hydrogels. Biomacromolecules 12:2137–2144
21. Abd El-Rehim HA, Hegazy ESA, Abd El-Mohdy HL (2004) Radiation synthesis of hydrogels to enhance sandy soils water retention and increase plant performance. J Appl Polym Sci 93:1360–1371
22. Baldrian P, Valaskova V (2008) Degradation of cellulose by basidiomycetous fungi. FEMS Microbiol Rev 32:501–521
23. Saboktakin MR, M.i Tabatabaei R (2015) Supramolecular hydrogels as drug delivery systems. Int J Bio Macro 75:426–436
24. Buwalda SJ, Kristel WM, Dijkstra PJ, Feijen J, Vermonden T, Hennink WE (2014) Hydrogels in a historical perspective: from simple networks to smart materials. J Controlled Release 190:254–273
25. Hassan CM, Peppas NA (2000) Structure and applications of polyvinyl alcohol hydrogels produced convestional cross-linking or by freeze/thawing method. Adv Polym Sci 153:37–65
26. Tsutsumi K, Takayama K, Machida Y, Ebert CD, Nakatomi I, Nagai T, Pharma STP (1994) Formulation of buccal mucoadhesive dosage form of ergotomine tartrate. Sci. 4:230
27. Dinu MV, Přádný M, Drăgan ES, Michálek J (2013) Morphogical and swelling properties of porous hydrogels based on poly(hydroxyethyl methacrylate) and chitosan modulated by ice-templating process and porogen leaching. Polym Res 20:285
28. Nho YC, Park JS, Lim YM (2014) Preparation of poly(acrylic acid) hydrogel by radiation crosslinking and its application for mucoadhesives. Polymers 6:890–898 [Open Access]
29. Saika S, Miyamoto T, Ohnishi Y (2003) Histology of anterior capsule opacification with a polyHEMA/HOHEXMA hydrophilic hydrogel intraocular lens compared to poly (methyl methacrylate), silicone, and acrylic lenses. J Cataract Refract Surg 29:1198
30. Kubinova S, Horak D, Kozubenko N, Vanecek V, Proks V, Price J, Cocks G, Sykova E (2010) The use of superporous Ac-CGGASIKVAVS-OH-modified PHEMA scaffolds to promote cell adhesion and the differentiation of human fetal neural precursors. Biomaterials 31:5966–5975
31. Kubinova S, Horak D, Sykova E (2009) Cholesterol-modified superporous poly(2-hydroxyethyl methacrylate) scaffolds for tissue engineering. Biomaterials 30:4601–4609
32. Kumar N, Ganapathy H, Kim J, Jeong YS, Jeong YT (2008) Preparation of poly 2-hydroxyethyl methacrylate functionalized carbon nanotubes as novel biomaterial nanocomposites. Eur Polym J 44:579–586
33. Rizzi S, Halstenberg S, Hubbell J (2001) Synthetic, enzymatically degradable extracellular matrices formed from recombinant protein-(poly)ethyleneglycol. Eur Cells Mat 2:82–83
34. Gong C, Shi S, Dong P-W, Kan B, Gou M, Wang XH, Li X-Y, Luo F, Zhao X, Wei Y-Q, Qian ZY (2009) Synthesis and characterization of PEG-PCL-PEG thermosensitive hydrogel. Int J Pharmaceutics 365:89–99
35. Lin C, Anseth K (2008) Cell–cell communication mimicry with poly(ethylene glycol) hydrogels for enhancing β-cell function. Pharmac Res 26:6380–6385
36. Kim B, Peppas N (2003) Poly(ethylene glycol)-containing hydro-gels for oral protein delivery applications. Biomed Microdevices 5:333–341
37. Lum L, Elisseeff J (2003) Ch. 4 In: Ashammakhi N, Ferretti P (eds) Topics in tissue engineering. University of Oulu, Oulu
38. Li J, Wang N, Wu X (1998) Poly(vinyl alcohol) nanoparticles prepared by freezing- thawing process for protein/peptide drug delivery. J Control Release 56:117–126
39. Ajji Z, Othoman I, Rosiak J (2005) Production of hydrogel wound dressings using gamma radiations. Nucl Instrum Methods Phys Res B 229:375–380

40. Ramires P, Miccoli M, Panzarini E, Dini L, Protopapa C (2005) In vitro and in vivo biocompatibility evaluation of a polyalkylimide hydrogel for soft tissue augmentation. J Biomed Mater Res B Appl Biomater 72:230–238
41. Zhang X, Yang Y, Yao J, Shao Z, Chen X (2014) Strong collagen hydrogels by oxidized dextran modification. ACS Sustainable Chem Eng 2:1318–1324
42. Choi B, Kim S, Lin B, Wu BM, Lee M (2014) Cartilaginous extracellular matrix- modified chitosan hydrogels for cartilage tissue engineering. ACS Appl Mater Interfaces 6:20110–20121
43. Park KM, Yang J-A, Jung H, Yeom J, Park JS, Park KH, Hoffman AS, Hahn SK, Kim K (2012) in situ supramolecular assembly and modular modification of hyaluronic acid hydrogels for 3D cellular engineering. ACS Nano 6:2960–2968
44. Ahmad N, Amin MCIM, Mahali SM, Ismail I, Chuang VTG (2014) Biocompatible and mucoadhesive bacterial cellulose-g-poly(acrylic acid) hydrogels for oral protein delivery. Mol Pharmaceutics 11:4130–4142
45. Wang J, Miao X, Fengzhao Q, Ren C, Yang Z, Wang L (2013) Using a mild hydrogelation process to confer stable hybrid hydrogels for enzyme immobilization. RSC Adv. 3:16739–16746
46. Hu X, Feng L, Xie A, Wei W, Wang S, Zhang J, Dong W (2014) Synthesis and characterization of a novel hydrogel: salecan/polyacrylamide semi-IPN hydrogel with a desirable pore structure. J Mater Chem B 2:3646–3658
47. Sui X, Feng X, Luca AD, van Blitterswijk CA, Moroni L, Hempenius MA, Vancso GJ (2013) Poly(N-isopropylacrylamide)–poly(ferrocenylsilane) dual-responsive hydrogels: synthesis, characterization and antimicrobial applications. Polym Chem 4:337–342
48. Li L, Ren L, Wang L, Liu S, Zhang Y, Tang L, Wang Y (2015) Effect of water state and polymer chain motion on the mechanical properties of a bacterial cellulose and polyvinyl alcohol (BC/PVA) hydrogel. RSC Adv 5:25525–25531
49. Qi X, Hu X, Wei W, Yu H, Li J, Zhang J, Dong W (2015) Investigation of Salecan/poly(vinyl alcohol) hydrogels prepared by freeze/thaw method. Carbo Polym 118:60–69
50. Selvam S, Pithapuram MV, Victor SP, Muthu J (2015) Injectable in situ forming xylitol–PEG-based hydrogels for cell encapsulation and delivery. Col Surf B Biointerfaces 126:35–43
51. Li D, Zhang X, Yao J, Zeng Y, Simon GP, Wang H (2011) Composite polymer hydrogels as draw agents in forward osmosis and solar dewatering. Soft Matter 7:10048–10056
52. Giammanco EG, Sosnofsky CT, Ostrowski AD (2015) Light-responsive iron(III)—polysaccharide coordination hydrogels for controlled delivery. ACS Appl Mater Interfaces 7:3068–3076
53. Ni T, Xu L, Sun Y, Yao W, Dai T, Lu Y (2015) Facile fabrication of reduced graphene oxide/polypyrrole composite hydrogels with excellent electrochemical performance and compression capacity. ACS Sustainable Chem Eng 3:862–870
54. Ananthoji R, Eubank JF, Nouar F, Mouttaki H, Eddaoudi M, Harmon JP (2011) Symbiosis of zeolite-like metal–organic frameworks (rho-ZMOF) and hydrogels: composites for controlled drug release. J Mater Chem 21:9587–9594
55. Li Z, Su Y, Xie B, Wang H, Wen T, He C, Shen H, Wuc D, Wang D (2013) A tough hydrogel–hydroxyapatite bone-like composite fabricated in situ by the electrophoresis approach. J Mater Chem B 1:1755–1764
56. Bakarich SE, Gorkin R, in het Panhuis M, Spinks GM (2014) Three-dimensional printing fiber reinforced hydrogel composites. ACS Appl Mater Interfaces 6:15998–16006
57. Qi X, Guan Y, Chen G, Zhang B, Ren J, Peng F, Sun R (2015) A non-covalent strategy for montmorillonite/xylose self-healing hydrogels. RSC Adv 5:41006–41012
58. Xia N, Zhang Y, Chang K, Gai X, Jing Y, Li S, Liu L, Qu G (2015) Ferrocene-phenylalanine hydrogels for immobilization of acetylcholinesterase and detection of chlorpyrifos. J Electroana Chem 746:68–74
59. Guarino V, Ambrosio L (2013) Thermoset composite hydrogels for bone/intervertebral disc interface. Mater Lett 110:249–252

60. Dumitrescu AM, Slatineanu T, Poiata A, Iordana AR, Mihailescu C, Palamaru MN (2014) Advanced composite materials based on hydrogels and ferrites for potential biomedical applications. Collo Surf A Physicochem Eng Aspects 455:185–194
61. Qu B, Chen C, Qian L, Xiao H, He B (2014) Facile preparation of conductive composite hydrogels based on sodium alginate and graphite. Mate Lett 137:106–109
62. Yang Q, Adrus N, Tomicki F, Ulbricht M (2011) Composites of functional polymeric hydrogels and porous membranes. J Mater Chem 21:2783–2811
63. Gaharwar AK, Avery RK, Assmann A, Paul A, McKinley GH, Khademhosseini A, Olsen BD (2014) shear-thinning nanocomposite hydrogels for the treatment of hemorrhage. ACS Nano 8:9833–9842
64. Wang H, Yi H, Chen X, Wang X (2014) One-step strategy to three-dimensional graphene/VO_2 nanobelt composite hydrogels for high performance supercapacitors. J Mater Chem A 2:1165–1173
65. Chang C-W, van Spreeuwel A, Zhang C, Varghese S (2010) PEG/clay nanocomposite hydrogel: a mechanically robust tissue engineering Scaffold. Soft Matter 6:5157–5164
66. Cheng C, Liu Z, Li X, Su B, Zhou T, Zhao C (2014) Graphene oxide interpenetrated Polymeric composite hydrogels as highly effective adsorbents for water treatment. RSC Adv 4:42346–42357
67. Baeissa A, Dave N, Smith BD, Liu Juewen (2010) DNA-functionalized monolithic hydrogels and gold nanoparticles for colorimetric DNA detection. ACS Appl Mater Interface 2:3594–3600
68. Gu X, Yang Y, Hu Y, Hu M, Wang C (2015) Fabrication of graphene-based xerogels for removal of heavy metal ions and capacitive deionization. ACS Sustainable Chem Eng 3:1056–1065
69. Xua M, Huanga Q, Wanga X, Suna R (2015) Highly tough cellulose/graphene composite hydrogels prepared from ionic liquids. Industrial Crops Products 70:56–63
70. Güler MA, Gök MK, Figenc AK, Özgümüş S (2015) Swelling, mechanical and mucoadhesion properties of Mt/starch-g-PMAA nanocomposite hydrogels. App Clay Sci 112–113:44–52
71. De Filpo G, Palermo AM, Munno R, Molinaro L, Formoso P, Nicoletta FP (2015) Gellan gum/titanium dioxide nanoparticle hybrid hydrogels for the cleaning and disinfection of parchment. Inter Biodeter Biodegrad 103:51–58
72. Isabel González-Sánchez M, Perni S, Tommasi G, Morris NG, Hawkins K, López-Cabarcos E, Prokopovich P (2015) Silver nanoparticle based antibacterial methacrylate hydrogels potential for bone graft applications. Mater Sci Eng C 50:332–340
73. Tóth IY, Veress G, Szekeres M, Illés E, Tombác E (2015) Magnetic hyaluronate hydrogels: preparation and characterization. J Magnet Magnetic Mater 380:175–180

Conductive Polymer Hydrogels

Damia Mawad, Antonio Lauto and Gordon G. Wallace

Abstract Combining electrical properties with synthetic scaffolds such as hydrogels is an attractive approach for the design of the ideal synthetic soft tissue, one that mimics the architecture of the native extracellular matrix and provides the electronic functionality needed for cell–cell communication. Conducting polymers (CPs) are carbon-based polymers that are electronically active and consequently are being investigated as the structural material for fabrication of electroactive hydrogels. CPs are attractive in that they could be processed in various forms, their chemistry could be modified to introduce different functionalities and most important is their capability to conduct electrons. In this chapter, electroconductive hydrogels (ECHs) fabricated from CP either as a single component or as an additive to conventional hydrogel networks are reviewed.

Keywords Conducting polymer · Hydrogel · Electroconductive · Single component · Hybrid

Abbreviations

3D	Three dimensional
AC	Alternating current
ADH	Adipoyl dihydrazide
APS	Ammonium persulfate

D. Mawad (✉)
Department of Materials, Imperial College London, London SW7 2AZ, UK
e-mail: damia.mawad@unsw.edu.au

D. Mawad
School of Materials Science and Engineering, UNSW, Sydney 2052, Australia

A. Lauto
Bioelectronics and Neuroscience (BENS) Research Group, University of Western Sydney, Sydney 2751, Australia

G.G. Wallace
Intelligent Polymer Research Institute, ARC Center of Excellence for Electromaterials Science, University of Wollongong, Wollongong 2500, Australia

S. Kalia (ed.), *Polymeric Hydrogels as Smart Biomaterials*,
Springer Series on Polymer and Composite Materials,
DOI 10.1007/978-3-319-25322-0_2

BF	Basic fuchsine
CCG	Chemically converted graphene
CNT	Carbon nanotube
CP	Conducting polymer
CTAB	Cetyl trimethyl ammonium bromide
DC	Direct current
DCC	*N,N′*-dicyclohexylcarbodiimide
DCI	1,1′-carbonyldiimidazole
ECH	Electroconductive hydrogel
Fmoc	*N*-fluorenylmethoxycarbonyl
FP	Phenylalanine
FTIR	Fourier transform infrared
Gd^{3+}	Gadolinium ion
GO	Graphene oxide
hMSC	Human mesenchymal stem cell
IC	Inhibitory concentration
LMWG	Low molecular weight gelator
MO	Sodium 4-[4′-(dimethylamino)phenyldiazo] phenylsulfonate
PANI	Poly(aniline)
PBS	Phosphate buffer solution
PCLF	Polycaprolactone fumarate
PEDOT	Poly(ethylenedioxythiophene)
PEG	Poly(ethylene glycol)
PMDIG	5,5′-(1,3,5,7-tetraoxopyrrolo[3,4-f]isoindole-2,6-diyl)diisophthalic acid
PPy	Polypyrrole
POWT	Poly(3-((*S*)-5-amino-5-carboxyl-3-oxapentyl)-2,5-thiophene) hydrochloride
PTAA	Poly(3-thiophene acetic acid)
PTh	Polythiophene
QCM	Quartz crystal microbalance
rGO	Reduced graphene oxide
ROS	Reactive oxygen species
SEM	Scanning electron microscopy
SWNT	Single wall nanotube

1 Introduction

The applicability of hydrogels across a range of biomedical applications such as biosensors, drug delivery and tissue engineering is driving researchers to develop 3D networks encompassing new tailored properties such as thermal, optical and electrical conductivities [1]. Electrically conductive hydrogels (ECHs) are attracting much interest in the field of biomaterial science due to their unique properties, combining a hydrated 3D structure while imparting electronic functionality.

Conducting polymers (CPs) are synthetic polymers that are characterised by their ability to conduct electrons, while providing flexibility and processability. Additionally, their organic nature facilitates their chemical modification to introduce different functionalities meeting specific needs required in the biomedical field. Consequently, CPs gained popularity in more recent years as components of complex systems designed to electrically communicate with physiological tissues such as nerve, brain, muscle and cardiac tissues [2]. These systems could be in the form of electrodes or implantable scaffolds. The capacity to manipulate the processability of CPs into various forms led to the design and fabrication of flexible bioelectronics applicable for a wide range of therapeutics. Processing CPs into hydrated 3D hydrogel scaffolds is an attractive approach to achieving synthetic soft tissue, one that matches the mechanical properties of the native extracellular matrix and preserves the electronic functionality needed for cell–cell communication.

Conversely, achieving this goal is rather challenging due to the rigorous requirements for fabricating a hydrogel. A 3D hydrogel network consists of crosslinked hydrophilic polymers with high water content, exhibiting elastic behaviour and porous internal structures [3]. CPs are inherently rigid due to the conjugated system in the backbone, and thus exhibit high stiffness [4]. The backbone is somewhat hydrophobic in nature due to the aromatic rings in the backbone, which cause π–π stacking of the chains [5]. Also the stiffness is often attributed to unwanted cross links. As such, it is counter intuitive that these polymers might be suitable precursors for the synthesis of conductive hydrogels. However, with the advent of fabrication techniques, composite formulations and development of smart chemistries, researchers have succeeded to overcome many of these limitations. Two main fabrication routes are being investigated for the development of conductive hydrogels. One is to grow the conducting polymer in a prefabricated hydrogel; these are referred to as hybrid systems. The other is to use the CP as the sole polymeric component (continuous phase) in the hydrogel network: this could be achieved either by self-assembly or by introducing water soluble and cross-linkable moieties in the backbone. Subsequently, conductive hydrogels are being developed and tested for a plethora of biomedical applications.

This chapter presents an overview of the current state of conducting polymer hydrogels, with emphasis on 3D-network in which the conducting polymer is the continuous phase. We present an overview of conjugated polymers and their charge transport mechanism. We then briefly introduce the first approach taken to develop electroconductive hydrogels (ECHs) in which the conducting polymer is grown within a prefabricated conventional hydrogel made from insulators. Owing to the emergence of excellent reviews that discuss ECHs, we focus on single component conducting polymer hydrogels. These are 3D-networks formed only from CPs and free of any insulating matrix. We present selected research papers in this field; recent advances and challenges will be highlighted to gain a better insight of various strategies employed. This section will be followed by introducing the role of other electrically active carbon-based materials such as graphene and carbon nanotubes (CNTs) in the fabrication of conductive hydrogels for biomedical applications. This is an emerging and rapidly growing field that warrants attention.

2 Conducting Polymers in 3D-Hydrogel Networks

2.1 Conducting Polymers and the Origin of Their Conductivity

The very first scientific paper reporting on organic CPs and their unique behaviour as semiconductors or metals was in 1977, published in the journal *Journal of the Chemical Society, Chemical Communications* and was co-authored by Shirakawa et al. [6]. The authors reported on the synthesis of polyacetylene and its doping by halogens to produce polymers with remarkable dc conductivity. Consequently, a new class of smart polymers has been introduced driving advances in material science, electronics and biomedical applications. This discovery was acknowledged by the awarding of the Chemistry Nobel Prize in 2000.

Halogenated polyacetylene was described to have electric conductivity due to mobile charge carriers introduced in the π complexes of the polymeric chain [6, 7]. CPs are also known as conjugated polymers because of the structure of their backbone that is formed from alternating single and double C–C bonds (C=C) (see Fig. 1a). Carbon has 3 sp^2 orbitals that lie in one plane and a *p*-orbital that is perpendicular to the plane (p_z) (see Fig. 1b).

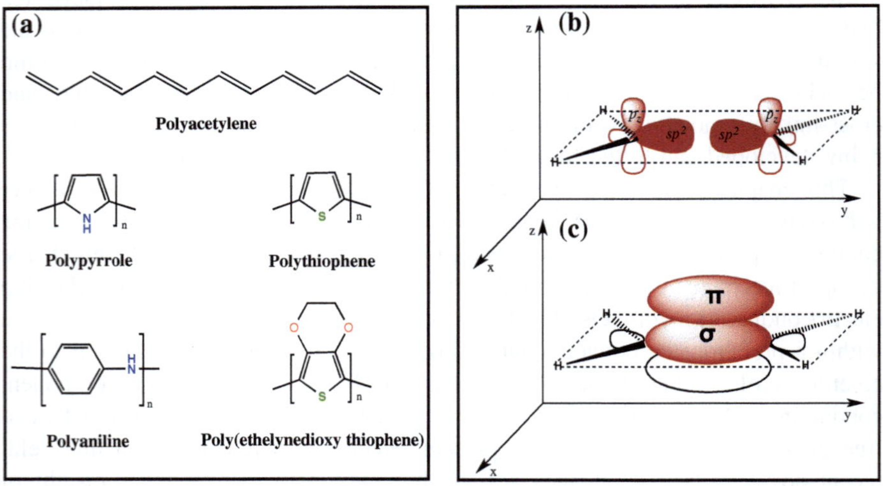

Fig. 1 **a** Chemical structures of conjugated polymers. **b** 2 sp^2 orbitals approach each other to form a σ bond and **c** the 2 p_z orbitals form a π bond which is in parallel to the σ bond plane

When two carbon atoms come close together, two of the sp^2 orbitals overlap forming a σ bond. This in turn brings two of the p_z orbitals in close proximity and they form π orbitals that are delocalized over the backbone chain (see Fig. 1c) [7]. The delocalised π orbitals can be either filled π-bonding orbital which forms the valence band or empty π*-anti-bonding orbital which is the conduction band [8]. These two states have degenerate energy levels [9] that upon excitation can allow charge mobility along the conjugated backbone and thus generating conductivity.

The term "doped" polymer refers to charge transfer to the backbone either by n-type (reduction) or p-type (oxidation). In case of n-type doping, an electron is introduced in the conduction band generating an "electron charge carrier," whereas for p-type doping an electron is abstracted from the valence band creating a "hole charge carrier." Therefore, doping is a mechanism by which an electron is either removed or added to the CP backbone creating unbound charge carriers. The mobility of these carriers is enhanced by the π orbital system of the conjugated backbone [10].

Thus conductivity in conjugated systems depends on the dopant, its type and efficiency to abstract or induce an electron within the backbone [8]. Additionally, conductivity is highly dependent on the mobility of these charge carriers that could occur intra or inter chain [10]. Hence any structural disorder in the conjugated polymer will cause deterioration in its value. Also, the packing of the polymeric chains highly affects the transport of the charges; too much disorder will hinder interchain hoping and dramatically lowers the conductivity.

This is particularly relevant and remains a challenge in the design and fabrication of conductive hydrogels. While electronic conductivity of a hydrogel in the dried state could be determined by standard techniques such as 2 or 4-probe measurements [11, 12], in their swollen state this task becomes more challenging. As the hydrogel swells, conducting polymer chains become further apart which might hinder electron transport [13]. Additionally, hydrogels for tissue engineering are swollen in a buffer at pH = 7.4. This introduces ions into the network, which contribute to ionic conductivity. As such, alternative techniques are sought such as AC impedance spectroscopy to discriminate between the components contributing to each of the electronic and ionic conductivity [14–16].

2.2 Electroconductive Hydrogels (ECHs)

ECH is the term used to describe a hybrid network fabricated from conventional insulating polymers combined with CPs. While the insulating polymer provides the 3D aqueous gel, the CP imparts electrical conductivity to the scaffold. Various synthetic approaches have been developed to fabricate hybrid networks. These include: (i) electro or chemical polymerization of a conducting monomer in a prefabricated hydrogel, and (ii) mixing the precursor monomers followed by simultaneous or step-wise polymerization to produce the ECH (Fig. 2). The end product is a hydrogel network in which the conducting polymer chains are physically entrapped within the conventional hydrogel matrix.

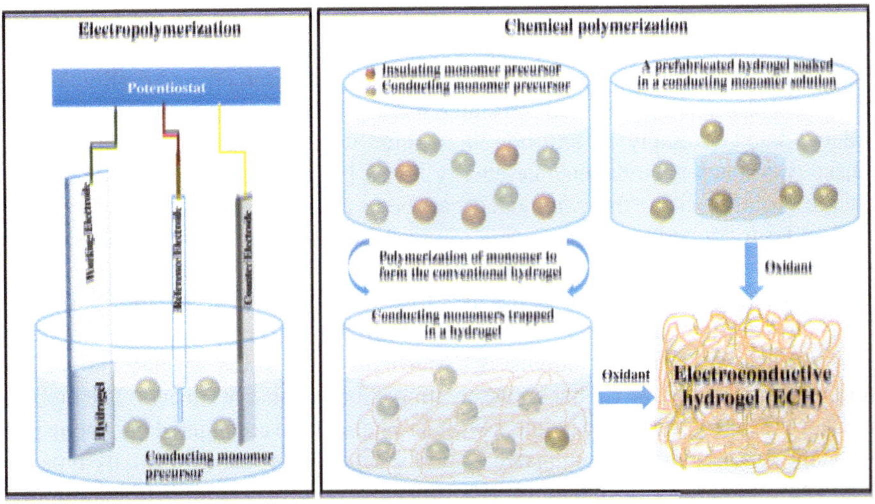

Fig. 2 Schematic representation of approaches developed to fabricate electroconductive hydrogels (ECHs). **Electropolymerization**: a prefabricated hydrogel around a working electrode such as ITO or gold is immersed in a monomer solution such as pyrrole or aniline. By applying either a voltage or a current, the conducting polymer grows in the pores of the hydrogel network. **Chemical polymerization**: this could be achieved in two ways; (i) the monomer precursors of both the insulator and the CP are mixed together, then polymerization occurs either simultaneously or step wise by the addition of the appropriate initiators/oxidants; (ii) a prefabricated hydrogel is soaked in a solution of the monomer precursor of the CP. The monomer diffuses into the pores of the hydrogel. Upon addition of an oxidant, the CP is chemically grown and gets physically entrapped within the network

The first reported ECH was by Gilmore et al. [17] describing the fabrication of a hybrid composite based on polypyrrole (PPy) directly electropolymerised on a preformed polyacrylamide hydrogel. Despite the introduction of the hydrophobic PPy polymer, the hybrid hydrogel retained its hydration/rehydration properties. Additionally, these gels were shown to be electroactive and conductive. This study has opened a new era in conducting polymer science. Owing to the many available hydrogels and a range of CPs, there have been a surge in the number of reports describing the fabrication of ECHs and their application in the biomedical field; these have been discussed in recent reviews [18–22]. Table 1 lists an up to date summary of ECHs reported in the past 10 years.

The ideal ECH network is one that combines both mechanical properties that are comparable to electroresponsive tissues (1–10 kPa) [48] and suitable conductivity for electrical stimulation. In a study by Ding et al. [34], biologically derived conductive hydrogel fabricated from methacrylate modified heparin and polyaniline (PANI) has been fabricated. In the first step, a heparin hydrogel network was formed by UV curing. The hydrogel was then soaked in an aniline solution to allow for the monomer to diffuse inside the porous structure. PANI was formed by oxidative polymerisation. By controlling the aniline monomer concentration, the electrical/ionic properties of the hydrogel could be tailored. Hydrogels formed from 1 and 0.1 M aniline solution

Table 1 Examples of ECHs: type of CP and insulator employed, fabrication technique and the application investigated

Conducting polymer	Insulating polymer	Fabrication route[a]	Application	Refs.
Polypyrrole	Agarose	Chemical	Self-healable electrodes	[23]
	Agarose	Electropolym	Patterned films	[24]
	Poly(hydroxyethylmethacrylate) (HEMA)	Electropolym	Implantable biosensors, bioelectronics	[25]
	Chitosan	Chemical		[26]
	Poly(HEMA)	Electropolym	Glucose responsive biotransducers	[27]
	Oligo(polyethylene glycol) fumarate (OPF)	Chemical	Tissue engineering	[28]
	Poly(HEMAco-PEGMA)	Electropolym	Coatings for implantable biosensors and neuronal prostheses	[29]
	Poly(acrylic acid)	Chemical	Drug delivery	[30]
	Poly(vinyl alcohol)	Electropolym	Drug delivery	[31]
Polyaniline	αCD-containing polyacrylamide (αCD-PAAm)	Chemical	Flexible supercapacitors	[32]
	iota-Carrageenan (ι-CGN)	Chemical	Tissue engineering Electrochemical capacitor	[33]
	Heparin	Chemical	Tissue engineering	[34]
	α-CD-based supramolecular hydrogel	Chemical	Biosensors	[35]
	Polyethyleneglycol diacrylate (PEGDA)	Chemical	Tissue engineering	[36]
	Poly(2-hydroxyethylmethacrylate-co-glycidylmethacrylate)	Chemical	Biosensors	[37]
	Chitosan	Chemical	Actuators	[38]

(continued)

Table 1 (continued)

Conducting polymer	Insulating polymer	Fabrication route[a]	Application	Refs.
PEDOT	Alginate	Chemical	Drug delivery	[39]
	Polyurethane (PU)	Electropolym	Tissue engineering	[40]
	Agarose	Electropolym	Autografts	[41]
	RGD-functionalized alginate	Electropolym	Drug delivery	[42]
	Poly(vinyl alcohol)/poly(acrylic acid)	Electropolym	Optogenetics	[43]
	Fibrin	Electropolym	Tissue engineering	[44]
	Agarose	Electropolym	Tissue engineering	[45]
	Alginate	Electropolym	Bioelectronic sensors	[46]
	Poly(ethylene glycol) methyl ether methacrylate (PPEGMA)/ poly(acrylic acid) (PAA)	Chemical	Soft strain sensors	[47]

[a]Fabrication route: the CP was either formed by chemical or electrochemical (Electropolym) polymerization

had an impedance of 900 and 2×10^4 Ω at 0.01 Hz, respectively. The storage modulus was in the order of 900 Pa. Additionally, the heparin/PANI hydrogel supported the growth, proliferation and differentiation of C2C12 muscle cells. Runge et al. [28] reported the synthesis of a hydrogel scaffold based on polycaprolactone fumarate (PCLF) in which polypyrrole was chemically grown. The hydrogels exhibited a conductivity of $\sim 6 \times 10^{-3}$ S/cm in their dried state. In comparison to PCLF, neural cells seeded on the PCLF/PPy hydrogels showed better cell morphology denoted by elongated cell bodies, extended neurites and a higher cell number. This hybrid hydrogel presents potential for neural regeneration.

An injectable conductive hydrogel based on collagen and infused with poly (ethylenedioxy thiopehene) (PEDOT) or PPy fibres has been reported by Sirivisoot et al. [49]. Prefabricated PPy or PEDOT fibres were added to a cell-containing collagen solution, followed by its gelation at physiological pH and temperature. The conductive scaffolds supported the growth and differentiation of PC-12 and human mesenchymal stem cells (hSMCs) up to 7 days. Furthermore, compared to the collagen control, the conductive scaffolds induced an increased neurite growth of PC-12 cells. This study presents a simple but elegant fabrication technique for injectable cell seeded conductive scaffolds that could be potentially applied in vivo for nerve, muscle and cardiac applications.

While many of the reported ECHs demonstrated relevant properties for tissue engineering such as appropriate mechanical strength, remarkable hydration and biocompatibility, some limitations remain with the fabrication approach used to develop ECHs. Typically, growing the conducting polymer in a prefabricated hydrogel network, whether by electropolymerization or chemical oxidation, leads to physically entrapped CP chains. Upon hydration in physiological media, these polymeric chains diffuse out of the network causing a drop in conductivity as well as possible cytotoxic effects [50]. When the hybrid network is formed from a negatively charged insulator such as alginate, poly(acrylic acid) or polyacrylamide, ionic interactions occur between the positively charged conducting polymer and the insulator. However, the potential application of these hydrogels in tissue engineering requires incubation in pH = 7.4 which leads to neutralisation of the charges on the CP backbone and consequently dissociation from the network [51].

2.3 Single Component CP Hydrogel

"Single component" CP hydrogel is a term used to refer to conductive hydrogels made from the conjugated polymer as the main continuous phase. The advantage of this approach is to overcome the aforementioned limitations of ECHs. However, this is a rather challenging approach because conjugated polymers lack water solubility, functional side chains and flexibility. Up to date, there are scarce reports on the fabrication of single components hydrogels. Fabrication approaches used include self-assembly of the CP chains, crosslinking either chemically, using metal ions, or by employing low molecular weight gelator (LMWG). We herein report the studies that investigated synthesis of single component conductive hydrogels and their characteristics.

2.3.1 Polythiophenes

The chemistry of the thiophene monomer is versatile; thiophene could be modified at α and β positions leading to a wide range of chemical structures. Consequently, a great diversity of thiophene-based materials has been prepared such as regioregular oligomers and polymers [52]. For biomedical applications in general and for the fabrication of a hydrogel network in particular, water solubility of the polymeric chains is a pre-requisite. As such, a range of water-soluble polythiophenes has been synthesised as described in a recent review by Das et al. [53]. Figure 3 highlights some of the chemical structures of water soluble polythiophenes used to fabricate conductive hydrogels.

The first reported polythiophene-based hydrogel was fabricated from poly (3-thiophene acetic acid) (PTAA) [54, 55]. The ionisable carboxylic groups on the side chains render PTAA water soluble. Additionally, by using appropriate crosslinkers, the functional carboxylic groups could be linked to form a 3D-network. Chen et al. [55] used adipoyl dihydrazide (ADH) as the crosslinker and *N,N'*-dicyclohexylcarbodiimide (DCC) as a condensation agent to form the PTAA gel. The authors showed that conformational changes of PTAA backbone occur in response to the carboxylic groups ionisation under different pH, despite the chemical crosslinking of the polymeric chains. They also reported the electric conductivity ($4 \times 10^{-3} - 2.0 \times 10^{-2}$ S/cm) of the gels doped with 60 wt% $HClO_4$ solution.

Similarly taking advantage of the carboxylic groups on the PTAA backbone, Mawad et al. [56] reported the fabrication of a chemically crosslinked single component PTAA hydrogel (Fig. 4a) using a more amiable crosslinking agent, 1,1′-carbonyldiimidazole (DCI). In contrast to DCC used previously by Chen [55], CDI

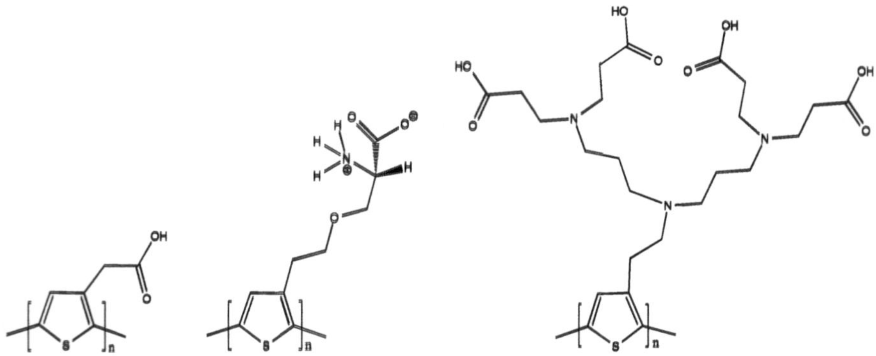

Fig. 3 Chemical structures of water-soluble polythiophenes. *PTAA* poly(3-thiophene acetic acid). *POWT* poly(3-((S)-5-amino-5-carboxyl-3-oxapentyl)-2,5-thiophene) hydrochloride

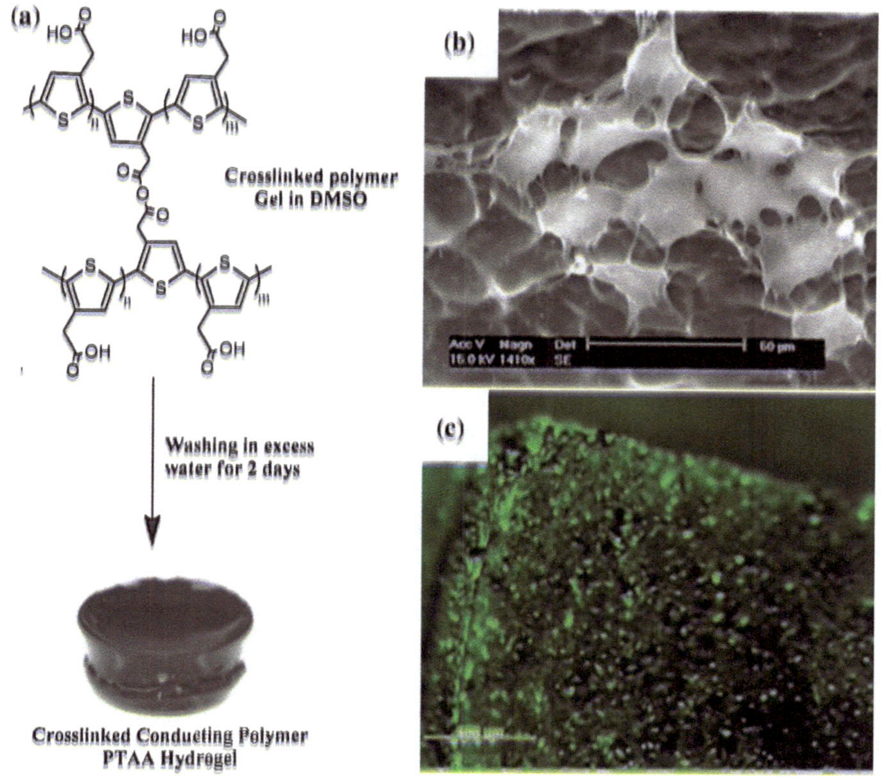

Fig. 4 **a** Schematic illustration of crosslinked PTAA polymeric chains. **b** Representative SEM images of myoblast cells adhered to the hydrogel substrates following 72 h incubation. *Scale bar* represents 50 μm. **c** Fluorescent images of myoblast cells on hydrogel substrates following 72 h incubation. Cells are stained with CalceinAM to visualise metabolically active cells. *Scale bar* represents 400 μm. Reprinted from [56]. Copyright 2012 with permission of John Wiley and Sons

and its decomposition products are water soluble and they could be easily washed from the network. The swelling ratio, internal porous structure and mechanical properties could be tuned by varying the ratio of the crosslinker (DCI) to the monomeric unit of PTAA. The hydrogels supported the adhesion and proliferation of C2C12 cells seeded on their surface (Fig. 4b, c). Of significance was the conductivity ($\sim 10^{-2}$ S/cm) and good electroactivity these hydrogels exhibited in physiological conditions. This is the only reported single component conductive hydrogel that has been tested as a scaffold for tissue engineering. In addition to having functional groups on its side chain, PTAA aqueous solution has been shown to have a half inhibitory concentration (IC_{50}) of 9.1 mg/mL [57], indicating a high degree of tolerance by the cells to this water-soluble polymer. These studies suggest that PTAA is a very promising CP candidate for biomedical applications.

A conductive hydrogel film based on water soluble zwitterionic polythiophene, poly(3-((S)-5-amino-5-carboxyl-3-oxapentyl)-2,5-thiophene) hydrochloride (POWT)

(Fig. 3), has been reported by Asberg et al. [58]. A thin film of POWT was deposited on a substrate from a 0.5 mg/mL POWT aqueous solution. While exposing the POWT film to buffer solutions of different pH, the hydration and stiffness of the film was monitored by quartz crystal microbalance (QCM). The authors reported that a hydrogel forms upon uptake of water and ions from the buffer. This hydrogel exhibited higher uptake of DNA in comparison to the compact POWT film. Since POWT is a conjugated polyelectrolyte, its polymeric chains change conformation after binding to the DNA strands. The authors suggested that by taking advantage of these conformational changes, POWT hydrogel could be applied as biochip for the specific binding and sensing of complementary DNA. Multifunctional hydrogels based on PT–COOH (Fig. 3) and DNA were fabricated by using gadolinium ions (Gd^{3+}) as a chelator. The hydrogel exhibited good swelling properties ($Q \sim 8$) and they were stable in phosphate buffer solution (PBS) for up to 60 h. Jurkat T cells were mixed with the DNA and PT–COOH solution followed by gelation using the Gd^{3+} ion. This is the first example of a conductive hydrogel that could form in situ allowing cell encapsulation prior to gelation. By irradiating the hydrogels with white light, the authors demonstrated that cells could be killed as a result of reactive oxygen species (ROS) produced by PT–COOH–Gd^{3+} complex in response to light.

2.3.2 Polyaniline

One approach to fabricate conducting polymer hydrogels is through the assembly of the polymeric chains with LMWG. LMWG are small organic molecules that can cause water or other solvents to form a network at very low concentrations [59]. PANI-based hydrogels have been prepared using LMWG such as 5,5′-(1,3,5,7-tetraoxopyrrolo[3,4-f]isoindole-2,6-diyl)diisophthalic acid (PMDIG) [60] and *N*-fluorenylmethoxycarbonyl (Fmoc) phenylalanine (FP) [61]. PMDIG-PANI hydrogels were fabricated by preparing a solution of anilinium chloride with the salt of PMDIG [60]. Upon addition of ammonium persulfate (APS), a stable hydrogel was formed after 24 h at 30 °C. FTIR spectra suggested that the supramolecular interactions between PANI and PMDIG are due to strong H-bonding between the two components as denoted by the shifts of both carbonyl and hydroxyl groups of the gelator in the hydrogel structure. The conductivity of the gel was 0.3×10^{-4} S/cm and the elasticity 14.59 kPa. In a similar fabrication approach, FP-PANI hydrogels were prepared and characterised [61]. Good electrical conductivity was achieved (1.2×10^{-2} S/cm) and the hydrogel exhibited a viscoelastic nature. Because of the Fmoc group on FP that can act as a weak acceptor while PANI acts as a donor, the photoconductivity of the FP-PANI hydrogel was investigated. The gel exhibited a reversible photoresponse under white light illumination. Such investigations open the way for the fabrication of responsive gels for applications such as optoelectronics and biosensors.

A conducting hydrogel fabricated from PANI chains crosslinked with phytic acid (Fig. 5a, b) has been prepared by Pan et al. [62]. The phytic acid was both the gelator and the dopant leading to a hydrogel with good electrical properties (conductivity value of 0.11 S/cm) as well as high water content (~92.6 % wt/wt). Due to the porous structure and high surface area of the 3D network, the hydrogel exhibited high specific capacitance (480 F g^{-1}), and excellent cycling stability (~83 % capacitance after 10,000 cycles). Furthermore, the fabrication approach adopted in this study allowed micropatterning of the gel into various dimensions (Fig. 5c, d). This hydrogel can serve as a promising material for the fabrication of high energy storage electrodes. In a similar study by Zhai et al. [63], the PANI-phytic acid hydrogel was casted as a thin film on a platinum surface, followed by dipping it into a solution of chloroplatinic acid (H_2PtCl_6). Using formic acid, H_2PtCl_6 was reduced to Pt introducing Pt nanoparticles in the 3D network. The Pt/hydrogel electrode was tested as a glucose biosensor and it exhibited ultra-high sensitivity, low detection limit (0.7 μm), and fast response time (3 s).

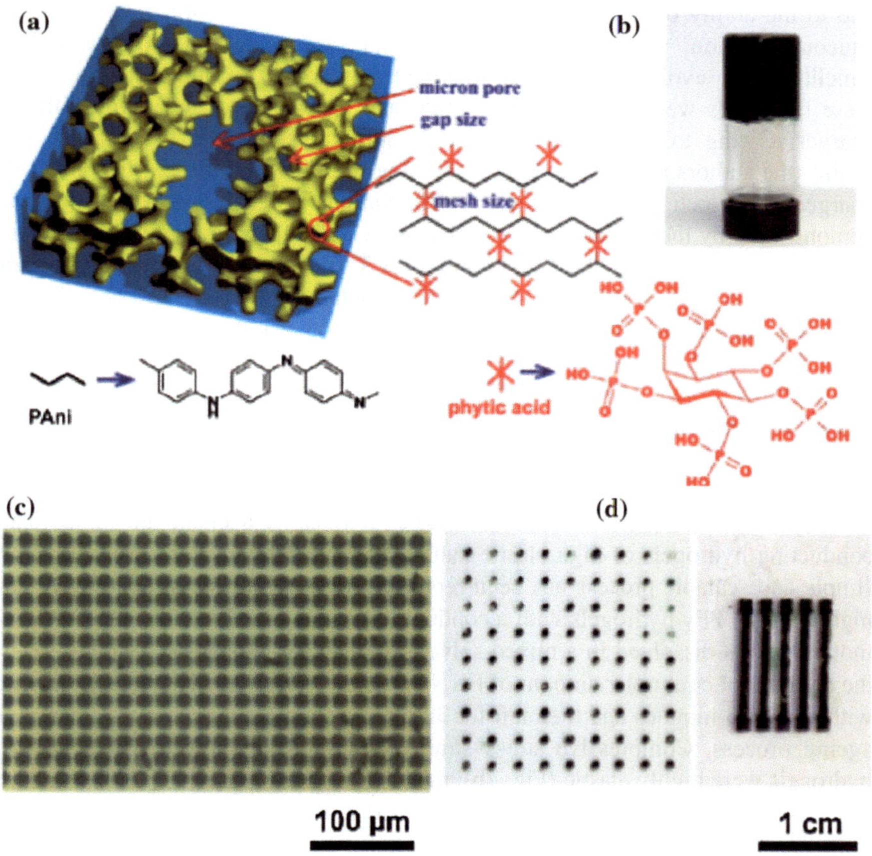

Fig. 5 Doped polyaniline hydrogels: **a** Schematic representation of the 3D network formed by crosslinking polyaniline with phytic acid. **b** Representative image showing the gel in a glass vial. **c** Ink jet printed hydrogel with dot dimension of 18 μm. **d** Patterned PANI hydrogels produced by mask-spray coating with either a dot diameter of 1 mm or line width of 2 mm. Reprinted from [62]. Copyright 2015, with permission of PNAS

2.3.3 Poly(Ethylenedioxythiophene) (PEDOT)

Amongst synthesised polythiophenes, PEDOT is of particular interest due to its high stability in aqueous solutions, high conductivity and versatility of side chains functionalization. To this effect, it has been explored for the fabrication of conductive hydrogels. Hydrogels based on alkoxysulfonate functionalised PEDOT have been fabricated in a one-step reaction by simply mixing the sulfonated EDOT (EDOT-S) monomer with an oxidant such as APS and $FeCl_3$ (Fig. 6a) [64]. Scanning electron microscopy (SEM) revealed that the internal 3D structure of the hydrogels could be controlled depending on the oxidant used (Fig. 6b) and its ratio. Additionally, the authors reported these gels to have conductivities in the range of 10^0–10^2 S/cm depending on the type of oxidant and monomer concentrations employed. The authors explained the formation mechanism of these hydrogels to be due to the ability of the amphiphilic EDOT-S monomers to form micelles in the aqueous solution, which upon addition of an oxidant fuse together and form lamellar sheets evolving into a 3D hydrogel network. In a follow up study [65], these hydrogels were tested for adsorption and desorption of dyes with different charges. Owing to the negatively charged sulfonic groups on the PEDOT side chains, the authors reported a high degree of adsorption (164 mg g^{-1}) of positively charged dyes such as basic fuchsine (BF). Similarly, desorption of the dye was demonstrated by the addition of the surfactant, cetyl trimethyl ammonium bromide (CTAB). By increasing the mass ratio of CTAB to the dried gel, desorption amounts up to 90 % of the adsorbed drug were obtained. These hydrogels exhibited favourable desorption efficiency in comparison to other reported systems [66].

2.3.4 Polypyrrole

The delocalised π-electron system along the backbone of conjugated polymers results in rigid polymeric chains; therefore, it remains a challenge to fabricate conducting hydrogels of high elastic nature. In a recent study by Lu et al. [67], a simple and scalable process has been reported for the synthesis of conductive and highly elastic PPy hydrogels with a conductivity value of 0.5×10^{-2} S/cm. Pyrrole monomer was dissolved in a mixed solvent of H_2O and ethanol (1:1) followed by the addition of equimolar amount of $Fe(NO_3)_3 \cdot 9H_2O$ (Fig. 7a). The gel was formed within several minutes and then left for 30 days at room temperature to undergo an ageing process. Compressive stress–strain measurements showed that aged PPy hydrogels were highly elastic (Fig. 7b); they could be compressed by ≥70 % with a full recovery to their original shape within 30 s. In contrast, PPy hydrogels that were not subject to the ageing process did not exhibit any elastic properties. The authors suggested that the origin of the elastic behaviour as a result of ageing is due to a secondary growth mechanism. Initially, polypyrrole forms immediately upon addition of the oxidant [68]. The insoluble polymers cluster together into spherical particles, then interconnect via π–π interactions leading to a 3D network. Over time, unreacted pyrrole monomers diffuse and oxidatively couple to the surface of these

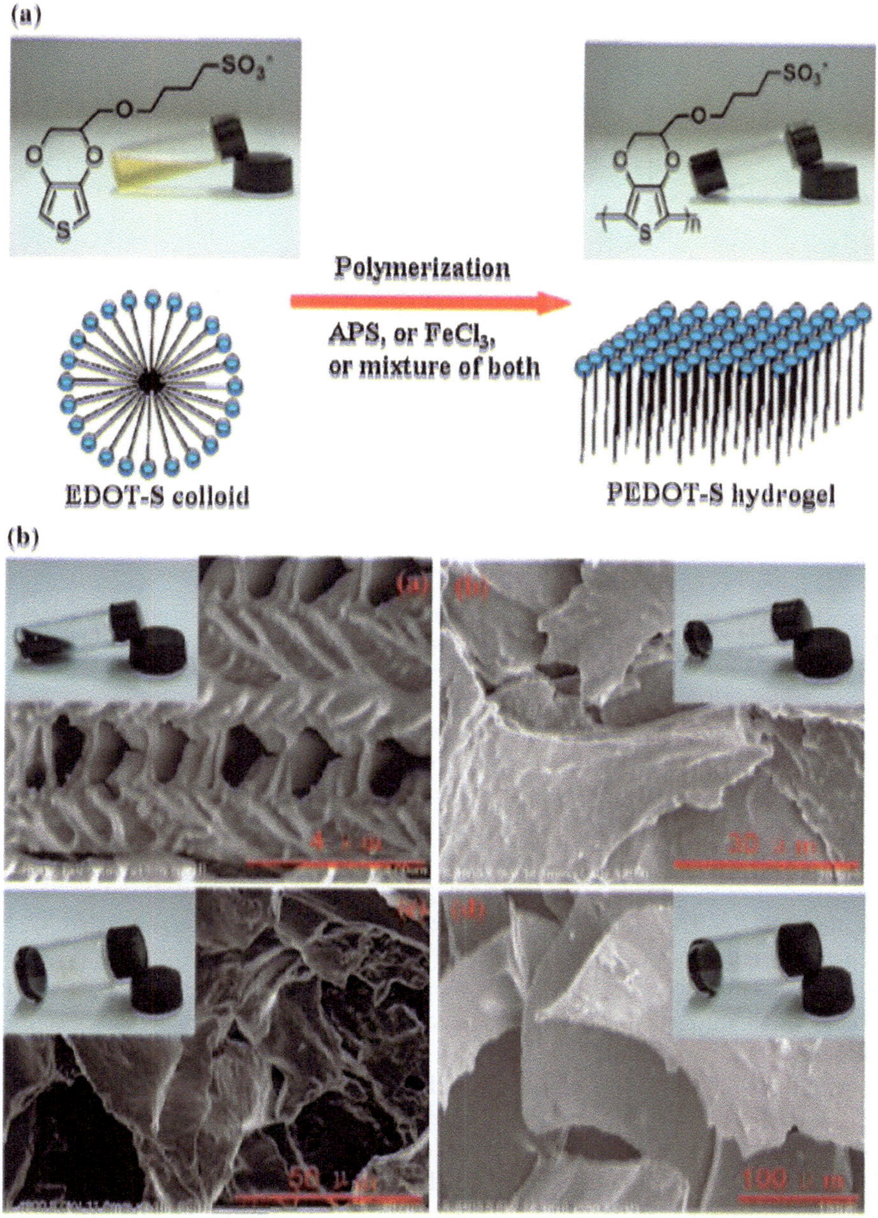

Fig. 6 **a** Schematic representation of the formation mechanism of PEDOT-S hydrogels. The amphiphilic monomers, EDOT-S, form micelles in solution, which upon addition of the oxidant convert into 3D hydrogel network. **b** SEM micrographs showing the internal structure of the gels synthesised by APS (a&b), $FeCl_3$ (**c**), and a mixture of APS and $FeCl_3$ (**d**). Reprinted from [64]. Copyright 2011, with permission of Royal Society of Chemistry

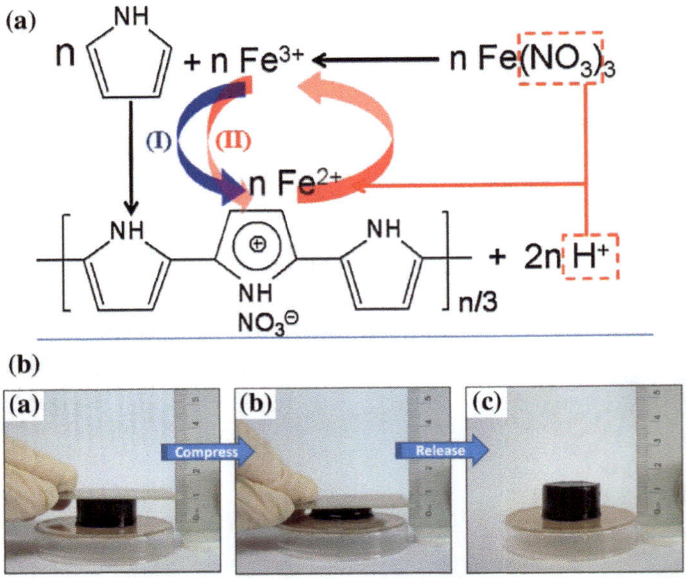

Fig. 7 **a** The polymerization process of polypyrrole by the addition of ferric nitrate. **b** Representative images showing the full recovery of the PPy hydrogels following compression by ≥70 %. Reprinted from [67]. Copyright 2014, with permission of Nature Publishing Group

particles forming protruded branches that eventually join together into building blocks. With ageing, a slow reaction process causes reinforcement of these joint branches, leading to a hydrogel that is less prone to fracturing and of high elasticity. The authors demonstrated the applicability of these elastic hydrogels for the efficient and fast removal of organic molecules from aqueous environments. Using methyl orange as a model dye, the hydrogel exhibited a high absorption efficiency of the dye (99.99 %) in just several seconds. Of significance is the ability to repeatedly reuse and refresh the hydrogel by simply treating it with 2 M NaOH at 80 °C. This highly elastic PPy hydrogel paves the way for the fabrication of elastic conducting hydrogels that have potential application in environmental engineering, biosensors and regenerative medicines.

Other PPy-based single component CP hydrogels have been fabricated by the self-assembly of PPy chains in the presence of sodium 4-[4′-(dimethylamino) phenyldiazo] phenylsulfonate (MO) [69, 70]. MO has a hydrophobic core with hydrophilic sulfonic groups at the side. The hydrophobic core stacks with neighbouring molecules causing MO to self-associate in aqueous media. Thus, MO forms cylindrical structure in water and can be used as a template for the polymerization of PPy microtubules [70]. Under static conditions, these microtubules can aggregate by van der Waals forces or chemical bonding resulting in a PPy conductive hydrogel [69]. The morphology and swelling/deswelling behaviour of these mesoscale networks have been studied in response to the type ($FeCl_3$, $Fe_2(SO_4)_3$, and $Fe(NO_3)_3$) and ratio of oxidant to the pyrrole monomer during synthesis. SEM micrographs showed clearly that the morphology is highly dependent on the type of oxidant employed, with $Fe(NO_3)_3$ resulting in a more coarse and granulated network. This in turn had an effect on improving the swelling/deswelling properties of the hydrogel with repeated cycles. Additionally, the electrical conductivities were correlated to the hydrated state of the hydrogel prepared using $Fe(NO_3)_3$ as an oxidant. When the gel was shrunk, it exhibited a conductivity of 1.7×10^{-2} S/cm, in comparison to a value of 8.2×10^{-3} S/cm when swollen. The authors suggested that the higher conductivity in the shrunk state is due to the closer proximity between the polymeric chains facilitating electron transport. Also, they demonstrated that the conductivity of the hydrogels could be further tailored by controlling the pH of the media.

3 Carbon-Based Conductive Hydrogels

Other organic-based conductive materials are being explored in biomedical applications. These include materials such as graphene and CNTs that combine both excellent electrical properties with superior mechanical characteristics. For tissue engineering, these two materials are being incorporated in 3D hydrogel networks to fabricate conductive scaffolds for various applications [71, 72].

3.1 Graphene

Graphene and its derivatives, graphene oxide (GO) and reduced graphene oxide (rGO), are emerging materials that have attracted much attention in research due to their unique properties such as high thermal [73] and electrical conductivity [74], remarkable mechanical properties [75] and superior optical transmittance [76]. As such, graphene is being investigated for a range of applications including biological fuel cells, electrochemical biosensors and tissue engineering. A recent review by Wallace et al. [77] presents a comprehensive overview of graphene and its application across the biomedical field. For the fabrication of ECHs, graphene has attracted much attention as filler improving elasticity and the mechanical strength of the hydrogel, while imparting electric conductivity in the system. Similar to CPs, conductive hydrogels based on graphene or its derivatives are either a hybrid or a "single component".

Hybrid hydrogels have been fabricated from graphene or its derivatives and insulating polymers such as polypropylene oxide (PPO)—polyethylene oxide [78], hydroxyapatite [79], polyvinyl alcohol [80], and chitosan [81]. Primarily, graphene has been added to enhance the mechanical properties of these hydrogels. "Single component" graphene-based hydrogels are formed via self-assembly of graphene sheets producing a 3D network [82–84]. The first graphene-based hydrogel reported by Xu et al. [82] was fabricated via a one-step hydrothermal method (Fig. 8).

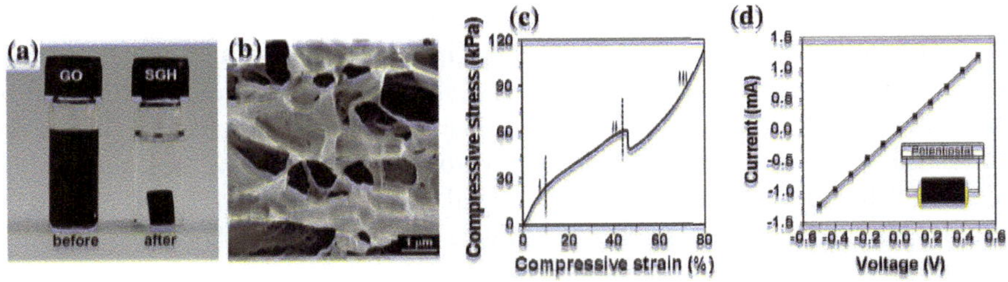

Fig. 8 **a** Fabrication of self-assembled graphene hydrogel (SGH) from a GO dispersion by photothermal reduction. **b** SEM micrograph of the porous internal structure of the SGH. **c** Compressive stress-strain curve of the SGH exhibiting 3 regions: (*I*) linear-elastic regime, (*II*) long plastic regime and (*III*) densification regime. **d** *I–V* curve of the SGH exhibiting an ohmic behaviour at room temperature, measured by the 2-probe technique (shown in the *Inset*). Reprinted from [82]. Copyright 2010, with permission of American chemical Society

A graphene oxide aqueous dispersion was heated up to 180 °C for a period of time causing the graphene sheets to self assemble into a mechanically robust 3D network hydrogel of high water content (97.4 % wt/wt) and elastic behaviour (Fig. 8c). By increasing the heating time from 1 to 12 h, the compressive modulus was varied from 29 ± 3 to 290 ± 20 kPa. Similarly, the conductivity (Fig. 8d) was increased from 0.23 ± 0.02 to 4.9 ± 0.2 mS/cm. In a similar mechanism, chemically converted graphene (CCG), prepared by chemical reduction of GO, has been shown to form a hydrogel film via a simple vacuum filtration process [83]. Gelation occurs during filtration as a result of the deposition of the CCG in a sheet-by-sheet fashion. The obtained product is a highly hydrated hydrogel film (92 % wt/wt) with remarkable conductivity of 0.87 S/cm in the swollen state. These graphene-based hydrogels combine critical properties such as mechanical strength, high electric conductivity and anisotropy due to sheets alignment.

3.2 *Carbon Nanotubes (CNTs)*

The rationale behind the incorporation of CNTs in hydrogel networks is to design engineered tissue constructs with properties closer to those of native electroresponsive tissue: mechanical integrity [85], nanofibrous architecture [86] and electric conductivity [87]. Subsequently, CNTs have been extensively investigated as additives into hydrogel networks; in particular, these hybrid hydrogels are being extensively investigated in cardiac tissue regeneration [88–90]. Gelatin hydrogels containing SWNTs [90] have been fabricated and tested both in vitro and in vivo as potential conductive hydrogels for the treatment of myocardium infarction (MI). SWNTs were mixed with gelatin solution and crosslinked with glutaraldehyde to produce the hydrogel. SEM revealed that the SWNTs were homogeneously distributed inside the porous structure. Despite the low conductivity of these hydrogels ($\sim 1 \times 10^{-6}$ S/cm), field stimulation revealed that neonatal cardiac cells seeded on the SWNT/gelatin hydrogel became more compact and in closer proximity in comparison to the control gelatin scaffold. For the first time, carbon nanotube-based hydrogels were tested in vivo for cardiac regeneration. The hydrogels fused with the myocardium infarct causing its regeneration and remodelling. A CNT/gelatin hydrogel exhibiting anisotropic electric conductivity has been fabricated by Ahadian et al. [91]. This could be achieved by vertically aligning the CNTs within the hydrogel matrix through the application of dielectrophoresis (DEP). This anisotropy led to significant increase in the myogenic genes and proteins of electrically stimulated C2C12 cells cultured on the gels.

While CNTs are showing promising properties in tissue engineering, their biocompatibility remains under debate with reports in the literature showing either positive or negative effects [92]. Their toxicity is mainly explained by their small size, which causes increased reactivity and potential inflammation and toxicity. However, the mechanisms by which they induce toxic effects remain to be fully investigated and confirmed [93]. Additionally, it appears that the dosage of the CNTs introduced in vivo plays an important role in the triggered inflammatory reaction and the dosage threshold remains to be identified [94]. However, the attractive properties of CNTs have pushed researchers to investigate ways to enhance their biocompatibility. Functionalization of CNTs with carboxylic groups appears to alleviate this issue. Carboxylated CNTs have been shown to degrade by the enzyme, myeloperoxidase (MPO) into non-inflammatory degradation products [95].

4 Applications of Conducting Polymer Hydrogels

The possibility to develop a crosslinked network that combines porosity, mechanical integrity, high water content with good electroactive properties is driving the use of conducting polymer hydrogels in a wide range of applications including biosensors, drug delivery and tissue engineering.

Electrochemical biosensors are based on an enzymatic reaction triggered on the surface of an electrode that functions as the sensor substrate. Since enzymes are redox active, the reaction either results in electron transfer across the double layer of the electrode producing a current or alters the double layer potential generating a voltage [96]. CPs have been used as the interface between the electrode and the analyte due to their redox active nature. Their doping/dedoping mechanism leads to a change in surface resistance, current or electrochemical potential that could be monitored as a response in relation to analyte concentrations [97]. Additionally, CPs combine both electronic and ionic conductivities lowering the impedance between the electrode and the analyte interface. Subsequently, both ECHs [98, 99] and single component CP hydrogels [62, 63] have attracted particular attention as the interface in electrochemical biosensors since they combine the electro-response of the CP and the ability of the hydrogel to retain bioactive molecules of interest. The CP facilitates the electron transport across the interface [100], the porous structure provides a large surface area and shorter diffusion length [101], and the high water content exhibited by the hydrogel component enhances the biocompatibility [1].

Conducting polymer hydrogels can serve as responsive drug delivery systems that undergo chemical or physical transitions under applied electric fields. By changing the redox state of the conducting polymer, the charge on the backbone can be varied as well as the volume of the CP scaffold. This in turn can be used to deliver drugs in a controlled manner. Anionic drugs could be easily loaded in the CP during synthesis. Following polymerization, CPs is in the oxidised form and the backbone is positively charged. These positive charges are balanced with anions to maintain electro-neutrality. Thus, negatively charged drugs can serve as the counterion and could be loaded during the polymerization process. Upon reduction, the CP backbone converts from a positively charged state to a neutral form; thus loosing the drug anion and causing its release [102]. Cationic drugs could also be loaded in CP networks. This could be achieved by electrostatic and hydrophobic interaction between the drug, the CP backbone and the anionic counter ion [103]. Alternatively, the cationic drug can be loaded post synthesis and following reduction of the CP network. Highest loading efficiencies could be achieved if the counterion has a large molecular weight. When the CP is reduced, it looses its positive charges allowing the cationic drugs to be attracted to the polymer backbone and to interact with the negatively charged counterion physically trapped inside the matrix [104, 105]. However, the application of conducting polymer hydrogels in delivery systems is limited by the low loading efficiency that could be achieved and the passive diffusion of the small counterions from the network.

One important application of conducting polymer hydrogel is tissue engineering. These systems are ideal candidates to serve as a 3D network for cell seeding and regeneration. The hydrogel component provides the hydrated environment needed, the porosity for exchange of nutrients, and the mechanical integrity to support the cells and promote their adhesion. On the other hand, the CP component provides the electronic communication required in electroresponsive tissues such as the heart, brain, muscle and nerve [20, 22]. Conductive hydrogels have shown promising results in regenerative medicine [106] and as highlighted throughout this chapter, many of the developed ECHs promote cell adhesion and proliferation; however, their long-term use remains limited by the degradation in their electronic properties after incubation in physiological conditions due to dopant loss.

5 Conclusion and Future Perspectives

In this chapter, the synthesis and use of conducting polymer hydrogels for biomedical applications have been reviewed. Two types of hydrogels were highlighted based on their fabrication routes. ECHs or hybrid networks are formed by incorporating the conducting material in insulating hydrophilic networks. Single component conducting hydrogels are formed by either self-assembly of the conjugated polymeric chains or by modifying the CP with water soluble and chemically crosslinkable moieties.

Until now, the incorporation of CPs in a prefabricated hydrogel network has dominated the field of conductive hydrogels. Advances in fabrication techniques have enabled smart multifunctional conductive hydrogels with tailored architectures to promote cell alignment, release of biomolecules under demand, and growth and differentiation in response to electrical stimulation. The synthesis of single component conducting polymer hydrogels offers the possibility of tissue-engineered scaffolds with enhanced electrical properties. Designing a network free of an insulating matrix but made from conjugated chains improves the efficiency of electron transport in the system. Additionally, chemical binding or self-assembly of the CPs chains results in a more stable network that is resilient to environmental changes such as pH. With the advent of smart chemistries, the possibility to design single component conductive scaffolds with tailored properties more suited to tissue engineering should be more feasible. The field will greatly benefit from conjugated polymers that are water soluble, regioregular, have good electron transport and contain side functional groups that could be either decorated with relevant biomolecules, used as crosslinking sites, or modified with biodegradable linkers.

The search for the ideal tissue-engineered scaffold remains on-going. Concerning the introduction of electronic properties, CPs along with graphene and CNTs are the most promising candidates due to their flexibility, processability and functionalization. However, this field is still in its infancy and much work is required to develop a practical biomedical device with superior electronics and one that could be used in the clinic.

Acknowledgements D.M. would like to acknowledge the Marie Curie International Incoming Fellowship for financial support. G.G.W. acknowledges the support of an ARC Australian Laureate Fellowship.

References

1. Annabi N, Tamayol A, Uquillas JA, Akbari M, Bertassoni LE, Cha C, Camci-Unal G, Dokmeci MR, Peppas NA, Khademhosseini A (2014) 25th anniversary article: Rational design and applications of hydrogels in regenerative medicine. Adv Mater 26:85–123
2. Bendrea AD, Cianga L, Cianga I (2011) Progress in the field of conducting polymers for tissue engineering applications. J Biomater Appl 26:3–84
3. Mawad D, Boughton EA, Boughton P, Lauto A (2012) Advances in hydrogels applied to degenerative diseases. Curr Pharm Des 18:2558–2575
4. Wijsboom YH, Patra A, Zade SS, Sheynin Y, Li M, Shimon LJ, Bendikov M (2009) Controlling rigidity and planarity in conjugated polymers: poly(3,4-ethylenedithioselenophene). Angew Chem Int Ed 48:5443–5447
5. Amrutha SR, Jayakannan M (2008) Probing the pi-stacking induced molecular aggregation in pi-conjugated polymers, oligomers, and their blends of p-phenylenevinylenes. J Phys Chem B 112:1119–1129

6. Shirakawa SR, Louis EJ, MacDiarmid AG, Chiang CK, Heeger AJ (1977) Synthesis of electrically conducting organic polymers: halogen derivatives of polyacetylene, $(CH)_x$. J Chem Soc Chem Commun 16:578–580
7. Heeger AJ, Kivelson S, Schrieffer JR, Su WP (1988) Solitons in conducting polymers. Rev Mod Phys 60:781–850
8. Malliaras G, Friend R (2005) An organic electronics primer. Phys Today 58:53–58
9. Harrison WA (1979) Solid state theory. Dover Publications Inc, New York
10. Dai L (1999) Conjugated and fullerene-containing polymers for electronic and photonic applications: advanced syntheses and microlithographic fabrications. J Macromol Sci Rev Macromol Chem Phys C 39:273–387
11. van der Pauw LJ (1958) A method of measuring specific resistivity and Hall effect of discs of arbitrary shape. Philips Res Rep 13:1–9
12. Smits FM (1957) Measurement of sheet resistivities with the four-point probe. Bell Syst Tech J 37:711–718
13. Giri G, Verploegen E, Mansfield SCB, Atahan-Evrenk S, Kim DH, Lee SY, Becerril HA, Aspuru-Guzik A, Toney MF, Bao Z (2011) Tuning charge transport in solution-sheared organic semiconductors using lattice strain. Nature 480:504–508
14. Lai W, Haile SM (2005) Impedance spectroscopy as a tool for chemical and electrochemical analysis of mixed conductors: a case study of ceria. J Am Ceram Soc 88:2979–2997
15. Rubinson JF, Kayinamura YP (2009) Charge transport in conducting polymers: insights from impedance spectroscopy. Chem Soc Rev 38:3339–3347
16. Bobacka J, Lewenstam A, Ivaska A (2000) Electrochemical impedance spectroscopy of oxidized poly(3,4-ethylenedioxythiophene) film electrodes in aqueous solutions. J Electroanal Chem 489:17–27
17. Gilmore K, Hodgson AJ, Luan B, Small CJ, Wallace GG (1994) Preparation of hydrogel/conducting polymer composites. Polym Gels Netw 2:135–143
18. Guiseppi-Elie A (2010) Electroconductive hydrogels: synthesis, characterization and biomedical applications. Biomaterials 31:2701–2716
19. Green RA, Baek S, Poole-Warren LA, Martens PJ (2010) Conducting polymer-hydrogels for medical electrode applications. Sci Technol Adv Mater 11:014107
20. Hardy JG, Lee JY, Schmidt CE (2013) Biomimetic conducting polymer-based tissue scaffolds. Curr Opin Biotechnol 24:847–854
21. Balint R, Cassidy NJ, Cartmell SH (2014) Conductive polymers: towards a smart biomaterial for tissue engineering. Acta Biomater 10:2341–2353
22. Molino PJ, Wallace GG (2015) Next generation bioelectronics: advances in fabrication coupled with clever chemistries enable the effective integration of biomaterials and organic conductors. APL Mater 3:014913
23. Hur J, Im K, Kim SW, Kim J, Chung DY, Kim TH, Jo KH, Hahn JH, Bao Z, Hwang S, Park N (2014) Polypyrrole/Agarose-based electronically conductive and reversibly restorable hydrogel. ACS Nano 8:10066–10076
24. Park S, Yang G, Madduri N, Abidian MR, Majd S (2014) Hydrogel-mediated direct patterning of conducting polymer films with multiple surface chemistries. Adv Mater 26:2782–2787
25. Kotanen CN, Wilson AN, Dong C, Dinu CZ, Justin GA, Guiseppi-Elie A (2013) The effect of the physicochemical properties of bioactive electroconductive hydrogels on the growth and proliferation of attachment dependent cells. Biomaterials 34:6318–6327
26. Huang H, Wu J, Lin X, Li L, Shang S, Yuen MC, Yan G (2013) Self-assembly of polypyrrole/chitosan composite hydrogels. Carbohydr Polym 95:72–76
27. Kotanen CN, Tlili C, Guiseppi-Elie A (2012) Bioactive electroconductive hydrogels: the effects of electropolymerization charge density on the storage stability of an enzyme-based biosensor. Appl Biochem Biotechnol 166:878–888
28. Runge MB, Dadsetan M, Baltrusaitis J, Ruesink T, Lu L, Windebank AJ, Yaszemski MJ (2010) Development of electrically conductive oligo(polyethylene glycol) fumarate-polypyrrole hydrogels for nerve regeneration. Biomacromolecules 11:2845–2853

29. Justin G, Guiseppi-Elie A (2009) Characterization of electroconductive blends of poly (HEMA-co-PEGMA-co-HMMA-co-SPMA) and poly(Py-co-PyBA). Biomacromolecules 10:2539–2549
30. Chansai P, Sirivat A, Niamlang S, Chotpattananont D, Viravaidya-Pasuwat K (2009) Controlled transdermal iontophoresis of sulfosalicylic acid from polypyrrole/poly(acrylic acid) hydrogel. Int J Pharm 381:25–33
31. Li Y, Neoh KG, Kang ET (2005) Controlled release of heparin from polypyrrole-poly(vinyl alcohol) assembly by electrical stimulation. J Biomed Mater Res A 73:171–181
32. Hao GP, Hippauf F, Oschatz M, Wisser FM, Leifert A, Nickel W, Mohamed-Noriega N, Zheng Z, Kaskel S (2014) Stretchable and semitransparent conductive hybrid hydrogels for flexible supercapacitors. ACS Nano 8:7138–7146
33. Vega-Rios A, Olmedo-Martínez JL, Farías-Mancilla B, Hernández-Escobar CA, Zaragoza-Contreras EA (2014) Synthesis and electrical properties of polyaniline/iota-carrageenan biocomposites. Carbohydr Polym 110:78–86
34. Ding H, Zhong M, Kim YJ, Pholpabu P, Balasubramanian A, Hu CM, He H, Yang H, Matyjaszewski K, Bettinger CJ (2014) Biologically derived soft conducting hydrogels using heparin-doped polymer networks. ACS Nano 8:4348–4357
35. Ma D, Zhang LM (2013) Novel biosensing platform based on self-assembled supramolecular hydrogel. Mater Sci Eng C Mater Biol Appl 33:2632–2638
36. Guarino V, Alvarez-Perez MA, Borriello A, Napolitano T, Ambrosio L (2013) Conductive PANi/PEGDA macroporous hydrogels for nerve regeneration. Adv Healthc Mater 2:218–227
37. Bayramoglu G, Altintas B, Arica MY (2013) Immobilization of glucoamylase onto polyaniline-grafted magnetic hydrogel via adsorption and adsorption/cross-linking. Appl Microbiol Biotechnol 97:1149–1159
38. Marcasuzaa P, Reynaud S, Ehrenfeld F, Khoukh A, Desbrieres J (2010) Chitosan-graft-polyaniline-based hydrogels: elaboration and properties. Biomacromolecules 11:1684–1691
39. Paradee N, Sirivat A (2014) Electrically controlled release of benzoic acid from poly (3,4-ethylenedioxythiophene)/alginate matrix: effect of conductive poly (3,4-ethylenedioxythiophene) morphology. J Phys Chem B 118:9263–9271
40. Sasaki M, Karikkineth BC, Nagamine K, Kaji H, Torimitsu K, Nishizawa M (2014) Highly conductive stretchable and biocompatible electrode-hydrogel hybrids for advanced tissue engineering. Adv Healthc Mater 3:1919–1927
41. Abidian MR, Daneshvar ED, Egeland BM, Kipke DR, Cederna PS, Urbanchek MG (2012) Hybrid conducting polymer-hydrogel conduits for axonal growth and neural tissue engineering. Adv Healthc Mater 1:762–767
42. Chikar JA, Hendricks JL, Richardson-Burns SM, Raphael Y, Pfingst BE, Martin DC (2012) The use of a dual PEDOT and RGD-functionalized alginate hydrogel coating to provide sustained drug delivery and improved cochlear implant function. Biomaterials 33:1982–1990
43. Lu Y, Li Y, Pan J, Wei P, Liu N, Wu B, Cheng J, Lu C, Wang L (2012) Poly (3,4-ethylenedioxythiophene)/poly(styrenesulfonate)-poly(vinyl alcohol)/poly(acrylic acid) interpenetrating polymer networks for improving optrode-neural tissue interface in optogenetics. Biomaterials 33:378–394
44. Nagamine K, Kawashima T, Sekine S, Ido Y, Kanzaki M, Nishizawa M (2011) Spatiotemporally controlled contraction of micropatterned skeletal muscle cells on a hydrogel sheet. Lab Chip 11:513–517
45. Sekine S, Ido Y, Miyake T, Nagamine K, Nishizawa M (2010) Conducting polymer electrodes printed on hydrogel. J Am Chem Soc 132:13174–13175
46. Kim DH, Wiler JA, Anderson DJ, Kipke DR, Martin DC (2010) Conducting polymers on hydrogel-coated neural electrode provide sensitive neural recordings in auditory cortex. Acta Biomater 6:57–62
47. Naficy S, Razal JM, Spinks GM, Wallace GG, Whitten PG (2012) Electrically conductive, tough hydrogels with pH sensitivity. Chem Mater 24:3425–3433

48. Engler AJ, Sen S, Sweeney HL, Discher DE (2006) Matrix elasticity directs stem cell lineage specification. Cell 126:677–689
49. Sirivisoot S, Pareta R, Harrison BS (2013) Protocol and cell responses in three-dimensional conductive collagen gel scaffolds with conductive polymer nanofibres for tissue regeneration. Interface Focus 4:20130050
50. Lin J, Tang Q, Wu J, Li Q (2010) A multifunctional hydrogel with high-conductivity, pH-responsive, and release properties from polyacrylate/polypyrrole. J Appl Polym Sci 116:1376–1383
51. Tang Q, Wu J, Sun H, Lin J, Fan S, Hu D (2008) Polyaniline/polyacrylamide conducting composite hydrogel with a porous structure. Carbohydr Polym 74:215–219
52. Barbarella G, Melucci M, Sotgiu G (2005) The versatile thiophene: an overview of recent research on thiophene-based materials. Adv Mater 17:1581–1593
53. Das S, Chatterjee DP, Ghosh R, Nandi AK (2015) Water soluble polythiophenes: preparation and applications. RSC Adv 5:20160–20177
54. Kim BS, Chen L, Gong JP, Osada Y (1999) Titration behavior and spectral transitions of water-soluble polythiophene carboxylic acids. Macromolecules 32:3964–3969
55. Chen L, Kim BS, Nishino M, Gong JP, Osada Y (2000) Environmental responses of polythiophene hydrogels. Macromolecules 33:1232–1236
56. Mawad D, Stewart E, Officer DL, Romeo T, Wagner P, Wagner K, Wallace GG (2012) A single component conducting polymer hydrogel as a scaffold for tissue engineering. Adv Funct Mater 22:2692–2699
57. Mawad D, Gilmore K, Molino P, Wagner K, Wagner P, Officer DL, Wallace GG (2011) An erodible polythiophenebased composite for biomedical applications. J Mater Chem 21:5555–5560
58. Asberg P, Bjork P, Hook F, Inganas O (2005) Hydrogels from a water-soluble zwitterionic polythiophene: dynamics under pH change and biomolecular interactions observed using quartz crystal microbalance with dissipation monitoring. Langmuir 21:7292–7298
59. Hanabusa K, Suzuki M (2014) Development of low-molecular-weight gelators and polymer-based gelators. Polym J 46:776–782
60. Bairi P, Chakraborty P, Shit A, Mondal S, Roy B, Nandi AK (2014) A co-assembled gel of a pyromellitic dianhydride derivative and polyaniline with optoelectronic and photovoltaic properties. Langmuir 30:7547–7555
61. Chakraborty P, Bairi P, Mondal S, Nandi AK (2014) Co-assembled conductive hydrogel of N-fluorenylmethoxycarbonyl phenylalanine with polyaniline. J Phys Chem B 118:13969–13980
62. Pan L, Yu G, Zhai D, Lee HR, Zhao W, Liu N, Wang H, Tee BC, Shi Y, Cui Y, Bao Z (2012) Hierarchical nanostructured conducting polymer hydrogel with high electrochemical activity. PNAS 109:9287–9292
63. Zhai D, Liu B, Shi Y, Pan L, Wang Y, Li W, Zhang R, Yu G (2013) Highly sensitive glucose sensor based on pt nanoparticle/polyaniline hydrogel heterostructures. ACS Nano 7:3540–3546
64. Du R, Xu Y, Luo Y, Zhang X, Zhang J (2011) Synthesis of conducting polymer hydrogels with 2D building blocks and their potential-dependent gel-sol transitions. Chem Commun 47:6287–6289
65. Du R, Zhang X (2012) Alkoxysulfonate-functionalized poly(3,4 - ethylenedioxythiophene) hydrogels. Acta Phys Chim Sin 28:2305–2314
66. Chang X, Chen D, Jiao X (2010) Starch-derived carbon aerogels with high-performance for sorption of cationic dyes. Polymer 51:3801–3807
67. Lu Y, He W, Cao T, Guo H, Zhang Y, Li Q, Shao Z, Cui Y, Zhang X (2014) Elastic, conductive, polymeric hydrogels and sponges. Sci Rep 4:5792
68. Myers RE (1986) Chemical oxidative polymerization as a synthetic route to electrically conducting polypyrroles. J Electron Mater 15:61–69
69. Wei D, Lin X, Li L, Shang S, Yuen MC, Yan G, Yu X (2013) Controlled growth of polypyrrole hydrogels. Soft Matter 9:2832–2836

70. Antonio J, Lira LM, Goncales VR, Cordoba de Torresi SI (2013) Fully conducting hydro-sponges with electro-swelling properties tuned by synthetic parameters. Electrochim Acta 101:216–224
71. Gao H, Duan H (2015) 2D and 3D graphene materials: preparation and bioelectrochemical applications. Biosens Bioelectron 65:404–419
72. Cirillo G, Hampel S, Spizzirri UG, Parisi OI, Iemma F (2014) Carbon nanotubes hybrid hydrogels in drug delivery: a perspective review. Biomed Res Int 2014, article ID 825017
73. Pop E, Varshney V, Roy AK (2012) Thermal properties of graphene: fundamentals and applications. MRS Bull 37:1273–1281
74. Wu ZS, Ren W, Gao L, Zhao J, Chen Z, Liu B, Tang D, Yu B, Jiang C, Cheng HM (2009) Synthesis of graphene sheets with high electrical conductivity and good thermal stability by hydrogen arc discharge exfoliation. ACS Nano 3:411–417
75. Lee C, Wei X, Kysar JW, Hone J (2008) Measurement of the elastic properties and intrinsic strength of monolayer graphene. Science 321:385–388
76. Li X, Zhu Y, Cai W, Borysiak M, Han B, Chen D, Piner RD, Colombo L, Ruoff RS (2009) Transfer of large-area graphene films for high-performance transparent conductive electrodes. Nano Lett 9:4359–4363
77. Thompson BC, Murray E, Wallace GG (2015) Graphite oxide to graphene. Biomaterials to bionics. Adv Mater. doi: 10.1002/adma.201500411
78. Lee Y, Bae JW, Hoang Thi TT, Park KM, Park KD (2015) Injectable and mechanically robust 4-arm PPO-PEO/graphene oxide composite hydrogels for biomedical applications. Chem Commun 51:8876–8879
79. Xie X, Hu K, Fang D, Shang L, Tranc SD, Cerruti M (2015) Graphene and hydroxyapatite self-assemble into homogeneous, free standing nanocomposite hydrogels for bone tissue engineering. Nanoscale 7:7992–8002
80. Du G, Nie L, Gao G, Sun Y, Hou R, Zhang H, Chen T, Fu J (2015) Tough and biocompatible hydrogels based on in situ interpenetrating networks of dithiol-connected graphene oxide and poly(vinyl alcohol). ACS Appl Mater Inter 7:3003–3008
81. Sayyar S, Murray E, Thompson BC, Chung J, Officer DL, Gambhir S, Spinks GM, Wallace GG (2015) Processable conducting graphene/chitosan hydrogels for tissue engineering. J Mater Chem B 3:481–490
82. Xu Y, Sheng K, Li C, Shi G (2010) Self-assembled graphene hydrogel via a one-step hydrothermal process. ACS Nano 4:4324–4330
83. Yang X, Qiu L, Cheng C, Wu Y, Ma ZF, Li D (2011) Ordered gelation of chemically converted graphene for next-generation electroconductive hydrogel films. Angew Chem Int Ed 50:7325–7328
84. Li D, Muller MB, Gilje S, Kaner RB, Wallace GG (2008) Processable aqueous dispersions of graphene nanosheets. Nat Nanotechnol 3:101–105
85. Chatterjee S, Lee MW, Woo SH (2009) Enhanced mechanical strength of chitosan hydrogel beads by impregnation with carbon nanotubes. Carbon 47:2933–2936
86. Lee CK, Shin SR, Mun JY, Han SS, So I, Jeon JH, Kang TM, Kim SI, Whitten PG, Wallace GG, Spinks GM, Kim SJ (2009) Tough supersoft sponge fibers with tunable stiffness from a DNA self-assembly technique. Angew Chem Int Ed 48:5116–5120
87. Ramón-Azcón J, Ahadian S, Estili M, Liang X, Ostrovidov S, Kaji H, Shiku H, Ramalingam M, Nakajima K, Sakka Y, Khademhosseini A, Matsue T (2013) Dielectrophoretically aligned carbon nanotubes to control electrical and mechanical properties of hydrogels to fabricate contractile muscle myofibers. Adv Mater 25:4028–4034
88. Shin SR, Jung SM, Zalabany M, Kim K, Zorlutuna P, Kim S, Nikkhah M, Khabiry M, Azize M, Kong J, Wan K, Palacios T, Dokmeci MR, Bae H, Tang X, Khademhosseini A (2013) Carbon-nanotube-embedded hydrogel sheets for engineering cardiac constructs and bioactuators. ACS Nano 7:2369
89. Liau B, Zhang D, Bursca N (2012) Functional cardiac tissue engineering. Regen Med 7:187–206

90. Zhou J, Chen J, Sun H, Qiu X, Mou Y, Liu Z, Zhao Y, Li X, Han Y, Duan C, Tang R, Wang C, Zhong W, Liu J, Luo Y, Xing M, Wang C (2013) Engineering the heart: evaluation of conductive nanomaterials for improving implant integration and cardiac function. Sci Rep 4:3733
91. Ahadian S, Ramon-Azcon J, Estili M, Liang X, Ostrovidov S, Shiku H, Ramalingam M, Nkjima K, Sakka Y, Bae H, Matsue T, Khademhosseini A (2014) Hybrid hydrogels containing vertically aligned carbon nanotubes with anisotropic electrical conductivity for muscle myofiber fabrication. Sci Rep 4:4271
92. Kostarelo K (2008) The long and short of carbon nanotube toxicity. Nat Biotechnol 26:774–776
93. Rodriguez-Yanez Y, Munoz B, Albores A (2013) Mechanisms of toxicity by carbon nanotubes. Toxicol Mech Methods 23:178–195
94. von Der Mark K, Park J, Bauer S, Schmuki P (2010) Nanoscale engineering of biomimetic surfaces: cues from the extracellular matrix. Cell Tissue Res 339:131–153
95. Kagan VE, Konduru NV, Feng W, Allen BL, Conroy J, Volkov Y, Vlasova II, Belikova NA, Yanamala N, Kapralov A, Tyurina YY, Shi J, Kisin ER, Murray AR, Franks J, Stolz D, Gou P, Klein-Seetharaman J, Fadeel B, Star A, Shvedova AA (2010) Carbon nanotubes degraded by neutrophil myeloperoxidase induce less pulmonary inflammation. Nat Nanotechnol 5:354–359
96. Grieshaber D, MacKenzie R, Voros J, Reimhult E (2008) Electrochemical biosensors—sensor principles and architectures. Sensors 8:1400–1458
97. Li L, Shi Y, Pan L, Shia Y, Yu G (2015) Rational design and applications of conducting polymer hydrogels as electrochemical biosensors. J Mater Chem B 3:2920–2930
98. Asberg P, Inganas O (2003) Hydrogels of a conducting conjugated polymer as 3-D enzyme electrode. Biosense Bioelectron 19:199–207
99. Mano N, Yoo JE, Tarver J, Loo YL, Heller A (2007) An electron-conducting cross-linked polyaniline-based redox hydrogel, formed in one step at pH 7.2, wires glucose oxidase. J Am Chem Soc 129:7006–7007
100. Singh RP (2012) Prospects of organic conducting polymer modified electrodes: enzymosensors. Int J Electrochem 2012, article ID 502707
101. Lupu S, Lete C, Balaure PC, Caval DI, Mihailciuc C, Lakard B, Hihn JY, del Campo FJ (2013) Development of amperometric biosensors based on nanostructured tyrosinase-conducting polymer composite electrodes. Sensors 13:6759–6774
102. Svirskis D, Travas-Sejdic J, Rodgers A, Garg S (2010) Electrochemically controlled drug delivery based on intrinsically conducting polymers. J Control Release 146:6–15
103. Thompson BC, Moulton SE, Ding J, Richardson R, Cameron A, O'Leary S, Wallace GG, Clark GM (2006) Optimising the incorporation and release of a neurotrophic factor using conducting polypyrrole. J Control Release 116:285–294
104. Zhou QX, Miller LL, Valentine JR (1989) Electrochemically controlled binding and release of protonated dimethyldopamine and other cations from poly(N-methylpyrrole)/ polyanion composite redox polymers. J Electroanal Chem 261:147–164
105. Barthus RC, Lira LM, Córdoba de Torresi SI (2008) Conducting polymer- hydrogel blends for electrochemically controlled drug release devices. J Braz Chem Soc 19:630–636
106. Guo B, Lei B, Li P, Ma PX (2015) Functionalized scaffolds to enhance tissue regeneration. Regen Biomater 2:47–57

Polysaccharide-Based Hydrogels as Biomaterials

Tejraj M. Aminabhavi and Anand S. Deshmukh

Abstract The ever-increasing interest to utilize renewable polysaccharide-based hydrogels as biomaterials has created renewed interest in many disciplines including biomedicine, bioengineering, pharmacy, chemistry, and materials science. The volume of literature published in this area is quite extensive as the diversity of these materials seeks novel applications. The polysaccharide-based hydrogels as smart biomaterials have attracted much interest in drug delivery, bioengineering, and electronics domain. Recent advances in micro- and nanobiotechnology have led to renewed interest for targeting drugs, genes, and other biotherapeutics like proteins, small interfering RNA (siRNA), and peptides. These applications have progressed exponentially due to their similarities with soft tissue body components as well as being responsive to external stimuli like temperature, pH, electric and magnetic fields. This chapter covers recent developments and advances in hydrogels derived from natural polysaccharides as biomaterials.

Keywords Polysaccharides · Nanotechnology · Drug delivery · Bioengineering · Hydrogels

Abbreviations

3D	Three-dimensional
AG	Agarose
ATDC5	Chondrocytic cell line
Az-chitosan	4-azidobenzamide grafted chitosan
BMP-2	Bone morphogenetic protein-2
bMSCs	Bovine mesenchymal stem cells
BMSCs	Bone marrow stem cells
BSA	Bovine serum albumin

T.M. Aminabhavi (✉) · A.S. Deshmukh
Department of Pharmaceutics, Shree Dhanvantary Pharmacy College, Near Kim Railway Station, Kim (E), Kudsad Road, Surat, Gujarat 394110, India
e-mail: aminabhavit@gmail.com; aminabhavi@yahoo.com

S. Kalia (ed.), *Polymeric Hydrogels as Smart Biomaterials*,
Springer Series on Polymer and Composite Materials,
DOI 10.1007/978-3-319-25322-0_3

CG	Carrageenan
CS	Chitosan
CT1	Connexin-43 carboxyl-terminus mimetic peptide
DD	Degree of deacetylation
DMF	Dimethyl formamide
DMSO	Dimethyl sulphoxide
DS	Dextran sulfate
ECM	Extracellular matrix
EE	Entrapment efficiency
EGFP	Enhanced green fluorescent protein
full-IPN	Full-interpolymeric network
GA	Glutaraldehyde
GAGs	Mammalian glycosaminoglycans
GAGs	Glycosaminoglycans
G-CSF	Granulocyte colony-stimulating factor
GFP	Green fluorescent protein
GM-CSF	Granulocyte-macrophage colony-stimulating factor
GRAS	Generally recognized as safe
HA	Hyaluronic acid
hASCs	Human adipose-derived stem cells
hNCs	Human nasal chondrocytes
HP	Heparin
k-carrageenan	*Kappa*-Carrageenan
MW	Molecular weight
MYO	*B*-galactosidase, myoglobin
NaAlg	Sodium alginate
NPs	Nanoparticles
PEG	Polyethylene glycol
PLGA	Poly(lactic-*co*-glycolic acid) copolymer
rhGH	Recombinant human growth hormone
semi-IPN	Semi-interpolymeric network
SLN	Solid lipid nanoparticles
US-FDA	United States Food and Drug Administration
VBL	Visible blue light

1 Introduction

In today's scientific scenario, researchers are trying to understand the importance of biocompatible and bioerodible materials of natural origin for designing and developing pharmaceuticals. Natural polysaccharides are the attractive candidates owing to their abundant availability at lower cost, amenability to modifications, structural diversity, biodegradability, and biocompatibility. Hydrogels, due to their structural similarity with extracellular matrix (ECM) of the biological system, offer excellent platform for proliferation, adhesion, and delivery of biologics and cells. Published literature on polysaccharide hydrogels as biomaterials has witnessed a tremendous growth globally. Despite rich flora and fauna, very limited numbers of polysaccharides have actually made their path from laboratory to clinical practice. Reasons for this drawback are many, though regulatory clearance of such materials has been the major hurdle. Researchers are aware of the directives of these governing bodies and are now expanding their horizons from basic to clinical practice. This chapter will cover a discussion on the most widely used polysaccharides that are used in clinical practice or have been accepted worldwide in formulation development.

2 Polysaccharides and Hydrogels—General Consideration

Polysaccharides are found in nature in a wide range of intricate structural arrangements formed by repeating saccharide units (such as mono-, di-, tri- or tetrasaccharides) and joined together by glycosidic bonds, which upon hydrolysis yield more than ten molecules of monosaccharides. Even though these structures are mostly linear, they may have different degree of branching. Polysaccharides are often heterogeneous and composed of repeating units of slightly modified monosaccharide units. These different structural arrays of their building blocks impart distinct physicochemical or mechanical properties. Furthermore, their solubility in water is largely governed by amorphous or crystalline molecular arrangements of the monosaccharide units. From the chemistry view point, polysaccharides have the general formula: $C_x(H_2O)_y$, where x is usually a large number varying between 200 and 2500, considering that repeating units in polymer backbone are often six-carbon monosaccharides. The general formula can also be represented as: $(C_6H_{10}O_5)_n$, where n may vary between 40 and 3000. If all monosaccharide units in a polysaccharide structure are of similar type, it is homopolysaccharide (e.g., starch, glycogen, cellulose, and pectin), but if more than one type of saccharide units are present, they are referred to as heteropolysaccharides (e.g., hyaluronic acid, chondrotin, alginate, xyloglucan, ghatti, etc.).

In nature, polysaccharide plays a very vital role in different arrays of life. For instance, cellulose provides basic structure for plants, chitin serves as exoskeleton of crawfish and shrimps, collagen helps in mechanical support in connective tissues, and silk in spider's webs, etc. [1, 2]. Principally, it is the versatility of their chemical structures that allows their crafting as advanced functionalized materials capable of meeting many requirements. The unparalleled properties of polysaccharides such as nontoxicity of monomer residues, high water solubility, and swelling capacity make them desired materials for pharmaceutical and medical applications [3]. Specifically, in biomedical research, their biodegradation into biocompatible physiological metabolites makes them excellent candidates in drug delivery and regenerative medicine. Hyaluronic acid (HA) [4, 5], cellulose [6, 7], carboxymethylcellulose [8], chitosan (CS) [9, 10], alginate [11, 12], starch [13], carrageenan (CG) [14], dextran (DEX), agarose (AG), and pullulan have received extensive technological acceptance due to their favorable characteristics and gelling nature.

Polysaccharides, being hydrophilic, are attractive alternatives to synthetic polymers, often used to develop hydrogels that are three-dimensional (3D) network structures capable of retaining a large quantity of water and remain insoluble due to their physical and/or chemical crosslinking matrix composition. Since their development in the early 1960s [15], innumerable studies have been made on hydrogels advocating their wide-ranging applications in pharmaceutical, agricultural, and bioengineering areas [16–19]. Hydrogels accommodate a large quantity of water into their water-filling porous structures and simulate soft network structure, resembling to natural ECM and minimize tissue irritation or cell adherence [20]. High loads of water-soluble therapeutically active proteins, peptides, siRNA, DNA, vaccines, etc., can be suitably encapsulated into these 3D networks. Unlike other drug delivery systems such as microparticles and microemulsions, where preparative conditions are detrimental to protein stability (due to the use of organic solvents and involvement of protein denaturing steps like homogenization, exposure to interfaces, shear forces, etc.), hydrogels can be prepared by simple approaches under mild water-based conditions to maintain protein's integrity.

Hydrogels can be produced either by chemical crosslinking or through physical entanglements. In chemically crosslinked structures, permanent junctions and/or covalent linkages are formed to build 3D network, where polymeric chains retain their shape irreversibly. In physically entangled network structure, transient junctions exist, arising from either chain entanglements or physical interactions (such as van der Waals interactions, ionic interactions, H-bond or hydrophobic interactions) and are reversible. Hence, they offer unique sol-to-gel reversibility and such structural variations of hydrogels are displayed in Fig. 1.

3 Classification of Hydrogels

The unique physicochemical properties of hydrogels arise out of their diverse internal structural arrangements [21], which can be further classified into different classes as outlined in Table 1. The first and foremost classification of hydrogels depends upon their origin or source, which falls broadly into two groups: natural and synthetic [22]. Furthermore, based on other parameters, they are classified into various classes, such as based on methods of preparation hydrogels are divided into three major categories, viz., homopolymeric, heteropolymeric, and copolymeric [23]. In homopolymeric hydrogels, crosslinked polymer networks are derived from single species of polymer, whereas in heteropolymeric hydrogels, two polymeric chains (natural, synthetic, or their combinations) constitute crosslinked structure. If both polymers are crosslinked, then it will be a full-interpolymer network (full-IPN)

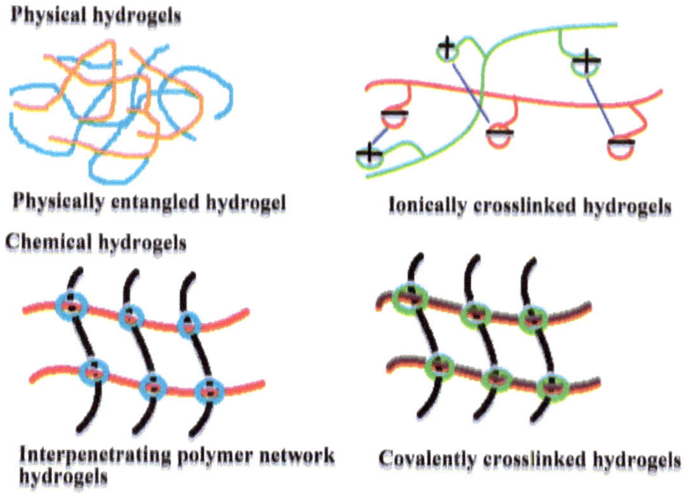

Fig. 1 Types of hydrogel 3D network formation structures

hydrogel and if only one polymer chain participates in crosslinking, while the other remains un-crosslinked, it is called as semi-IPN hydrogel [24].

In copolymeric hydrogels, two polymer chains are connected covalently as blocks, in a random fashion, in alternate order and as grafts [25]. Hydrogels are usually classified based on their sensitivity toward environmental stimulus. For instance, physical stimuli sensitive (such as temperature, electric or magnetic field, light, pressure, and sound) or chemical stimuli sensitive (like pH, solvent composition, ionic strength, and molecular species) are the governing parameters. Such hydrogels are also called "smart materials" as they exhibit dramatic changes in swelling and shrinking behavior with changes in stimuli level [26]. Hydrogels can also be classified into four groups based on the presence or absence of electrical charges located on crosslinked network such as nonionic (contains no charge), ionic (carry either anionic or cationic charges), amphoteric (carrying both acidic and

Table 1 Classification of hydrogels based on various parameters

S. no.	Parameters	Type of hydrogels
1.	Source of origin	Natural and synthetic polymers
2.	Polymeric composition	Homopolymeric hydrogels, heteropolymeric hydrogels and copolymeric hydrogels
3.	Network structure	Physical and chemical hydrogels
4.	Sensitivity to stimulus	Physical and chemical stimuli sensitive hydrogels
5.	Charge of polymeric network	Ionic, nonionic, zwitterion, and amphoteric hydrogels
6.	Physical appearance	Micro/nanoparticle, matrix, film, and gels
7.	Configuration	Noncrystalline, semicrystalline, and crystalline

alkali groups such as gelatin), and zwitterionic, where both anionic and cationic groups are present on repeating units of a polymer. Hydrogels can be further classified based on their structural configurations or molecular arrangements on the basis of degree of crystallinity such as amorphous, semicrystalline, and crystalline. Finally, hydrogels can be also classified based on their physical appearance, i.e., final dosage forms such as microspheres, films, sheets, matrix, etc., which in turn depend upon the processing parameters and the method of preparation [27].

In the recent past, research on the utilization of polysaccharide-based hydrogels in biomedical areas has greatly increased. The varying role of hydrogels from simple swelling structure to smart delivery device capable of swelling and shrinking at predetermined rates according to environmental stimuli has greatly revolutionized and extended their applications in biopharmaceutical and bioengineering fields. In particular, development of novel controlled release (CR) systems that maintain therapeutic plasma concentration in the surrounding tissues, provides longer blood circulation time and release cargos through a highly regulated feedback mechanism, this has led to novel studies in this area.

4 Polysaccharide-Based Hydrogels as Biomaterials

Naturally occurring polysaccharides are used in the CR of therapeutics due to their versatile functionality, formation of conjugates or complexes with proteins and peptides that can offer considerable advantages in terms of resemblance to biological macromolecules (both from chemical and physical viewpoints) as these are recognized by cell surface receptors, affecting adhesion, spreading, and proliferation [3, 28]. In view of their similarity to ECM, polysaccharides avoid stimulation of chronic inflammation or immunological reactions and toxicity, which is often seen with several synthetic materials. This chapter will cover the details on the members of polysaccharides that are being widely investigated including those of sodium alginate (NaAlg), CG, AG, dextran sulfate (DS), HA, and CS. These in combination with other polymers (synthetic or natural) offer desired chemical and biological advantages [29]. This chapter will also cover discussions on polysaccharide-based hydrogels as biomaterials in tissue engineering, organ replacement, implants, blood substitutes, wound dressings, bone replacement composite materials, scaffolds, and semipermeable membranes (Table 2).

4.1 Chitosan

CS is a copolymer of β-(1 → 4)-linked 2-acetamido-2-deoxy-D-glucopyranose and 2-amino-2-deoxy-D-glucopyranose, obtained by alkaline deacetylation of chitin and is the main component of exoskeleton of crustaceans like shrimps and crabs. The molecular weight (MW) and degree of deacetylation (DD), which represents the

Table 2 Structure and applications of biomedical polysaccharides

S. no.	Polysaccharide	Structure	Major biomedical applications
1.	Chitosan	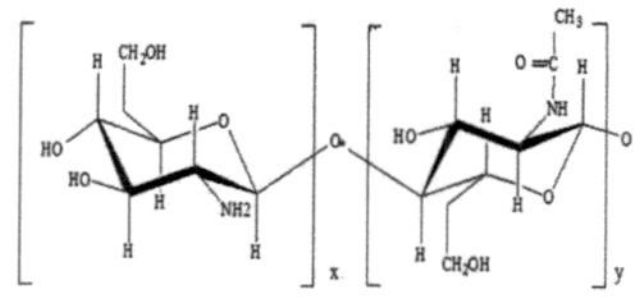	• Wound healing bandage [41] • Bioadhesives in peripheral neurosurgeries [43] • Tissue adhesive and hemostatic material [44]
2.	Sodium alginate	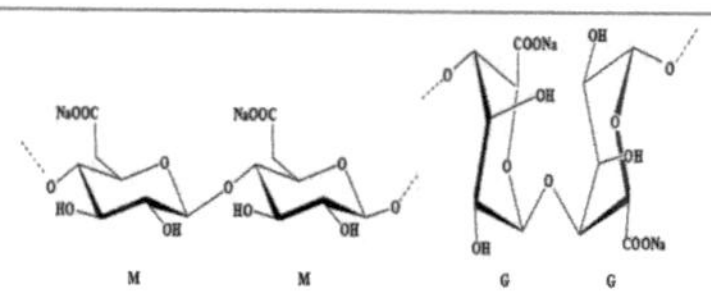	• Wound dressing scaffold [66] • 3D printed organs [68] • Prolonged release of therapeutic proteins [74]
3.	Carrageenan	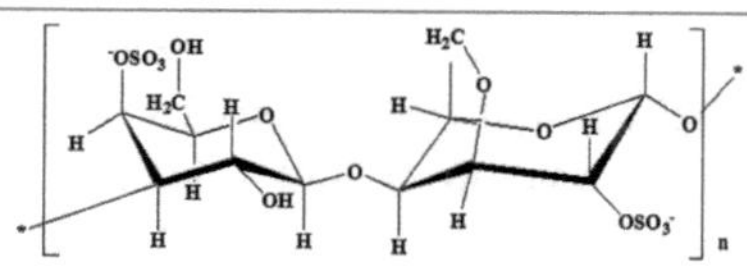	• Drug or growth factor delivery systems [79] • Cartilage regeneration [84] • Tissue engineering [85]
4.	Agarose	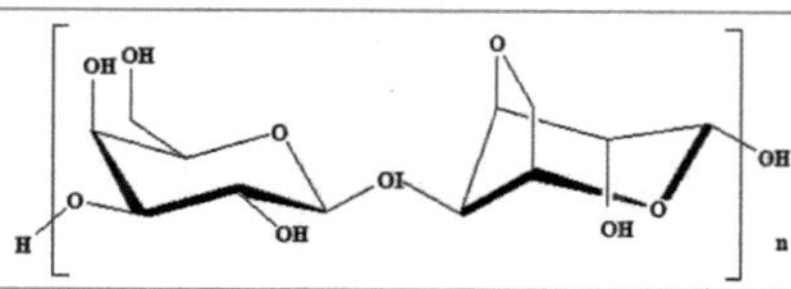	• Temporary scaffold for bony cells [94] • 3D scaffold for neural engineering [97]
5.	Hyaluronic acid	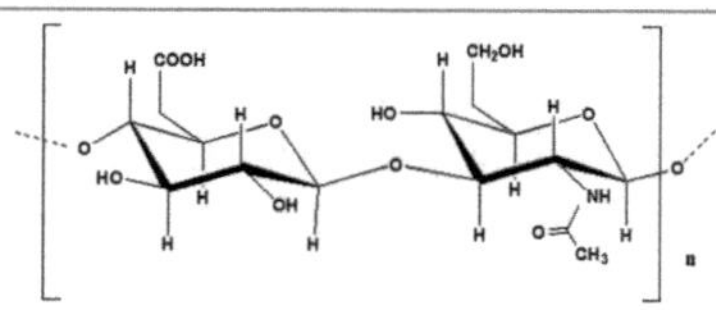	• Stem cell differentiation in vitro [104] • Nanofiber mimiking fibrous tissue architectures [114] • Bone regeneration implants [121]
6.	Dextran	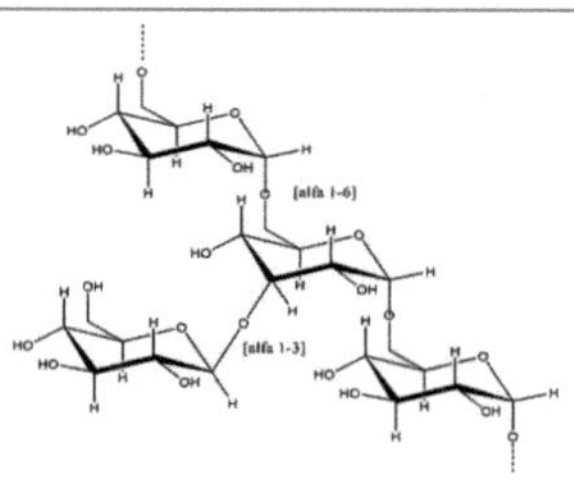	• Gene therapy [125] • Wound dressing, hemostatics and embolisation materials [127] • Cell proliferation [132] • Encapsulation of siRNA [136]

proportion of deacetylated units, greatly influences the characteristics of CS. The CS has many interesting properties like ease of modification, cationic charge at physiological pH, biocompatibility, bioadhesivity, and biodegradability. It is metabolized by certain human enzymes, especially by lysozyme, chitotriodidase, di-*N*-acetylchitobiase, and *N*-acetyl-β-D-glucosamineidase [30].

4.1.1 Chitosan-Based Hydrogels as Biomaterials

CS is widely studied in tissue engineering and drug delivery. The CS-based hydrogels are used as scaffolds for hepataocyte attachments and enzyme immobilization, as films for fabrication of amperometric glucose biosensor, as a material for supporting nerve repair, wound dressing, as implants and blood substitutes [31–36]. Techniques like ionic gelation, chemical crosslinking, and in situ gelation are used to produce hydrogels of CS. The cationic amine group of CS retains stability of hydrogels under mild physiological conditions (when injected into body) either by ionic gelation or by thermo reversible gelling [37]. This ability is utilized [38] to develop thermal and pH-responsive in situ hydrogels using inorganic phosphate salts as acid neutralizer and a gelling agent. This combination produces a cytocompatible hydrogel network capable of encapsulating MC3T3-E1 mouse osteoblast-like cells, which proves feasibility of CS hydrogel as a potential stem cell carrier.

The CS hydrogels are safer and cytocompatible materials but the solubility of CS at lower pH is a major challenge in terms of cell and growth factor encapsulation. To overcome these difficulties, a water-soluble carboxymethyl chitosan (CMCS) was proposed for drug delivery and tissue engineering [39]. To eliminate toxic crosslinkers, dextran dialdehyde was used for crosslinking CMCS hydrogel via imine bond to make it a useful scaffold for in vitro encapsulation of growth factors and stem cells in tissue regeneration [40]. The CS is also ideal wound dressing material to maintain moist environment at the wound interface, providing a cooling sensation, easy gaseous exchange, biodegradability and biocompatibility, allowing the absorption of wound exudates and maintaining the barrier to microorganisms. One of its major drawbacks is that in alkaline to neutral pH, the CS fails to act as a barrier to microorganisms and show poor antibacterial activity due to its positive charge; it has led to the modify the structure of CS.

Composite hydrogels of CS loaded with zinc oxide nanoparticles (Fig. 2) were prepared [41] for effective wound dressing to enhance wound healing and for fast re-epithelialization and collagen deposition at the wound site occurred due to burn, chronic wounds, and diabetic foot ulcers. Recently, enzyme-sensing hydrogels of CS modified with fluorogenic substrate as enzyme-selective reporting moiety are reported [42], wherein CS-based hydrogel was used for sensitive, fast, and effective detection of enzymes (for indirect detection of bacteria) compatible with infection-sensing wound dressings.

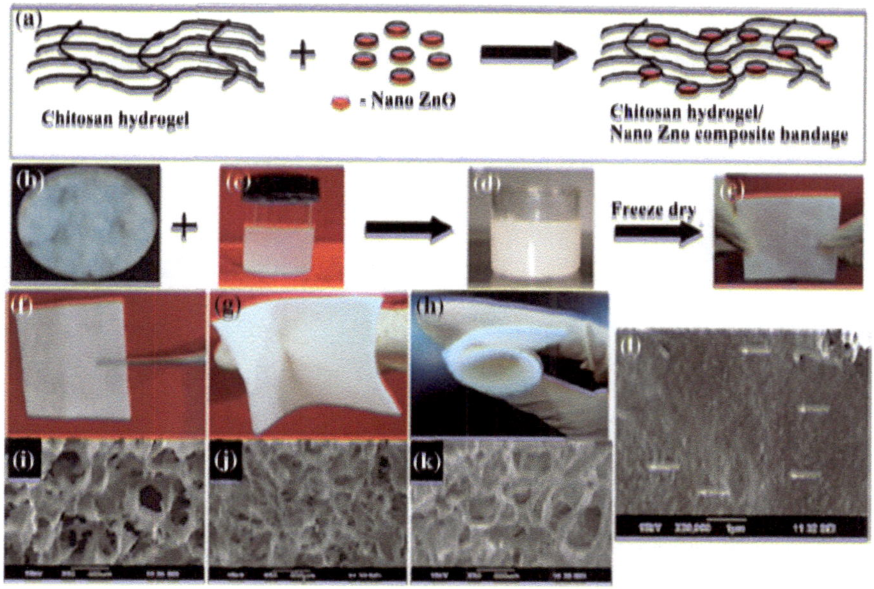

Fig. 2 **a** Schematic representation of CS hydrogel/nano ZnO composite bandage; **b**, **c**, **d** photographs of CS hydrogel, nano ZnO suspension, and CS hydrogel/nano ZnO mixture, respectively; **e**, **f**, **g**, **h** photographs of CS hydrogel/nano ZnO composite bandage; **i**, **j**, **k** SEM images of CS control, CS + 0.01 % nano ZnO, and CS + 0.005 % nano ZnO composite bandages, respectively; **l** SEM image of CS + 0.01 % nano ZnO composite bandage; *white arrows* indicate nano ZnO particles. Reprinted from [41]. Copyright © 2012, with permission from American Chemical Society

Developing surgical bioadhesive of CS for rapid wound healing needs modification of CS as noted in a study by Rickett et al. [43] who prepared UV photo-crosslinkable CS derivative, viz., 4-azidobenzamide grafted CS (Az-CS) for use as adhesive hydrogels for the treatment of peripheral nerve anastomosis. Similarly, CS/pluronic blend temperature-sensitive hydrogels were developed [44] as tissue adhesives and haemostatic materials showing solgel transition under physiological conditions. The CS was conjugated with multiple catechol groups and crosslinked with terminally thiolated Pluronic F-127 (Fig. 3) and this unique composite remained viscous at ambient temperature, but transformed into cross-linked hydrogel exhibiting excellent mechanical properties and stability (both in vitro and in vivo) at the body temperature and physiological pH. These data are undoubtedly leading to new applications of CS as antibleeding materials.

The CS-based hydrogels are also ideal scaffolds due to their resemblance to water-filled biological tissues. The CS/PLGA blend microsphere functionalized through heparin immobilization via attachment of osteoblast-like MC3T3-E1 cells proliferate the porous structure of the scaffolds for cell differentiation [45]. The scaffolds of CS-based hydrogels can reduce extracellular Ca^{2+} concentration to

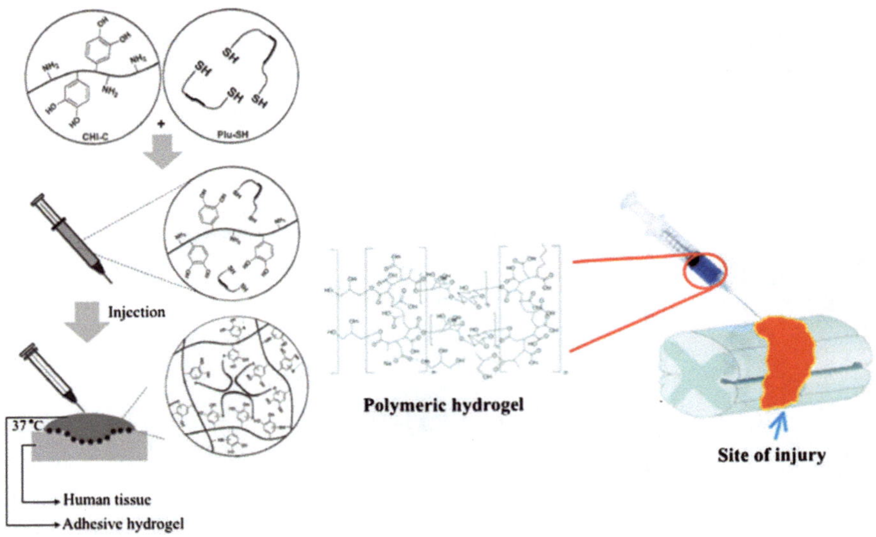

Fig. 3 Chitosan-pluronic-based hydrogels as antibleeding biomaterials. Reprinted from [44]. Copyright © 2011, with permission from American Chemical Society

reduce the secondary spinal cord injury cascades. Composite hydrogels of alginate, CS, and genipin were developed to interact with extracellular Ca^{2+} to initiate in situ gelation for producing a scaffold to maintain elastic modulus that is similar to the native spinal cord (≈1000 Pa). This property facilitates the CR of therapeutics to astrocyte inside an acutely injured spinal cord to regulate both astrocyte behavior and prevention of Ca^{2+} related secondary neuron damage during acute spinal cord injury [46].

The CS hydrogels are used in cartilage regeneration. In normal biological system, cartilaginous ECM components (type II collagen and chondroitin sulfate) play a crucial role in cartilage regeneration and use CS hydrogels as scaffolds for cartilage tissue engineering, which can be unsuccessful due to rapid enzymatic degradation. To improve regeneration of cartilage tissue, a novel photo-crosslinked injectable hydrogel of CS was prepared [47] by exposing CS to visible blue light (VBL) in the presence of riboflavin (Fig. 4) into which type II collagen and chondroitin sulfate can be incorporated to increase proliferation and deposition of cartilaginous ECM to promote cartilage regeneration.

The CS hydrogels are used in colon delivery of drugs [48] because at intestinal pH, CS deswells and causes insufficient delivery of therapeutics. This has opened up ways to develop more swellable polyelectrolyte complex of CS in combination with pectin and sodium alginate (NaAlg) along with water-soluble polyionic species [49, 50]. The nanoparticles of CS-NaAlg blend (850 nm size) produced by ionotropic gelation exhibit association efficiency of ≈81 % for insulin without changing its conformation (α-helix and β-sheet content of insulin) [51].

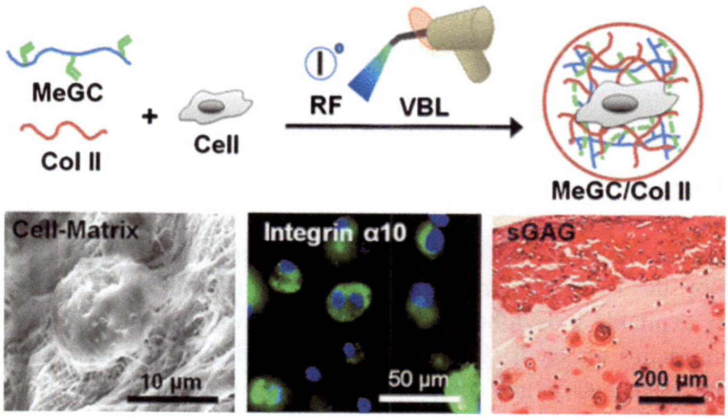

Fig. 4 ECM incorporated CS hydrogels for tissue repair. Reprinted from [47]. Copyright © 2014, with permission from American Chemical Society

Polyelectrolyte complexes of CS with other biopolymers [50, 52] such as folic acid to produce conjugated nanoparticles for targeted delivery of 5-aminolevulinic acid (5-ALA) and with tamarind kernel powder (interpolymer complex films) are used for colon targeting [53, 54]. CS upon crosslinking with polyethylene glycol (PEG) produces system swellable hydrogel in both acidic and alkaline pH media [55] and also using genipin as crosslinker [56].

Nanohydrogels of CS appended with ligand/aptamer are suitable for encapsulating negatively charged siRNA (via inter-polyelectrolyte complex phenomenon). The crosslinked CS nanoplexes (size <150 nm) were prepared [57] by ionic crosslinking at CS to siRNA mass ratio of 50:1 for encapsulating siRNA and delivering to lungs. Also, inter-polyelectrolyte complex of CS with siRNA showed a rapid uptake ($\approx$1 h) of Cy5-labeled nanoparticles into NIH 3T3 cells in about 24 h [58] that knocked down endogenous enhanced green fluorescent protein (EGFP) in both H1299 human lung carcinoma cells and murine peritoneal macrophages. Guanidinylated chitosan (GCS) hydrogel nanoplexes ($\sim$100 nm) with plasmid DNA at physiological pH exhibit lower cytotoxicity and higher transfection efficiency with about eightfold increase in cellular uptake [59, 60]. These systems after incorporating siRNA could deliver to lungs.

4.2 Seaweed-Based Polysaccharides (Alginate, Carrageenan, and Agarose) as Biomaterials

Algae-based polysaccharides like sodium alginate (NaAlg), AG, and CG are being extensively studied in biomedicine due to their free availability and ease of their fabrication [61] and these are gaining impetus in biomedical area due to the presence of functional groups like carboxylic acid and sulfate. This chapter will briefly discuss on the three major seaweeds, viz., alginate, CG, and AG hydrogels as biomaterials.

4.2.1 Alginate

Alginate is a naturally occurring anionic water-soluble polysaccharide first described by the British chemist E.C.C. Stanford in 1881. It is one of the most abundantly available biosynthesized materials, derived primarily from brown seaweed (comprising up to 40 % of dry matter) and bacteria. It is located in the intercellular matrix as a gel containing salts like sodium, calcium, magnesium, strontium, and barium. Because of its abilities like retention of water, gelling property, viscous nature, and stabilizing properties, alginate is widely used in biomedicine. Chemically, alginate contains blocks of (1–4)-linked β-D-mannuronic acid (M) and α-L-guluronic acid (G) monomers. Typically, these blocks are composed of three different forms of polymer segments: (a) consecutive G residues, (b) consecutive M residues, and (c) alternating MG residues [62].

Sodium Alginate-Based Hydrogels as Biomaterials

NaAlg is a widely used biopolymer in biomedical field as a supporting matrix for tissue repair and regeneration due to its nonantigenicity and chelating ability [63]. NaAlg hydrogels beads enclosing periosteum-derived chondrogenesis has been reported as superior scaffold capable of promoting the cartilaginous constructs useful for treatment of articular cartilage defects. The enclosed periosteal explants showed hyaline-like in appearance after 6 weeks in vitro (Fig. 5), suggesting its in vivo application as filling gels for treatment of partial- or full-thickness defects in articular cartilage. Further, in drug delivery area, NaAlg is used for its anionic property and used for targeting drugs at lower intestine since they exhibit pH responsivity due to presence of functional carboxyl groups and its high swelling is observed at increasing pH values due to chain expansion. Modification of synthetically derived NaAlg hydrogels has expanded its utility in therapeutic applications [64].

NaAlg-based biomaterials are particularly used in tissue engineering and as wound dressings like sponges, hydrogels, and electrospun mats that are promising substrates for wound healing due to advantages like hemostatic capability and gel-forming ability upon absorption of wound exudates. Similar to CS, the NaAlg possesses many critical elements desirable to develop wound dressing material. For instance, NaAlg dressings (like Kaltostat®) enhance wound healing through selective stimulation of monocytes to produce elevated levels of cytokines such as interleukin-6 and tumor necrosis factor-α [65]. Such wound dressings can prevent wound bed from drying out, giving better cosmetic repair of wounds. NaAlg-based

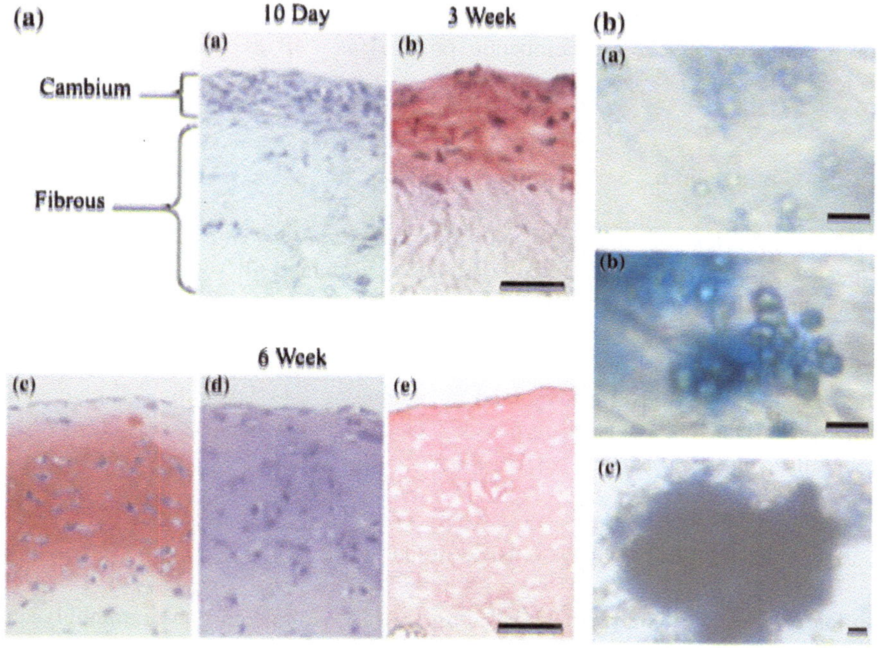

Fig. 5 **a** Periosteal explants cultured in alginate gel system in the presence of TGF-b1 after 10 days (*a*), 3 weeks (*b*), 6 weeks (*c–e*) in vitro culture; *a–c* safranin-O stain; *d* hematoxylin and eosin stain; *e* immunostain for type II collagen. Scale bar: 50 mm; **b** light micrographs of chondrocyte/alginate gel cultures fixed and stained with *alcian blue* after 4 days (*a*), 15 days (*b*), and 42 days (*c*) in vitro culture; *c* high concentration of cells is visualized as *dark region* in the center of the image. Scale bar: 20 mm. Reprinted from [63]. Copyright © 2004, with permission from Elsevier

dressings can avoid the secondary injury when dressing is peeled off, and is thus a popular alternative for the existing wound management dressings.

Many hybrid composites of NaAlg (tailored alginate) are produced by mixing with other materials with improved adhesion properties [66, 67]. Currently, silk-alginate-based hydrogels of defined molecular composition are used to support stem cell survival, its differentiation, and to mimic the host environment [68]. Since NaAlg is an unbranched polysaccharide consisting of 1-4′-linked β-D-mannuronic acid (M) and α-L-guluronic acid (G) moieties in varying compositions [69], it can form strong hydrogels by simple addition of metal ions to its aqueous solution, which is suitable for encapsulating living cells and study their CR properties. NaAlg hydrogels have the ability to regenerate tissues and simultaneously release growth factors or cytokines [70]. Insulin can be delivered through oral route using pH-responsive NaAlg-based nanoparticles prepared by spray drying, ionic crosslinking, electrohydrodynamic spraying, and solvent diffusion methods [71, 72]. Biodegradable PVA and NaAlg electrospun composite nanofiber-based transmucosal patches are used for sublingual delivery of insulin or BSA [57, 73].

Heparin incorporated photo-crosslinkable NaAlg hydrogels (HP-NaAlg) are used for prolonged delivery of therapeutic proteins as well as affinity-based growth factor [74].

4.2.2 Carrageenan

Different species of red seaweeds (family *Rhodophycae)* produce different polysaccharides such as CG, furcellaran, and agar, which fill voids within cellulose structure of the plant. The main species of *Rhodophycae* used in commercial production of CG are *Euchema cottonii* and *E. spinosum*. CG is a high MW linear polysaccharide comprising of repeating galactose units and 3,6-anhydrogalactose (both sulfated and nonsulfated) joined by alternating α-(1,3) and β-(1,4) glycosidic links [75]. The three different types of CG are lambda, kappa, and iota and are generally considered safe as per US-FDA.

Carrageenan-Based Hydrogels as Biomaterials

Sulfate groups in CG and chemical affinity with mammalian glycosaminoglycans are important to exhibit antiviral, anticoagulant, antioxidant, and anticancer activities [76]. CGs are soluble in water above 60 °C, which gel upon cooling to around 30–40 °C and are regarded as thermosensitive physical hydrogels [77]. CG hydrogels are widely used in drug or growth factor delivery [78, 79], immobilization of enzymes [80], various types of pharmaceuticals [81] and in cartilage regeneration [82]. The *kappa*-carrageenan (k-carrageenan) can be used [83] for encapsulating human adipose-derived stem cells (hASCs), human nasal chondrocytes (hNCs), and chondrocytic cell line (ATDC5). Popa et al. [84] developed simple ionically crosslinked CG hydrogels by adding cations for in situ cell matrix delivery. The photo-crosslinkable methacrylated derivative of k-carrageenan (MA-k-CA) with good elastic moduli can be used in tissue engineering [85]. The k-carrageenan hydrogel beads have been used for sustained release of growth factors [79], scaffold [86], and for the CR of parenteral formulations [87]. The oligosaccharides derived from both k-carrageenan and λ-carrageenan inhibit the growth of new vessels in chicken chorioallantoic membrane model [88].

4.2.3 Agarose

AG is a polysaccharide originated from sea algae, composed of repeating units of 3,6-anhydrous-L-galactose and D-galactose, and these produce "physical gels" due to presence of H-bonds. These hydrogels are porous reticulums that appear as a pyrogenic, colorless, and transparent at temperatures above 45 °C.

Agarose-Based Hydrogels as Biomaterials

AG hydrogels are used in numerous biocompatibility tests such as mutagenesis and sensitivity in subcutaneous implants [89], cytotoxity, genotoxicity, and as a substrate for cell growth [90, 91]. AG is used as a vehicle in a wide range of preclinical applications, such as bioengineering as a substrate for 3-D tissue growth [92], in gene therapy, and in the delivery of therapeutics [93]. AG is also used as a temporary scaffold for bony cells such as in orthopedic and oral–maxillofacial surgery as a biocompatible substrate enriched with osteoconductive particles for bone grafting/augmentation procedures as well as a bone spacer in guided tissue regeneration [94, 95]. An extrudable in situ hydrogel of AG was developed for subcutaneous implantation that compared well with established hydrogels of collagen and HA [96] and these are used as scaffolds in regenerative medicine. The in vitro cell study on the blends AG and CS [97] revealed increased neuron adhesion.

AG is widely studied for cartilage tissue engineering [98]. The articular chondrocytes seeded onto AG demonstrated enhanced chondrogenic matrix elaboration when cultured under physiological deformation loading, which accelerated the formation of a cartilage-like tissue [99]. The ability of AG to promote and maintain chondrogenic phenotype of bone marrow stem cells (BMSCs) with deposition of cartilage was also demonstrated [100]. Alginate and AG blend hydrogels are used to encapsulate cells and tissues in for protection from the host immune system [101]. The AG hydrogel in droplets can be suspended in liquid paraffin and gelated by cooling, followed by gelation of alginate in $CaCl_2$ solution. The mammalian cells enclosed in subsieve-size alginate–AG capsules (Fig. 6) showed mitochondrial

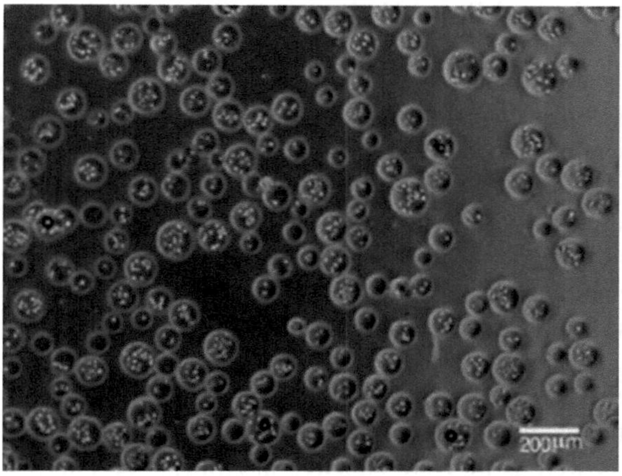

Fig. 6 SEM image of subsieve-size cell-enclosing noncoated alginate–agarose particles; white particles are seen in spherical vehicles (ca. 100 μm in dia) which represents individual cells. Reprinted from [101]. Copyright © 2006, with permission of Elsevier

activity in 27 days. Similarly, combination of AG with mesenchymal stem cells including human adipose-derived stem cells (hASCs) and bovine mesenchymal stem cells (bMSCs) are also investigated for cartilage repair.

5 Miscellaneous Types of Polysaccharides

5.1 Hyaluronic Acid

Among the glycosaminoglycans (GAGs), HA is a unique biopolymer produced and secreted from cells as a linear polymer unattached to a polypeptide. HA is distributed natively as a component of ECM in connective, epithelial, and neutral tissues in the body to help in regulating lubrication, water content for retaining structural properties, inter/intracellular communication, pathological processes, and shock absorption in synovial fluid [102]. Structurally, it comprised of alternating α-1,4-D-glucuronic acid and β-1,3-*N*-acetyl-D-glucosamine units linked together by β-1,3 linkages. Typically, HA MWs are above 1 million Da (10^7 Da) and are degraded by both reactive oxygen intermediates as well as enzyme hyaluronidase; HA is synthesized by macrophages, fibroblasts, and endothelial cells. HA has many important physicochemical and biological properties such as lubricity, viscoelasticity, water retention, biocompatibility, cell proliferation, morphogenesis, inflammation, and wound repair properties as well as specific signal transduction and cellular interactions through cell surface receptors such as CD44, CD54, and CD168 [103–106].

5.1.1 Hyaluronic Acid-Based Hydrogels as Biomaterials

HA is currently investigated as a bioactive material as plastic fillers to eliminate facial wrinkles, carrier to deliver stem cells, bioactive materials to treat specific diseases and scaffold for tissue engineering of bone, cartilage, blood vessel, and nerves. Since HA is produced and secreted from the cells as a linear polymer unattached to a polypeptide, its production is easy to adapt for genetic engineering areas using microbial fermentation technique [107]. These recombinant HAs are nonimmunogenic and are available in a wide range of well-defined MWs. Hydrogels based on HA have an interconnected porous structure that allows both transportation of (nutrition) and penetration of (cells, nerve fibers, and blood vessels). Therefore, HA-based hydrogel scaffolds are used as implants to enhance the neural regeneration. Implantation of HA scaffolds help to reduce glial scar formation [108] as well as they are effective in reducing scar formation to enhance neural regeneration in peripheral nervous system (PNS) [109] and also in central nervous system [110].

One of the major drawbacks of HA is its inability to adhere to biological membrane and hence, various modifications of its structure are attempted to improve its functionality. The abundance of primary and secondary hydroxyl groups, glucoronic acid's carboxylic acid, and *N*-acetyl group on HA are amenable for modification with reactive groups to form hydrogels. Generally, HA modification produces two types of HA: monolithic and living. Monolithic type HA is "terminally modified" form of HA, which may not form new chemical bonds in presence of added cells or molecules, but living HA derivatives can form new covalent bonds in presence of cells, tissues, and small or large molecules. Due to this feature, living HA derivatives are generally used in clinical and preclinical uses in 3D cell culture and in vivo cell delivery [111–114].

Heparin, a naturally occurring sulfated polysaccharide, isolated from animal tissues is most commonly used for incorporation into HA hydrogels due to its high degree of sulfation [115]. Researchers have attempted sulfation of HA directly through nucleophilic substitution of primary hydroxyl hydrogens with SO_3 in organic solvent [116–118] to improve its binding affinity. Recently, thiolated HA is gaining much importance as living hydrogels as they allow for easy incorporation of cells [119]. Due to viscoelastic properties of HA, it helps to reduce the friction between bones [120]. The photopolymerized HA hydrogels and HA–tyramine conjugates crosslinked by disulfide bond formation are used in delivering proteins and peptides [121]. HA is readily available in the market as an intraocular viscoelastic agent (Alcon Co., Texas, USA) under the trade name "Viscoat", and as a viscosupplementation for treatment of arthritic joints (Seikagaku Co., Tokyo, Japan). Fidia Pharmaceutical (USA), a subsidiary of Italian pharmaceutical giant, is also marketing HA with moderate molecular size as a viscosupplement.

Conjugation of HA with synthetic polymers can lead to a variety of assembled structures such as micelles, nanoparticles, and nanogels. Lee et al. [122] studied conjugation of HA and grafted HA with poly(lactic-*co*-glycolic acid) (PLGA) in the form of nanoparticles to deliver anticancer drugs. Other efforts include conjugation of Pluronic F127 di-acrylate with methacrylated HA via photo-crosslinking to produce thermosensitive hydrogels that release human growth hormone for over 13 days and plasmid DNA for over 10 days by inducing in vitro transfection. Grafting of amine-functionalized Pluronic® F127 onto methacrylated HA by carbodiimide coupling and subsequent photo-crosslinking with acrylated cell adhesion domains gave hydrogels, which when encapsulated with chondrocytes increased production of ECM proteins in cell culture studies [122].

5.2 Dextran

DEX is one of the most important polysaccharides used in biomedical area, which is produced by bacterial strains. DEX exists in nature without any relevant imperfection and shows a narrow MW distribution, which is advantageous for its chemical modification. It belongs to the family of neutral polysaccharides consisting of α-1,6 linked D-glucose main chain with varying proportions of linkages and branches, depending on the type of bacteria used. The α-1,6 linkages in DEX may vary between 97 and 50 % of the total glycosidic bonds and the balance represents α-1,2, α-1,3, and α-1,4 linkages [123]. Due to good solubility in water and various other solvents (DMSO, formamide, etc.,), DEX is used in many biomedical applications [124].

5.2.1 Dextran-Based Hydrogels as Biomaterials

Hydrogels prepared by homogeneous esterification of DEX with unsaturated carboxylic acids are used in drug delivery and as protective encapsulants for viruses used in gene therapy [125, 126]. Maleate esters of DEX are easily soluble in common organic solvents (DMSO, DMF, *N*-methyl-2-pyrrolidone and DMA), which can be converted into hydrogels by irradiation under long-wave UV light (≈365 nm). The DEX hydrogels are transparent and are useful as adhesion inhibitors when used as tissue adhesives and wound dressings [127].

Du et al. [128] and Tomme et al. [129] reviewed on DEX-based hydrogels for protein delivery. Crosslinking with peptide resulted in enzyme-dependent degradation controlled by cell-secreting enzymes, thus mimicking degradation of natural ECM [130]. The self-assembled nanoparticles of quaternized CS (*N*-(2-hydroxyl) propyl-3-trimethyl ammonium CS chloride (HTCC)) and DEX sulfate prepared by ionic gelation method showed rapid internalization of nanoparticles into Caco-2 cells without loss of cell viability as demonstrated by a fast release of therapeutics in pH 7.4 compared to slow release in pH 1.4 media [131]. DEX-based hydrogel nanoparticles (20–170 nm) have shown excellent Nb_2-11 cell proliferation [132, 133] confirming no protein aggregation or loss of bioactivity.

Delgado et al. [134] investigated a multicomponent delivery system based on DEX, protamine, and solid lipid nanoparticles (SLN) containing pCMS-EGFP plasmid for clathrin/claviolae-dependant transfection efficiency. These surface-modified SLNs of DEX are suitable for erythrocyte interaction and potential agglutination. Also, insulin-loaded nanoparticles (500 nm size) produced by complexation of DEX and CS [135] showed good stability at optimal composition of DEX:CS (mass ratio of 1.5:1) and these oral insulin delivery systems did not release insulin in pH 4.8 for up to 24 h, but released in pH of 6.8. The cationic character can be imparted to DEX by conjugating spermine to oxidized DEX by reductive amination [136]; these nanoparticles when encapsulated with *CXCR4*-siRNAs significantly downregulated *CXCR4* expression as tested in colorectal cancer metastasis in Balb/c mice through *CXCR4* silencing.

6 Application Areas of Polysaccharide-Based Biomaterials

The unique properties of polysaccharides make them suitable for the design of biomaterials in view of their inherent advantages over the synthetic polymeric materials. The presence of functional groups like carboxylic, amine, hydroxyl, and sulfhate, presents much wider scope for their structure modifications for better cell attachments (such as in case of sulfated HA), biodegradability, and tissue adhesion. Few of the promising areas of application of polysaccharide-based hydrogels are discussed here.

6.1 *Scaffold*

The 3D scaffolds help in cell adhesion, migration, differentiation, and proliferation providing guidance for the formation of new tissues. Hydrogels are useful scaffolding biomaterials because of their close resemblance to natural tissues. The aqueous environment of hydrogels mimics natural conditions of cells in the body and finds a special place in tissue engineering, especially in repair of cartilage, tendon, ligament, skin, blood vessels, and heart valves. Polysaccharides like AG, alginate, CS, collagen, fibrin, gelatin, and HA are widely investigated for designing scaffolds. Of all these, CS is the most widely used polymer due to its net positive charge, which readily interacts with anionic biological membranes, cells, and tissues. Polysaccharides such as silk-alginate-based hydrogels are also used to support stem cell survival, its differentiation, and to mimic the host environment.

6.2 *Cell Encapsulation*

Cell technology provides a promising therapeutic modality for diabetes, hemophilia, cancer, and renal failure [137, 138]. While using hydrogel-based cell encapsulation therapy, key features of hydrogels like biocompatibility, microporous structure, and minimal surface irritation within the surrounding tissues are important. These can be designed with the required porosity that can selectively restrict entrance of immune cells while allowing stimuli, oxygen, nutrients, and waste transfer through the pores. The major challenge in cell encapsulation therapy is selection of a suitable biomaterial membrane. Natural polysaccharide-based hydrogel is efficient as cell encapsulating biomaterial. The major issue using such crosslinked hydrogels is toxic residues of unreacted crosslinker, which causes the death of enclosed cells. To circumvent this, living type HA derivatives can be synthesized to form new covalent bonds in the presence of cells, tissues, and small or large molecules and these can be used in clinical and preclinical 3D cell cultures s well as in vivo cell delivery.

6.3 Wound Dressing Material

An ideal wound dressing material helps to maintain moist environment at the wound interface, providing a cooling sensation, easy gaseous exchange, biodegradability, and biocompatibility, allows the absorption of wound exudates, and maintains the barrier property to microorganisms. Many polysaccharide-based hydrogels meet these requirements as discussed in earlier parts of this chapter.

7 Conclusions

Promising results derived from natural polysaccharide-based hydrogels as biomaterials encouraged many researchers to exploit the natural resources for a wide range of biomedical applications. The biodegradability and biocompatibility of these materials are the major advantages with their natural tendency to retain water in their hydrogel network. The safety, better therapeutic outcomes, and protection of drug(s) are the most conducive environments for cell growth and proliferation, substrate ability for tissue and bone regeneration, etc. These unique features are important for a natural choice of these polymers as biomaterials. There are tremendous opportunities for utilization of these materials in the near future.

Acknowledgments Authors are grateful to the Board of Research in Nuclear Science [BRNS, Grant No.: 34(1)14/35/2014-BRNS], Mumbai, India for financial support.

References

1. Rudzinski WE, Aminabhavi TM (2010) Chitosan as a carrier for targeted delivery of small interfering RNA. Int J Pharm 399:1–11
2. Pasqui D, Cagna MD, Barbucci R (2012) Polysaccharide-based hydrogels: the key role of water in affecting mechanical properties. Polymer 4:1517–1534
3. Mundargi RC, Babu VR, Rangaswamy V, Patel P, Aminabhavi TM (2008) Nano/micro technologies for delivering macromolecular therapeutics using poly(d, l-lactide-co-glycolide) and its derivatives. J Control Release 125:193–209
4. Robert L, Robert AM, Renard G (2010) Biological effects of hyaluronan in connective tissues, eye, skin, venous wall role in aging. Pathol Biol 82:187–198
5. Cowman MK, Matsuoka S (2010) Experimental approaches to hyaluronan structures. Carbohydr Res 340:791–809
6. Luo Y, Wang S, Shen M, Qi R, Fang Y, Guo R, Cai H, Cao X, Tomás H, Zhu M, Shi X (2013) Carbon nanotube-incorporated multilayered cellulose acetate nanofibers for tissue engineering applications. Carbohydr Polym 91(1):419–427
7. Hoo SP, Loh QL, Yue Z, Fu J, Tan TTY, Choong C, Chan PPY (2013) Preparation of a soft and interconnected macroporous hydroxypropyl cellulose methacrylate scaffold for adipose tissue engineering. J Mater Chem. B 1(24):3107–3117
8. Ramli NA, Wong TW (2011) Sodium carboxymethylcellulose scaffolds and their physicochemical effects on partial thickness wound healing. Int J Pharm 403:73–82

9. Hirano S, Zhang M, Chung BS, Kim SK (2000) The N-acylation of chitosan fibre and the N-deacetylation of chitin fibre and chitin cellulose fibre at a solid state. Carbohydr Polym 41:175–179
10. Hirano S, Nagamura K, Zhang M, Kim SK, Chung BG, Yoshikawa M, Midorikawa T (1999) Chitosan staple fibres and their chemical modification with some aldehydes. Carbohydr Polym 38:293–298
11. Yu J, Du KT, Fang Q (2010) The use of human mesenchymal stem cells encapsulated in RGD modified alginate microspheres in the repair of myocardial infarction in the rat. Biomaterials 31:7012–7020
12. Man Y, Wang P, Guo Y, Xiang L, Yang Y, Qu Y, Gong P, Deng L (2012) Angiogenic and osteogenic potential of platelet-rich plasma and adipose-derived stem cell laden alginate microspheres. Biomaterials 33:8802–8811
13. Xu R, Feng X, Xie X, Xu H, Wu D, Xu L (2011) Grafted starch-encapsulated hemoglobin (GSEHb) artificial red blood cells substitutes. Biomacromolecules 12(5):1935
14. Nishinari K, Takahashi R (2003) Interaction in polysaccharide solutions and gels. Curr Opin Colloid Interface Sci 8:396–400
15. Wichterle O, Lim D (1960) Hydrophilic gels for biological use. Nature 185:117–118
16. Kim JI, Lee BS, Chun C, Cho JK, Kim SY, Song SC (2012) Long-term theranostic hydrogel system for solid tumors. Biomaterials 33(7):2251–2259
17. Hamidi M, Azadi A, Rafiei P (2008) Hydrogel nanoparticles in drug delivery. Adv Drug Deliv Rev 60:1638–1649
18. Rudzinski WE, Dave AM, Viashnav UH, Kumbar SG, Kulkarni AR, Aminabhavi TM (2002) Hydrogels as controlled release devices in agriculture. Des Monomers Polym 5:39–65
19. Ganguly K, Aminabhavi TM, Kulkarni AR (2011) Colon targeting of 5-fluorouracil using polyethylene glycol crosslinked chitosan microspheres enteric coated with cellulose acetate phthalate. Ind Eng Chem Res 50(21):11797–11807
20. Kashyap N, Kumar N, Ravi Kumar MNV (2005) Hydrogels for pharmaceutical and biomedical applications. Crit Rev Ther Drug Carrier Syst 22(2):107–150
21. Hacker MC, Mikos AG (2011) Synthetic polymers, principles of regenerative medicine, 2nd edn. Academic press, San Diego
22. Zhao W, Jin X, Cong Y, Liu Y, Fu J (2013) Degradable natural polymer hydrogels for articular cartilage tissue engineering. J Chem Technol Biotechnol 88(3):327–339
23. Takashi L, Hatsumi T, Makoto M, Takashi I, Takehiko G, Shuji S (2007) Synthesis of porous poly(N-isopropylacrylamide) gel beads by sedimentation polymerization and their morphology. J Appl Polym Sci 104(2):842
24. Yang L, Chu JS, Fix JA (2002) Colon-specific drug delivery: new approaches and in vitro/in vivo evaluation. Int J Pharm 235:1–15
25. Maolin Z, Jun L, Min Y, Hongfei H (2000) The swelling behaviour of radiation prepared semi-interpenetrating polymer networks composed of poly-NIPAAm and hydrophilic polymers. Radiat Phys Chem 58:397–400
26. Jinsub S, Paul VB, Wonmok L (2010) Fast response photonic crystal pH sensor based on templated photopolymerized hydrogel inverse opal. Sens Actuat B: Chem 150(1):183–190
27. Ahmed EM (2013) Hydrogel: preparation, characterization and applications. J Adv Res 6 (2):105–121
28. Kumar A, Sahoo B, Montpetit A, Behera S, Lockey R, Mohapatra S (2007) Development of hyaluronic acid-Fe_2O_3 hybrid magnetic nanoparticles for targeted delivery of peptides. Nanomed NBM 3:132–137
29. Kadokawa J (2011) Precision polysaccharide synthesis catalyzed by enzymes. Chem Rev 111:4308–4345
30. Berger J, Reist M, Mayer JM, Felt O, Peppas NA, Gurny R (2004) Structure and interactions in covalently and ionically crosslinked chitosan hydrogels for biomedical applications. Eur J Pharm Biopharm 57:19–34

31. Soppimath KS, Kulkarni AR, Aminabhavi TM, Rudzinski WE (2001) Microspheres as floating drug delivery systems to increase gastric retention of drugs. Drug Metab Rev 33 (2):149–160
32. Ramesh Babu V, Patel P, Mundargi RC, Rangaswamy V, Aminabhavi TM (2008) Developments in polymeric devices for oral insulin delivery. Expert Opin Drug Deliv 5:403–415
33. Chaturvedi K, Ganguly K, Nadagouda MN, Aminabhavi TM (2013) Polymeric hydrogels for oral insulin delivery. J Controlled Release 165:129–138
34. Aminabhavi TM, Nadagouda MN, More UA, Joshi SD, Kulkarni VH, Noolvi MN, Kulkarni PV (2014) Controlled release of therapeutics using interpenetrating polymeric networks. Expert Opin Drug Deliv 12(4):669–688
35. Ganguly K, Chaturvedi K, More UA, Nadagouda MN, Aminabhavi TM (2014) Polysaccharide-based micro/nanohydrogels for delivering macromolecular therapeutics. J Controlled Release 193:162–173
36. Aminabhavi TM, Nadagouda MN, Joshi SD, More UA (2014) Guar Gum as a platform for the oral controlled release of therapeutics. Expert Opin Drug Deliv 11:753–766
37. Bhattarai N, Ramay HR, Gunn J, Mastan FA, Zahn MJ (2005) PEG-grafted chitosan as an injectable thermosensitive hydrogel for sustained protein release. J Cont Release 103:609–624
38. Nair LS, Laurencin CT (2007) Biodegradable polymers as biomaterials. Prog Polym Sci 32 (8):762–798
39. Jayakumar R, Prabaharan M, Nair S, Tokura S, Tamura H, Selvamurugan N (2010) Novel carboxymethyl derivatives of chitin and chitosan materials and their biomedical applications. Progr Mater Sci 55:675–709
40. Cheng Y, Nada AA, Valmikinathan CM, Lee P, Liang D, Yu X, Kumbar SG (2014) In situ gelling polysaccharide-based hydrogel for cell and drug delivery in tissue engineering. J Appl Polym Sci 131:39934
41. Sudheesh PT, Lakshmanan VK, Anilkumar TV, Ramya C, Reshmi P, Unnikrishnan AG, Nair SV, Jayakumar R (2012) Flexible and microporous chitosan hydrogel/nano ZnO composite bandages for wound dressing: in vitro and in vivo evaluation. Appl Mater Interfaces 4:2618–2629
42. Ebrahimi MMS, Schönherr H (2014) Enzyme-Sensing chitosan hydrogels. Langmuir 30:7842–78501
43. Rickett TA, Amoozgar Z, Tuchek CA, Park J, Yeo Y, Shi R (2011) Rapidly photo-cross-linkable chitosan hydrogel for peripheral neurosurgeries. Biomacromolecules 12:57–65
44. Ryu JH, Lee Y, Kong WH, Kim TG, Park TG, Lee H (2011) Catechol-functionalized chitosan/pluronic hydrogels for tissue adhesives and hemostatic materials. Biomacromolecules 12:2653–2659
45. Jiang T, Khan Y, Nair LS, Abdel-Fattah WI, Laurencin CT (2010) Functionalization of chitosan/poly (lactic acid-glycolic acid) sintered microsphere scaffolds via surface heparinization for bone tissue engineering. J Biomed Mater Res A 93(3):1193–1208
46. McKay CA, Pomrenke RD, McLane JS, Schaub NJ, DeSimone EK, Ligon LA, Gilbert RJ (2014) An injectable, calcium responsive composite hydrogel for the treatment of acute spinal cord injury. Appl Mater Interfaces 6:1424–1438
47. Choi B, Kim S, Lin B, Wu BM, Lee M (2014) Cartilaginous extracellular matrix-modified chitosan hydrogels for cartilage tissue engineering. Appl Mater Interfaces 6:20110–20121
48. Zhang H, Alsarra IA, Neau SH (2002) An in vitro evaluation of a chitosan containing multiparticulate system for macromolecule delivery to the colon. Int J Pharm 239:197–205
49. Bigucci F, Luppi B, Cerchiara T, Sorrenti M, Bettinetti G, Rodriquez L, Zecchi V (2008) Chitosan/pectin polyelectrolyte complexes: Selection of suitable preparative conditions for colon-specific delivery of vancomycin. Eur J Pharm Sci 35:435–441

50. Assaad E, Wang YJ, Zhu XX, Mateescu MA (2011) Polyelectrolyte complex of carboxymethyl starch and chitosan as drug carrier for oral administration. Carbohydr Polym 84:1399–1407
51. Sarmento B, Ferreira DC, Jorgensen L, Van de Weert M (2007) Probing insulin's secondary structure after entrapment into alginate/chitosan nanoparticles. Eur J Pharm Biopharm 65:10–17
52. Du J, Dai J, Liu J, Dankovich T (2006) Novel pH-sensitive polyelectrolyte carboxymethyl konjac glucomannan-chitosan beads as drug carriers. React Funct Polym 66:1055–1061
53. Li P, Wang Y, Zeng F, Chen L, Peng Z, Kong LX (2011) Synthesis and characterization of folate conjugated chitosan and cellular uptake of its nanoparticles in HT-29 cells. Carbohydr Res 346:801–806
54. Thakker SP, Rokhade AP, Abbigerimeth SS, Iliger SR, Kulkarni VH, More UA, Aminabhavi TM (2014) Inter-polymer complex microspheres of chitosan and cellulose acetate phthalate for oral delivery of 5-fluorouracil. Polymer Bull 71:2113–2131
55. Kulkarni AR, Hukkeri VI, Sung HW, Liang HF (2005) A novel method for the synthesis of the PEG-crosslinked chitosan with a pH independent swelling behaviour. Macromol Biosci 5:925–928
56. Chen SC, Wu YC, Mi FL, Lin YH, Yu LC, Sung H (2004) A novel pH-sensitive hydrogel composed of N, O-carboxymethyl chitosan and alginate cross-linked by genipin for protein drug delivery. J Control Release 96:285–300
57. Sharma A, Gupta A, Rath G, Goyal A, Mathur RB, Dhakate SR (2013) Electrospun composite nanofiber-based transmucosal patch for anti-diabetic drug delivery. J Mater Chem B 1(27):3410–3418
58. Howard KA, Rahbek UL, Liu X, Damgaard CK, Glud SZ, Andersen MØ, Hovgaard MB, Schmitz A, Nyengaard JR, Basenbacher F, Kjems J (2006) RNA interference in vitro and in vivo using a chitosan/siRNA nanoparticle system. Mol Ther 14(4):476–484
59. Zhai X, Sun P, Luo Y, Ma C, Xu J, Liu W (2011) Guanidinylation: a simple way to fabricate cell penetrating peptide analogue-modified chitosan vector for enhanced gene delivery. J Appl Polym Sci 121:3569–3578
60. Luo Y, Zhai X, Ma C, Sun P, Fu Z, Liu W, Xu J (2012) An inhalable β2-adrenoceptor ligand-directed guanidinylated chitosan carrier for targeted delivery of siRNA to lung. J Control Release 162:28–36
61. Ko HF, Sfeir C, Kumta PN (2010) Novel synthesis strategies for natural polymer and composite biomaterials as potential scaffolds for tissue engineering. Philos Trans R Soc A 368:197–198
62. Narayanan RP, Melman G, Letourneau NJ, Mendelson NL, Melman A (2012) Photodegradable iron(III) cross-linked alginate gels. Biomacromolecules 13:2465–2471
63. Stevens MM, Qanadilo HF, Langer R, Shastri VP (2004) A rapid-curing alginate gel system: utility in periosteum-derived cartilage tissue engineering. Biomaterials 25:887–894
64. Popa GE, Gomes ME, Rui LR (2011) Cell delivery systems using alginate–carrageenan hydrogel beads and fibers for regenerative medicine applications. Biomacromolecules 12 (11):3952–3961
65. Balakrishnan B, Mohanty Umashankar M, Jayakrishnan PR (2005) Evaluation of an in situ forming hydrogel wound dressing based on oxidized alginate and gelatin. Biomaterials 26:6335–6342
66. Bidarra SJ, Barrias CC, Fonseca KB, Barbosa MA, Soares RA, Granja PL (2011) Injectable in situ crosslinkable RGD-modified alginate matrix for endothelial cells delivery. Biomaterials 32:7897–7904
67. Wayne JS, McDowell CL, Shields KJ, Tuan RS (2005) In vivo response of polylactic acid-alginate scaffolds and bone marrow-derived cells for cartilage tissue engineering. Tissue Eng 11:953–963
68. Ziv K, Nuhn H, Ben-Haim Y, Sasportas LS, Kempen PJ, Niedringhaus TP, Hrynyk M, Sinclair R, Barron AE, Gambhir SS (2014) A tunable silk–alginate hydrogel scaffold for stem cell culture and transplantation. Biomaterials 35(12):3736–3743

69. Tonnesen HH, Karlsen J (2002) Alginate in drug delivery systems. Drug Dev Ind Pharm 28:621–630
70. Rabbany SY, Pastore J, Yamamoto M, Miller T, Rafii S, Aras R, Penn M (2010) Continuous delivery of stromal cell-derived factor-1 from alginate scaffolds accelerates wound healing. Cell Transplant 19:399–408
71. Bowey K, Swift BE, Flynn LE, Neufeld RJ (2013) Characterization of biologically active insulin-loaded alginate microparticles prepared by spray drying. Drug Dev Ind Pharm 39:457–465
72. Suksamran T, Opanasopit P, Rojanarata T, Ngawhirunpat T, Ruktanonchai U, Supaphol P (2009) Biodegradable alginate microparticles developed by electrohydrodynamic spraying techniques for oral delivery of protein. J Microencapsulation 26:563–570
73. Moore K, Amos J, Davis J, Gourdie R, Potts JD (2013) Characterization of polymeric microcapsules containing a low molecular weight peptide for controlled release. Microsc Microanal 19:213–226
74. Jeon O, Powell C, Solorio LD, Krebs MD, Alsberg E (2011) Affinity-based growth factor delivery using biodegradable, photocrosslinked heparin-alginate hydrogels. J Control Release 154:258–266
75. Anderson NS, Dolan TCS, Rees DA (1965) Evidence for a common structural pattern in the polysaccharide sulfates of the Rhodphyceae. Nature 205:1060–1062
76. Vera J, Castro J, Gonzalez A, Moenne A (2011) Seaweed polysaccharides and derived oligosaccharides stimulate defence responses and protection against pathogens in plants. Mar Drugs 9:2514–2525
77. Nunez-Santiago MC, Tecante A, Garnier C, Doublier JL (2011) Rheology and microstructure of k-carrageenan under different conformations induced by several concentrations of potassium ion. Food Hydrocoll 25:32–41
78. Rocha PM, Santo VE, Gomes ME, Reis RL, Mano JF (2011) Encapsulation of adipose-derived stem cells and transforming growth factor-β1 in carrageenan-based hydrogels for cartilage tissue engineering. J Bioact Compat Polym 26(5):493–507
79. Santo VE, Frias AM, Carida M, Cancedda R, Gomes ME, Mano JF, Reis RL (2009) Carrageenan-based hydrogels for the controlled delivery of pdgf-bb in bone tissue engineering applications. Biomacromolecules 10:1392–1401
80. Desai PD, Dave AM, Devi S (2004) Entrapment of lipase into K-carrageenan beads and its use in hydrolysis of olive oil in biphasic system. J Mol Catal B-Enzym 31(4):143–150
81. Sipahigil O, Dortunc B (2001) Preparation and in vitro evaluation of verapamil HCl and ibuprofen containing carrageenan beads. Int J Pharm 228(1):119–128
82. Pereira Rui C, Scaranari M, Castagnola P, Grandizio M, Azevedo HS, Reis RL, Cancedda R (2009) Novel injectable gel (system) as a vehicle for human articular chondrocytes in cartilage tissue regeneration. J Tissue Eng Regen Med 3(2):97–106
83. Silva TH, Alves A, Popa EG, Reys LL, Gomes ME, Sousa RA, Silva SS, Mano JF, Reis RL (2012) Marine algae sulfated polysaccharides for tissue engineering and drug delivery approaches. Biomatter 2:278–289
84. Popa EG, Rodrigues MT, Santo VE, Goncalves AI, Reis RL, Gomes ME (2011) Magnetic-responsive hydrogels for cartilage tissue engineering. Biomacromolecules 12 (11):3952–3961
85. Mihaila SM, Gaharwar AK, Reis RL, Marques AP, Gomes ME, Khademhosseini A (2013) hotocrosslinkable kappa -carrageenan hydrogels for tissue engineering applications. Adv Healtc Mater 2:895–907
86. Liang W, Mao X, Peng X, Tang S (2014) Effects of sulfate group in red sea-weed polysaccharides on anticoagulant activity and cytotoxicity. Carbohydr Polym 101:776–785
87. Silva FRF, Dore CMPG, Marques CT, Nascimento MS, Benevides NMB, Rocha HAO, Leite EL (2010) Anticoagulant activity, paw edema and pleurisy induced carrageenan: action of major types of commercial carrageenans. Carbohydr Polym 79(1):6–33
88. Yao Z, Wu H, Zhang S, Du Y (2014) Enzymatic preparation of kappa-carrageenan oligosaccharides and their anti-angiogenic activity. Carbohyder Polym 101:359–367

89. Scarano A, Carinci F, Piattelli A (2009) Lip augmentation with new filler (agarose gel): a 3-year follow-up study. Oral Surg Oral Med Oral Pathol Oral Radiol Endod 108:e11–e15
90. Sheehy EJ, Buckley CT, Kelly DJ (2011) Chondrocytes and bone marrow-derived mesenchymal stem cells undergoing chondrogenesis in agarose hydrogels of solid and channelled architectures respond differentially to dynamic culture conditions. J Tissue Eng Regen Med 5(9):747–758
91. Cregg JM, Wiseman SL, Pietrzak-Goetze NM, Smith MR, Jaroch DB, Clupper DC, Gilbert RJ (2010) A rapid, quantitative method for assessing axonal extension on biomaterial platforms. Tissue Eng Part C: Methods 16:167–172
92. Fernandez-Cossio S, León-Mateos A, Sampedro FG, Oreja MT (2007) Biocompatibility of agarose gel as a dermal filler: histologic evaluation of subcutaneous implants. Plast Reconstr Surg 120:1161–1169
93. Rossi F, Chatzistavrou X, Perale G, Boccaccini AR (2012) Synthesis and degradation of agar-carbomer based hydrogels for tissue engineering applications. J Appl Polym Sci 123:398–408
94. Suzawa Y, Funaki T, Watanabe J, Iwai S, Yura Y, Nakano T, Umakoshi Y, Akashi M (2010) Regenerative behaviour of biomineral/agarose composite gels as bone grafting materials in rat cranial defects. J Biomed Mater Res A 93:965–975
95. WatanabeJ Kashii M, Hirao M, Oka K, Sugamoto K, Yoshikawa H, Akashi M (2007) Quick-forming hydroxyapatite/agarose gel composites induce bone regeneration. J Biomed Mater Res A 83:845–852
96. Varoni E, Tschon M, Palazzo B, Nitti P, Martini L, Rimondini L (2012) Agarose gel as biomaterial or scaffold for implantation surgery: characterization, histological and histomorphometric study on soft tissue response. Connect Tissue Res 53(6):548–554
97. Cao Z, Gilbert RJ, He W (2009) Simple agarose-chitosan gel composite system for enhanced neuronal growth in three dimensions. Biomacromolecules 10:2954–2959
98. Benya PD, Shaffer JD (1982) Dedifferentiated chondrocytes reexpress the differentiated collagen phenotype when cultured in agarose gels. Cell 30:215–224
99. Mauck RL, Soltz MA, Wang CCB et al (2000) Functional tissue engineering of articular cartilage through dynamic loading of chondrocyte-seeded agarose gels. J Biomech Eng 122:252–260
100. Bougault C, Paumier A, Aubert-Fouche E, Mallein-Gerin F (2009) Investigating conversion of mechanical force into biochemical signaling in three-dimensional chondrocyte cultures. Nat Protoc 4:928–938
101. Sakai S, Hashimoto I, Kawakami K (2006) Development of alginate–agarose subsieve-size capsules for subsequent modification with a polyelectrolyte complex membrane. Biochem Eng J 30:76–81
102. Kogan G, Soltés L, Stern R, Gemeiner P (2007) Hyaluronic acid: a natural biopolymer with a broad range of biomedical and industrial applications. Biotechnol Lett 29:17–25
103. Choi S, Choi W, Kim S, Lee S, Noh I, Kim C (2014) Purification and biocompatibility of fermented hyaluronic acid for its applications to biomaterials. Biomaterials Research 18 (6):1–10
104. Burdick JA, Prestwich GD (2011) Hyaluronic acid hydrogels for biomedical applications. Adv Mater 23(12):41–56
105. Choh SY, Cross D, Wang C (2011) Facile synthesis and characterization of disulfide-cross-linked hyaluronic acid hydrogels for protein delivery and cell encapsulation. Biomacromolecules 12(4):1126–1136
106. Gojgini S, Tokatlian T, Segura T (2011) Utilizing cell-matrix interactions to modulate gene transfer to stem cells inside hyaluronic acid hydrogels. Mol Pharm 8(5):1582–1591
107. Yamada T, Kawasaki T (2005) Microbial synthesis of hyaluronan and chitin: new approaches. J Biosci Bioeng 99:521–528
108. Hou S, Tian W, Xu Q, Cui F, Zhang J, Lu Q, Zhao C (2006) The enhancement of cell adherence and inducement of neurite outgrowth of dorsal root ganglia co-cultured with

hyaluronic acid hydrogels modified with Nogo-66 receptor antagonist in vitro. Neuroscience 137:519–529
109. Ozgenel GY (2003) Effects of hyaluronic acid on peripheral nerve scarring and regeneration in rats. Microsurgery 23:575–581
110. Lin CM, Lin JW, Chen YC, Shen HH, Wei L, Yeh YS, Chiang YH, Shih R, Chiu PL, Hung KS, Yang LY, Chiu WT (2009) Hyaluronic acid inhibits the glial scar formation after brain damage with tissue loss in rats. Surg Neurol 72(Suppl. 2):S50–S54. doi:10.1016/j.wneu.2009.09.004
111. Guvendiren M, Burdick JA (2012) Stiffening hydrogels to probe short- and long-term cellular responses to dynamic mechanics. Nat Commun 3:792
112. Khetan S, Guvendiren M, Legant WR, Cohen DM, Chen CS, Burdick JA (2013) Three-dimensional hydrogels. Nat Mater 12:458–465
113. Purcell BP, Elser JA, Mu KB, Margulies Burdick JA (2012) Synergistic effects of SDF-1α chemokine and hyaluronic acid release from degradable hydrogels on directing bone marrow derived cell homing to the myocardium. Biomaterials 33:7849–7857
114. Kim IL, Khetan S, Baker BM, Chen CS, Burdick JA (2013) Fibrous hyaluronic acid hydrogels that direct MSC chondrogenesis through mechanical and adhesive cues. Biomaterials 34:5571–5580
115. Nie T, Baldwin A, Yamaguchi N, Kiick KL (2007) Production of heparin-functionalized hydrogels for the development of responsive and controlled growth factor delivery systems. J Controlled Release 122:287–296
116. Nagira T, Nagahata-Ishiguro M, Tsuchiya T (2007) Effects of sulfated hyaluronan on keratinocyte differentiation and Wnt and Notch gene expression. Biomaterials 28:844–850
117. Hintze V, Moeller S, Schnabelrauch M, Bierbaum S, Viola M, Worch H, Scharnweber D (2009) Modifications of hyaluronan influence the interaction with human bone morphogenetic protein-4 (hBMP-4). Biomacromolecules 10:3290–3297
118. Yamada T, Sawada R, Tsuchiya T (2008) The effect of sulfated hyaluronan on the morphological transformation and activity of cultured human astrocytes. Biomaterials 29:3503–3513
119. Prestwich GD (2011) Hyaluronic acid-based clinical biomaterials derived for cell and molecule delivery in regenerative medicine. J Controlled Release 155:193–199
120. Girish KS, Kemparaju K (2007) The magic glue hyaluronan and its eraser hyaluronidase: a biological overview. Life Sci 80(21):1921–1943
121. Hahn SK, Oh EJ, Miyamoto H, Shimobouji T (2006) Sustained release formulation of erythropoietin using hyaluronic acid hydrogels crosslinked by michael addition. Int J Pharm 322:44–51
122. Lee H, Park TG (2009) Photo-crosslinkable, biomimetic, and thermo-sensitive pluronic grafted hyaluronic acid copolymers for injectable delivery of chondrocytes. J Biomed Mater Res A 88:797–806
123. Taylor C, Cheetham NWH, Walker GJ (1985) Application of high-performance liquid chromatography to a study of branching in dextrans. Carbohydr Res 137:1–12
124. Debelder AN (1996) Medical applications of dextran and its derivatives. In: Dimitriu S (ed) Polysaccharides in medicinal applications. Marcel Dekker, New York
125. Kim SH, Won CY, Chu CC (2000) Dextran-maleic acid monoesters and hydrogels based thereon. WO Patent No. 2000012619
126. Kim SH, Won CY, Chu CC (1999) Synthesis and characterization of dextran-maleic acid based hydrogel. J Biomed Mater Res 46:160–170
127. Yamaoka T, Tanihara M, Mikami H, Kinoshita H (2003) Temperature-responsive material and composition comprising the same. JP Patent No. 2003252936
128. Du YZ, Weng H, Yuan HuFQ (2010) Synthesis and antitumor activity of stearate-*g*-dextran micelles for intracellular doxorubicin delivery. ACS Nano 4:6894–6902
129. Van Tomme SR, Hennink WE (2007) Biodegradable dextran hydrogels for protein delivery applications. Expert Rev Med Dev 4:147–164

130. Lévesque SG, Shoichet MS (2007) Synthesis of enzyme-degradable, peptide-cross-linked dextran hydrogels. Bioconjug Chem 18:874–885
131. Shu S, Zhang X, Wu Z, Wang Z, Li C (2011) Delivery of protein drugs using nanoparticles self-assembled from dextran sulfate and quaternized chitosan. J Control Release 152:e170–e172
132. Yuan W, Hu Z, Su J, Wu F, Liu Z, Jin T (2012) Preparation and characterization of recombinant human growth hormone–Zn^{2+}dextran nanoparticles using aqueous phase–aqueous phase emulsion. Nanomedicine NBM 8:424–427
133. Wu F, Zhou Z, Su J, Wei L, Yuan W, Ji (2013) Development of dextran nanoparticles for stabilizing delicate proteins. Nanoscale Res Lett 8:197
134. Delgado D, Gascón AR, Del Pozo-Rodríguez A, Echevarría E, Ruiz de Garibay AP, Rodríguez JM, Solinís MA (2012) Dextran-protamine-solid lipid nanoparticles as a non-viral vector for gene therapy: in vitro characterization and in vivo transfection after intravenous administration to mice. Int J Pharm 425:35–43
135. Sarmento B, Ribeiro A, Veiga F, Ferreira D (2006) Development and characterization of new insulin containing polysaccharide nanoparticles. Colloids Surf B Biointerfaces 53:193–202
136. Abedini F, Hosseinkhani H, Ismail M, Domb AJ, Omar AR, Chong PP, Hong PD, Yu DS, Farber IY (2012) Cationized dextran nanoparticle-encapsulated CXCR4-siRNA enhanced correlation between CXCR4 expression and serum alkaline phosphatase in a mouse model of colorectal cancer. Int J Nanomed 7:4159–4168
137. Shu XZ, Ahmad S, Liu Y, Prestwich GD (2006) Synthesis and evaluation of injectable, in situ crosslinkable synthetic extracellular matrices for tissue engineering. J Biomed Mater Res, Part A 79(4):902–912
138. Miura S, Teramura Y, Iwata H (2006) Encapsulation of islets with ultra-thin polyion complex membrane through poly(ethylene glycol)-phospholipids anchored to cell membrane. Biomaterials 27:5828–5835

Protein-Based Hydrogels

Alexandra L. Rutz and Ramille N. Shah

Abstract Protein-based hydrogels are composed of isolated or enriched proteins from natural extracellular matrix. Inherent and controllable bioactivity makes these hydrogels promising candidates as smart biomaterials for drug delivery, tissue engineering and regenerative medicine, and other applications. Desirable characteristics for these applications include natural cell binding, cell degradable, and growth factor-binding sequences. This chapter covers the unique properties of a variety of proteins (collagen, gelatin, fibrin, silk, elastin, keratin, and decellularized, tissue-specific extracellular matrix) as well as hydrogel synthesis, fabrication, modification, and established applications. Conditions of solubility and the mechanism of the sol–gel transition are discussed. Since each protein presented undergoes self-assembly to form a gel network, gelation parameters that affect this assembly and subsequently the gel ultrastructure are specifically presented. Emerging applications and technologies for protein-based hydrogels are also briefly mentioned.

Keywords Proteins · Extracellular matrix · Hydrogels · Biomaterials · Tissue engineering · Drug delivery

A.L. Rutz
Biomedical Engineering, Northwestern University, Evanston, IL, USA

R.N. Shah (✉)
Materials Science and Engineering, Northwestern University, Evanston, IL, USA
e-mail: ramille-shah@northwestern.edu

R.N. Shah
Surgery (Transplant Division), Northwestern University, Chicago, IL, USA

A.L. Rutz · R.N. Shah
Simpson Querrey Institute for BioNanotechnology, Northwestern University, Chicago, IL, USA

S. Kalia (ed.), *Polymeric Hydrogels as Smart Biomaterials*,
Springer Series on Polymer and Composite Materials,
DOI 10.1007/978-3-319-25322-0_4

1 Introduction

Extracellular matrix (ECM) is a complex mixture of proteins and carbohydrates that are not only responsible for the structure of tissues and organs, but also cell signaling in both normal and wounded or diseased states. Generally, cells secrete and build ECM by synthesizing monomers that then aggregate. Either monomers or oligomers are excreted to the extracellular space in which a change occurs that induces their assembly into fibers. Researchers have harnessed this phenomenon to synthetically fabricate ECM protein hydrogels in vitro, enabled by understanding the conditions of protein solubility and the mechanisms of sol–gel transition. More recently, gelation parameters that can change and tune the gel ultrastructure have been discovered and utilized. This chapter will discuss each protein in detail, and cover topics of protein structure and properties as well as hydrogel synthesis, fabrication, and application. Particularly for in vivo applications, these protein hydrogels typically possess inherent desirable characteristics: biocompatibility, biodegradability, and bioactivity (such as growth factors or growth factor-binding sequences).

A hydrogel is defined as an interconnected polymer network that holds over 90 % water. The fact that this material holds such little polymer and so much water enables passive diffusion of solutes throughout the material. Hydrogels are the only biomaterial that allows cell encapsulation since required nutrients and wastes can freely diffuse across the material to sustain cell viability. The polymer network of the hydrogel also serves as an extracellular matrix mimic by providing a completely 3D growth environment. Protein hydrogels in particular are ideal for cell encapsulation. Firstly, gelation is typically physical (assembly of fibers) and is therefore, gentle enough for maintaining viable cells. Secondly, protein hydrogels contain sequences for cell adhesion and degradation without further modification. In addition to cell therapies and tissue engineering, hydrogels can also provide controlled drug release. Protein hydrogels are ideal for drug delivery applications in which the implant must degrade over a short period of time (weeks to months). Gentle encapsulation methods and ability to be remodeled by cells make protein hydrogels especially ideal for controlled release of biomolecules for regenerative medicine. Finally, hydrogels also serve as injectable biomaterials for minimally invasive therapies by preparing the polymer solution and coaxing the sol–gel transition either in the body or immediately before injection. These unique properties of hydrogels, and especially those specific to protein hydrogels, make these biomaterials valuable to the development of new therapies for drug delivery, tissue engineering and regenerative medicine. The future may hold use of protein hydrogels in other promising applications, particularly with the emerging technologies of genetic engineering, biomanufacturing, and microfabrication.

2 Collagen

2.1 *Structure, Solubilization, and Assembly*

Collagen is the most abundant protein in mammals, and not surprisingly, has become the most widely used natural biomaterial. Collagen exists in a variety of forms, 28 different types [1]. The most typically occurring collagens are fibrous collagens, types I, II, III, V, and XI [2]. Type I makes up 90 % of the protein in connective tissue, and therefore, is the most readily available and most commonly used type, typically from murine and bovine sources [3]. A collagen molecule consists of three polypeptide chains assembled into a triple helix structure that is 300 nm in length and 1.5 nm in diameter (Fig. 1). A repeating amino acid sequence (Gly-X-Y, X and Y most often proline and hydroxyproline) is responsible for the helical conformation. At the ends of the helix, telopeptides assist in assembling the molecules to form cross-striated fibrils with a characteristic 67 nm banding. Cross-links occur between molecules to stabilize fibrils. Tissue can be processed to break some of these cross-links and extract soluble collagen monomers (or even multiple monomers that are still together such as dimers, trimers, etc.) in an acidic solution [2]. Extraction is more easily achieved with the addition of proteolytic enzymes (pepsin), which help cleave telopeptide regions, and is particularly needed for tissues with densely cross-linked collagen [3]. This soluble solution is the starting material for collagen-based biomaterials research. When the solubilized

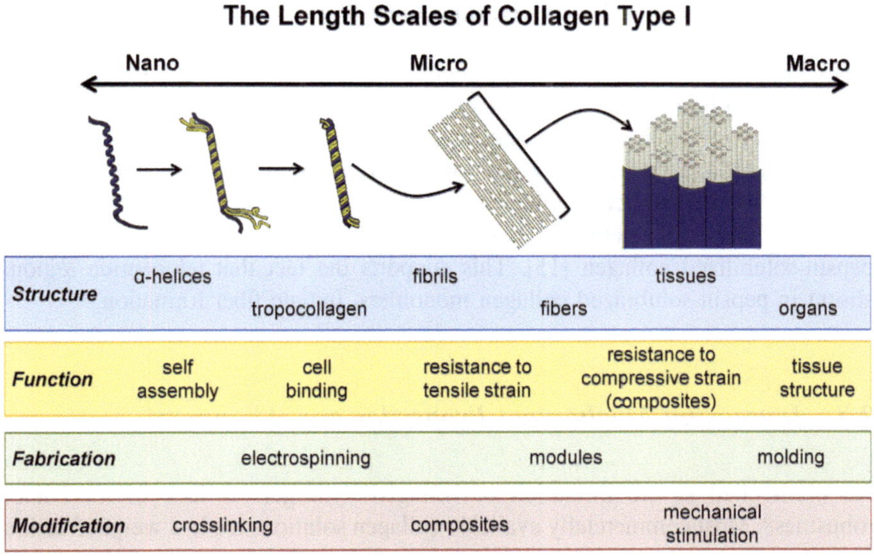

Fig. 1 The assembly of collagen. Three alpha helix collagen molecules form a triple helix. These triple helices stagger to form collagen fibrils that then bundle to form tissue. Reprinted from [10], Copyright 2014, with permission from Elsevier

collagen is neutralized and the temperature is raised (20–37 °C), collagen monomers assemble into fibers and form a physically cross-linked hydrogel. Because of the abundance of collagen in tissues, collagen hydrogels have been most often used in medical and tissue engineering applications of all types including skin, bone, tendon, ligament, etc. [4–9]. As well, the physical cross-linking of collagen is desirable for cell-encapsulating biomaterials. Of important note is that collagen can also be processed into a variety of forms including fibers via electrospinning, sheets, and sponges, but these are not considered hydrogels and therefore will not be discussed further in this chapter.

2.2 Gelation Parameters and Influence on Gel Ultrastructure

There are many factors during gelation that influence the characteristics of the resulting fibrous polymer network. Increasing temperatures increases the rate of gel formation/collagen assembly, and therefore, this results in less collagen bundling (smaller fiber diameters, smaller mesh pores) and less order [3]. Comparing gels formed at a variety of temperatures (4–37 °C), gels formed at lower temperatures exhibited numerous, desirable properties, including larger pore sizes for enhanced cellular responses, higher strength, and prolonged degradation over those formed at higher temperatures [11–13]. Similar to forming gels at higher temperatures (37 ° C), increasing the pH or decreasing the ionic strength of the collagen solution also speeds fibrillogenesis and therefore, leads to smaller diameter fibers and pore sizes [3]. Specifically for cell-encapsulating hydrogels, the pH must be restricted to a narrow range, 7.4–8.4, during gel formation to ensure cell viability. Heterogeneous properties can be instilled in collagen gels by forming regions of varying gelation temperatures or collagen concentrations using microfluidics [14]. Comparing pepsin-solubilized (typically bovine and porcine sources) and acid-solubilized (typically murine sources) collagens, acid-solubilized collagens gel faster and therefore, the gel network has shorter fibers and smaller pores than pepsin-solubilized collagen [15]. This supports the fact that telopeptide regions, absent in pepsin-solubilized collagen monomers, initiate fiber formation.

2.3 Increasing Mechanical Properties and Alignment

An often cited severe limitation of collagen hydrogels is lack of mechanical robustness. Most commercially available collagen solutions are low weight fractions (less than 5 mg/mL). Even if a high concentration of collagen is produced, high weight fraction collagen gels (greater than 20 mg/mL) may not be suitable for cell encapsulation since dense polymer matrices restrict cell spreading and migration as well as nutrient and waste diffusion [3]. The most typical resource for increasing mechanical properties of protein biomaterials is chemical cross-linking. Ubiquitous protein cross-linking strategies are discussed in later sections of this chapter. There, however, are other strategies for controlling mechanical properties of collagen. First, collagen gels can be mechanically conditioned. For example, plastic compression may be performed to expel water from the gel. This can be done either with an unconfined compressive load and/or by blotting with an absorbent material [10]. The gel significantly shrinks (more than 100 fold), and the collagen concentration increases by 7–10 fold [16, 17]. The resulting collagen sheets have been used in many tissue engineering applications [4]. Second, collagen gels can be formed with aligned fibers. Aligned fibers are not only desirable for increased mechanical properties, but also for anisotropic mechanical properties (particularly increased tensile mechanical properties). Cells respond to this anisotropic environment through alignment, and cell alignment is desirable for many tissue engineering targets, especially for load-bearing musculoskeletal tissues [18–20]. As discussed above, fibril assembly and order can be influenced by temperature, pH, and ionic strength, but there are also other influential factors. For instance, cells along with the presence of a mechanical stimulus (either tensile or compressive strains or shear strains from fluid flow) can remodel and direct the assembly of aligned collagen fibrils [21, 10]. This process, however, is slow, and furthermore, without cells, alignment has only been achieved with high strain (30–50 %) [22, 23]. Applying strong magnetic fields (greater than 6 T), either during or after fibrillogenesis, can also align fibrils [24–26]. Weak magnetic fields (as low as 10^{-4} T) can result in alignment with the incorporation of magnetic particles into the collagen solution and simply placing a magnetic bar on the solution container [27]. With isoelectric focusing, aligned collagen bundles (up to 7 cm in length) have been produced in solutions exposed to an electric current to create a pH gradient [28]. Others have shown that molecular crowding with hyaluronic acid or poly(ethylene glycol), mimetic of conditions during embryonic development, can also induce aligned and planar fibrillar lamellae, strikingly similar to that seen in the ECM of native tissue [29].

2.4 Other Fabrication Methods

In addition to bulk collagen gels, collagen can be easily fabricated into other shapes by utilizing the solution phase prior to gelation. Collagen can be fabricated into microspheres by simply pipetting and incubating collagen droplets on parafilm or through controlled fluid flow and droplet formation using microfluidics [30]. These microspheres could be used for cell-encapsulating gels as injectable cell therapies [31] or microspheres may be fused together with additional collagen or through cell–cell adhesions to form a scaffold that can even have multiple layers of different cell types [32, 33]. Additionally, collagen gels can be formed in a variety of shapes or surface features by filling micromolds with a collagen solution and subsequently gelling [34, 35]. Microfabrication techniques may favor high polymer fraction gels for desirable mechanical properties in order to remove the gel from the mold, yet these high concentrations (2 % or greater) may not promote the appropriate cell behavior when encapsulated [36]. Alternatively, collagen can be cast into a sacrificial mold that would later diffuse away [37, 38]. Tubes or microfluidic channels can be fabricated and later perfused and lined with endothelial cells to achieve engineered vessel-like structures of all physiologically relevant sizes (μm to multi-mm scales) [12, 39]. Collagen has even been 3D printed by inducing gelation layer-by-layer by applying heat or nebulized sodium bicarbonate [40–42]. Molding also allows hydrogels to be cast into geometries relevant for traditional mechanical testing such as "dogbones" for tensile testing [43], cylindrical disks for compression testing, and between cone plate and parallel plate fixtures for rheological testing [44]. These molding and microfabrication techniques are not just restricted to collagen, but can be applied to all protein hydrogels discussed ahead. Since hydrogel fabrication utilizes a sol–gel transition, the solution phase of the hydrogel can be used to generate different structural architectures.

3 Gelatin

3.1 Structure, Solubilization, and Assembly

Gelatin is a heterogeneous mixture of polypeptides derived from collagen. Key amino acid sequences, such as cell adhesion sites (RGD) and cell degradation sites (MMP sensitive), are retained, and therefore, gelatin retains the biocompatibility and biodegradability of collagen. As well, gelatin still possesses GXY repeats that form triple helices. Although gelatin is degraded to such a degree that it no longer retains the ability to form collagen fibrils, small regions between polypeptides link and form physical cross-links that are responsible for gelation (Fig. 2). Lack of this fibrous structure is why gelatin is transparent whereas fibrous collagen hydrogels are opaque. Gelation of gelatin is thermoreversible and occurs in the opposite

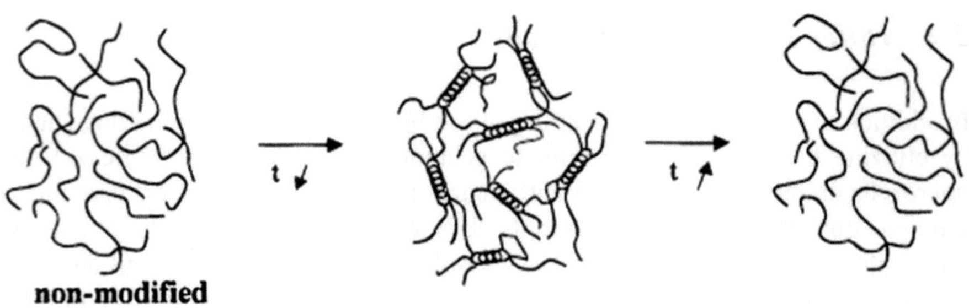

Fig. 2 Gelatin is a solution at temperatures above ~35 °C. When cooled, local regions of triple helices form physical cross-links producing a gel. If not further cross-linked, the gelation is reversible. Reprinted from [48], Copyright 2000, with permission from American Chemical Society

fashion as collagen hydrogels, which are soluble in only cold acid but gel at or above room temperature at physiological pH. Gelatin solutions are simply prepared by dissolving gelatin powder in warm water or buffer. The ease of isolation, solubilization, and processing as well as low costs have resulted in the use of gelatin across many industries and uses, ranging from medical (plasma expanders [45], tissue engineering [46], drug delivery), pharmaceutical (drug capsules), cosmetic, and food [47] industries.

Interestingly, gelatin can be isolated with a variety of isoelectric points (IEPs). Gelatin is derived from collagen by two main methods that result in two types of gelatin, type A and type B. Before extraction, collagen is pretreated with either an acidic (type A) or basic (type B) processing [49]. Alkaline processing (liming) converts amide groups of asparagine and glutamine to carboxylic acid groups, thus converting some positive charges to negative charges (at physiological pH). Acidic processing does not largely affect any specific amino acid functional groups. Therefore, type A gelatin has an IEP similar to collagen (positive, IEP 5.0) whereas type B is lowered (negative, IEP of 9.0).

Gelatin is a solution above approximately 35 °C and gels below. Therefore, gelatin is a solution at physiological conditions and requires chemical cross-linking for cell culture and in vivo study. Gelatin solutions are used for coating tissue culture substrates, and have even been used in monitoring traction forces of moving cells [50]. Warm gelatin solutions can be cast into molds and cooled to room temperature, and subsequently cross-linked [51]. Alternatively, an aqueous cross-linker can be mixed into the gelatin solution prior to casting [52]. The temperature during cross-linking is also an important parameter that can affect gel properties. If cross-linking occurs below the sol–gel transition, the physical cross-links of triple helix regions may be stabilized with chemical cross-links. After cross-linking and swelling, gelatin hydrogels can also be freeze-dried to instill microporosity, depending on the method of freezing [53]. Of note is that in addition to cast hydrogels, micro- and nanoparticles of gelatin can be formed using a variety of methods including desolvation, coacervation-phase separation, emulsification-solvent evaporation, reverse phase microemulsion, nanoprecipitation, self-assembly, and layer-by-layer coating [54, 55]. Gelatin can also be 3D printed into porous scaffolds that are subsequently cross-linked [56]. Alternatively, gelatin can be 3D co-printed with other materials and used as a sacrificial material that dissolves away when incubated at 37 °C [40].

3.2 *Drug Delivery and Tissue Engineering*

Both gelatin hydrogel scaffolds and nano- and microparticles can be loaded with a drug payload for sustained release. Proteins, polysaccharides, and plasmids may be absorbed by polyion complexation using a gelatin with an IEP matched to the biomolecule of interest [49]. Delivery is not just restricted to biomolecules, but also has been performed with small molecule drugs including cancer chemotherapeutics [54].

Particularly for biomolecules, the main desired interaction is largely electrostatic. Although other interactions, such as physical entrapment, covalent conjugation, hydrogen bonding, and hydrophobic interactions, may also be present. Without favorable interactions, biomolecules quickly diffuse from hydrogels due to the heavily hydrated environment and thus do not provide any enzymatic or immunological protection to the payload. Absorbed biomolecule release from gelatin could occur through two methods: (1) release from gelatin in a high ionic strength environment or (2) degradation of gelatin that releases gelatin fragments with the absorbed molecules. The latter is the accepted mechanism in vivo [49].

Loading of particles or scaffolds is achieved by rehydrating freeze-dried gelatin in a solution of the biomolecule [52]. This provides safe loading for sensitive payloads, particularly proteins. Growth factors have been safely loaded into freeze-dried scaffolds to coax tissue regeneration and vascularization [57–60]. Alternatively, particles may be loaded by direct incorporation of the drug into the gelatin solution during particle preparation [54]. These microspheres can be used alone or loaded in another hydrogel or scaffold of interest to provide controlled release [61–63]. In either particles or scaffolds, the degree of cross-linking affords a means to control the release rate [64, 65, 52].

4 Fibrinogen

4.1 Structure, Solubilization, and Assembly

Fibrin is the material that comprises a blood clot and provides a matrix for recruited cells to aid in tissue repair. A fibrinogen molecule is a 45 nm long, 340 kDa glycoprotein with a dumbbell shape and a dimer of three polypeptide chains held together by disulfide bonds (Fig. 3) [66]. A central globular domain E is flanked by two globular regions, D, connected with three-stranded alpha helix coiled-coils. In the last steps of the coagulation cascade, the active protease thrombin cleaves fibrinopeptides of fibrinogen, circulating in the blood, to produce a fibrin monomer (Fig. 4). The fibrinopeptides (A and B) are on the E domain, and once cleaved, change the electrostatic charges of fibrinogen that facilitate fibrin monomer interaction for lateral fibril polymerization. Fibrinopeptides A are cleaved first that allow assembly into protofibrils, and further cleavage of fibrinopeptides B allow assembly into the full fibril [67]. Fibrin monomers associate in a half-staggered fashion, with the center of one molecule associating with the end of another. The fibrils are stabilized with Factor XIII, transglutaminase, which links glutamine and lysine residues.

Fibrin hydrogels are produced in the same fashion as a blood clot by simply mixing isolated thrombin and fibrinogen in the presence of calcium, which aids in fibrinopeptide cleavage. Fibrin gels are widely used in the clinic as a glue or sealant to stop blood flow (hemostatic agent) and close tissue (used in place of sutures)

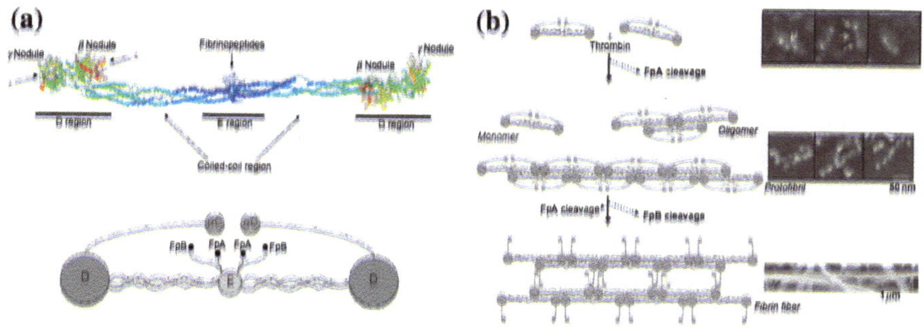

Fig. 3 **a** Fibrinogen has a dumb-bell shape and is a dimer of three polypeptide chains held together by disulfide bonds. A central globular domain E is flanked by two globular regions, D, connected with three-stranded alpha helix coiled-coils. **b** Activated protease thrombin cleaves fibrinopeptides of fibrinogen to produce fibrin monomers. Fibrinopeptides A are cleaved first that allow assembly into protofibrils, and further cleavage of fibrinopeptides B allow assembly into the full fibril. Monomers link with the center of one molecule associating with the end of another. Reprinted from [66], Copyright 2011, with permission from Elsevier

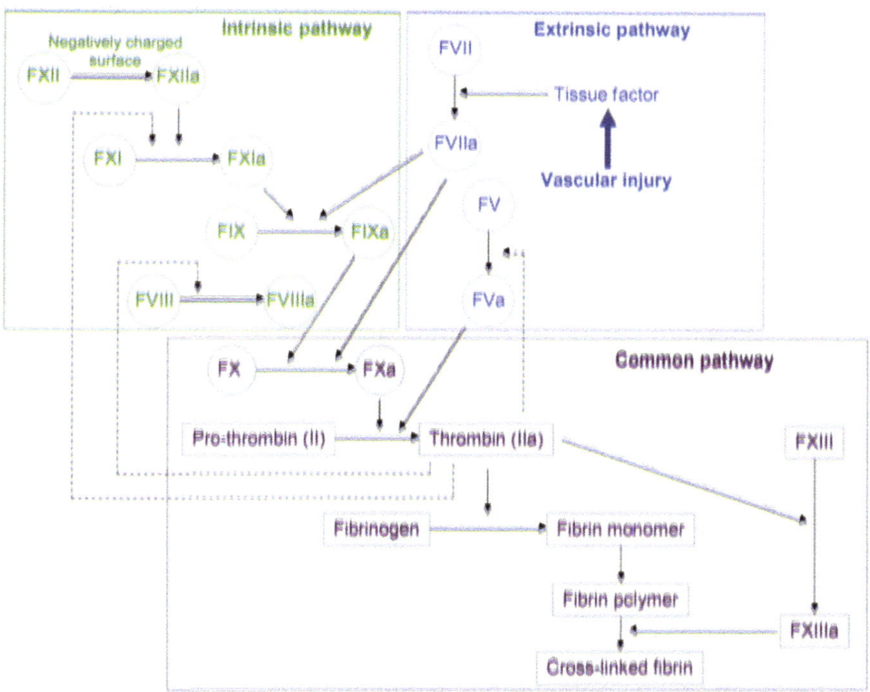

Fig. 4 The coagulation cascade of the formation of a blood clot. Activated thrombin cleaves fibrinopeptides of fibrinogen to produce the fibrin monomer. Fibrin monomers assemble and form fibrils that are then cross-linked with Factor XIII/transglutaminase, which links glutamine and lysine residues. Reprinted from [66], Copyright 2011, with permission from Elsevier

[68]. In order to quickly gel, as is needed for these applications, fibrin glue is prepared at a high concentration of both fibrinogen (up to 60 mg/mL) and thrombin (up to 300 IU/mL). Fibrinogen is easily isolated by simply processing blood, either from humans or other mammals. This means that as long as a patient does not suffer from any clotting disorders, fibrinogen can be isolated from a patient's serum for autologous therapy [69]. The resulting enriched fibrinogen typically contains other components such as growth factors and fibronectin, transglutaminase, and fibrinolytic inhibitors [70, 71]. There is no added cost to using autologous versus commercially available fibrin, and in fact, can even be cheaper [72]. Although fibrin glues cannot be used for patient-specific emergency care, autologous products can be used for planned procedures. Autologous fibrin gels could also even be used for culture, expansion, and differentiation of a patient's cells to be influenced by their own matrix and bioactive factors [73]. Although fibrin has long been utilized as a biomaterial and the clotting cascade has been studied for decades, it was only until very recently that the crystal structure of human fibrinogen was identified [74]. Furthermore, there is still much unknown about its structure, properties, and role in physiology and diseases caused by both genetic and environmental factors.

4.2 Gelation Parameters and Influence of Gel Ultrastructure

Fibrin fibrils assemble up to 600 nm in length, after which point either begin to branch or become thicker diameters [66]. The most common branch point type is the junction of three fibers [75]. Many conditions in vitro influence the gel network structure of a fibrin gel [76, 75]. Higher thrombin concentrations result in faster gelation, as used in fibrin glues. Low thrombin concentrations lead to thicker diameter fibers and therefore, less branching and smaller pore sizes [67]. Incorporation of PEGylated peptides synthesized to interact with fibrin monomers has also shown modulation of gel ultrastructure, particularly to increase porosity for cell encapsulation and infiltration [77, 78]. Fibrinogen contains many calcium-binding sites, and calcium has been shown to influence fibrin polymerization, prevent denaturation caused by pH and heat changes, as well as help prevent disulfide bond reduction [66]. At nonphysiological concentrations of calcium, varying the calcium concentrations resulted in different diameter fibrils: increasing the calcium concentration increased fibril diameter (from 150 to 510 nm) [79]. The presence of calcium during gelation has also been shown to be critical for in vivo stability [80]. Likewise, in fibrin gels prepared with varying sodium chloride concentrations ranging from 0 to 4.45 % w/v, both fiber network and mechanical properties change. The density of the network and mechanical properties increase with increasing sodium chloride until 2.6 % w/v [81]. Another group noted slowed gelation when sodium chloride was included in gel formulations [82]. Varying pH, calcium concentration, and fibrinogen concentration together led one group to find a condition that produced a transparent fibrin gel with enhanced mechanical properties and prolonged degradation in vivo [83]. In addition to chemical additions and similar to collagen, mechanical loading of fibrin hydrogels has been used to align fibers, and can be used for aligning cells in culture [84–86]. Alignment has also been achieved by gelling in the presence of a strong magnetic field [79] or by magnetically guiding fibrin polymerization by covalently attaching thrombin to magnetic microbeads, which were then arranged in a pattern on a substrate [87]. Fibrin gels have also been inkjet printed into defined, cell-encapsulating hydrogels [88, 89]. This has been performed by applying thrombin droplets to a fibrinogen solution to cause localized gelation.

4.3 *Bioactivity*

Mechanical properties, bioactivity, and cell and in vivo-friendly cross-linking have made fibrin gels attractive candidates for tissue engineering and growth factor delivery. Fibrin gels are used for growth factor delivery, either by entrapment or binding [90, 91], as well as for cell culture, both on and within (encapsulating) the hydrogel. Fibrinogen and cells can be injected in vivo to subsequently form gels. Just like collagen and gelatin, fibrinogen can also be fabricated into microbeads, including those for cell-encapsulating applications and those for drug and gene delivery [67, 92, 91]. Fibrinogen has many specific binding sequences, those for cell adhesion, heparin binding, ECM protein binding, and growth factor binding, enabling either direct or indirect binding (via heparin or other ECM proteins) of growth factors. Gels loaded with growth factors can release the payload for up to a few weeks [67]. Especially for encapsulated growth factors, high fibrinogen concentrations are preferred to ensure a slower diffusion rate through a smaller mesh. As mentioned above, isolated fibrinogen may also come enriched with certain factors that contribute to fibrin's bioactive properties aside from its inherent sequences [71, 70].

4.4 *Mechanical Properties and Controllable Degradation*

Fibrin gels also possess amazing extensibility (can be stretched more than three times length before breaking), due to unique properties at many scales: protein folding and unfolding on the molecular level as well as movements within fiber and network structures [84]. These remarkable mechanical properties of fibrin clots are physiologically important for a clot to withstand the shear forces of blood flow [93]. The bioactivity and unique mechanical properties have led to the increasing exploration of many tissue engineering applications, including adipose, bone, cardiac, muscle, cartilage, and nerve repair [72, 94]. Furthermore, fibrin gels have been shown to promote and support angiogenesis and neurite extension, and this is significant since vascularization and innervation are critical components to a successful tissue engineered implant [95, 96]. However, a drawback to fibrin gels is that although fibrin is quite extensible, the gels are still weak. To improve gel stiffness, Tranquillo et al. has reported a cell-friendly method in which fibrin can be photocrosslinked with blue light using ruthenium as a catalyst to couple tyrosines [97, 98]. Another significant limitation to fibrin gels is rapid degradation since the gels are cell degradable by both plasmin and matrix metalloproteases (MMPs). The degradation rate can be slowed with the incorporation of fibrinolytic inhibitors such as aprotinin [80]. The Hubbell group has demonstrated methods to control degradation rates as well as enhance fibrin's bioactivity. Bi-domain proteins or peptides containing both a bioactive region and a Factor XIII substrate are added to the fibrin mixture and are covalently attached to the fibrin matrix during gel formation [99]. This has been demonstrated with RGD [99], laminin [100], N-cadherin [100], and fibronectin domain (including integrin binding site capable of binding VEGF, PDGF, BMP-2) peptides [101], as well as fusion proteins of the Factor XIII substrate and growth factors [102–104] or fibrinolysis inhibitors (aprotinin) [102]. This technique provides site-specific cross-linking for enhanced retention of bioactivity (i.e., decreased risk of denaturing or reacting at an active site with nonspecific cross-linking). Since growth factors or peptides are covalently attached, release is cell mediated from degradation by proteases. Furthermore, adding fibrinolysis inhibitors provides a means to control degradation and thereby, control release rate.

5 Silk

5.1 *Structure, Solubilization, and Assembly*

Silk is the most widely studied nonmammalian protein biomaterial. It is produced from species within the class of Arachnida, and is most commonly utilized from silkworms (*Bombyx mori*) and orb-weaving spiders (*Nephila clavipes*) [105]. Because of silk's heavy use in the textile industry, silk is easily acquired in many forms from domesticated *B. mori*. Spider silks are not as widely available as that from silkworms. Large populations of spiders cannot be maintained and furthermore, yield is lower [105]. Therefore, large-scale production of spider silk and subsequent biomaterial use will come from biomanufacturing/genetic engineering strategies [106–109]. Particularly for silk-based hydrogel research (derived from sequences of any species), silk-like and silk-elastin-like polymers dominate the field [105].

Silk fibroin fibers for *B. mori* are 10–25 μm in diameter and are coated with hydrophilic proteins called sericins, which are removed during material processing.

Silk proteins have large hydrophobic domains of short-chain amino acids (e.g., glycine, alanine) with short hydrophilic domains in-between. The intermingled hydrophilic domains assist in the assembly of the silk proteins that form tightly packed sheets from hydrogen bonding and hydrophobic interactions of the hydrophobic domains. The soluble form of silk is in alpha helix and random coil conformations, and by transitioning to beta sheets, silk is rendered insoluble. Silk's most unique and desirable property as a protein biomaterial is its incredible mechanical properties [105]. Silk fibers have a tensile modulus of few-tens GPa, tensile strength up 4.8 GPa, and yet is still elastic. Silk also has high thermal and chemical stability, especially relative to other protein-based materials [110].

The first medical use of silk was using weaved silk fibers as sutures. The long use of silk sutures has established silk as a biocompatible material, yet silk carries a risk of an immune and inflammatory response. Silk is generally regarded as a biocompatible material having only a mild inflammatory response and low adhesion of immune-competent cells [105]. Although retention of sericins has been shown to increase cell adhesion and proliferation in cell culture experiments, these proteins have caused hypersensitivity (an exaggerated immune response) in vivo and therefore, are typically removed. Long-term compatibility and compatibility of degradation products are still in need of evaluation [111]. Like other proteins, silk is degradable via proteases, but the degradation rate is heavily dependent on the processing [105]. For example, silk fibroin fibers are classified as nondegradable. Applications of silk biomaterials include tissue engineering [111, 105] (musculoskeletal [112]), drug delivery [113, 114, 105, 115, 110] (genes, small molecules, and biologics by covalent coupling or adsorption), biosensors [105], and most recently, optics, photonics, and electronics [116–118].

5.2 *Gelation Parameters and Influence on Gel Ultrastructure*

Like collagen, silk can be processed into a variety of forms including mats, films, electrospun fibers, porous sponges, micro- and nanoparticles, and hydrogels [105]. Depending on the way silk is processed, there are different resulting cell behaviors and in vivo host responses [112]. Silk is solubilized by first "de-gumming", which is performed by either boiling cocoons in an alkaline solution or exposing to enzymes that produce sericin-free fibers (Fig. 5). The fibers are then dissolved in a highly concentrated chaotropic salt solution (commonly 9 M LiBr) and then dialyzed against water to obtain an aqueous solution [105]. In the aqueous solution, silk proteins aggregate as random coils. The presence of calcium brings proteins together through hydrophilic domains, and thus facilitates gelation. The proteins transition to beta sheets through hydrophobic interactions, and hydrogen bonding eventually forms a percolating network that forms the gel. Thus, increasing silk concentration or temperature facilitates more rapid interaction and therefore,

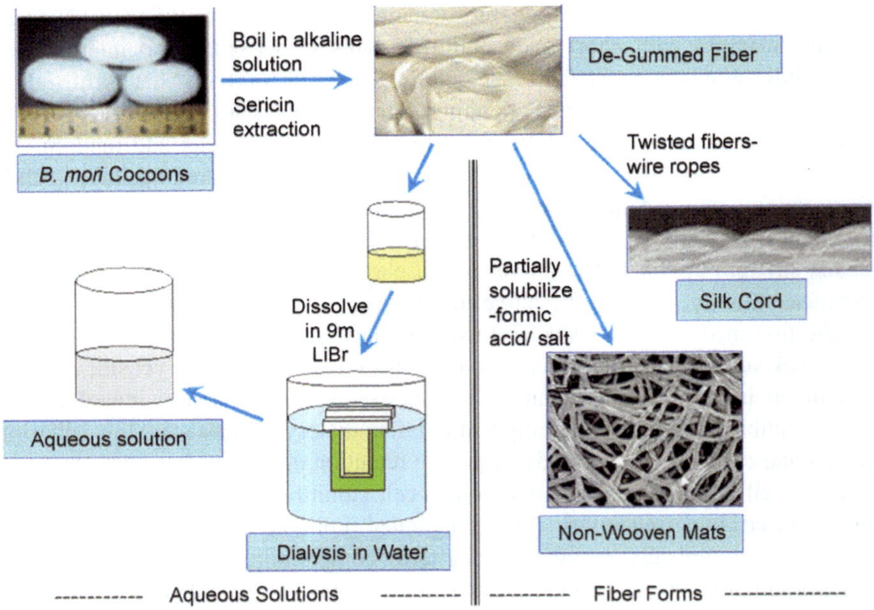

Fig. 5 Silk fibers are "de-gummed" (rid of sericins) by boiling *B.mori* cocoons in an alkaline solution. The fibers are then dissolved in a highly concentrated chaotropic salt solution and dialyzed against water to obtain an aqueous solution. Reprinted from [105], Copyright 2007, with permission from Elsevier

quicker gelation [119, 120]. Higher temperatures cause hydrophobic domains to be more accessible; the hydrophobic domains can subsequently aggregate by either less solvation or from greater exposure of these domains after temperature-induced unfolding [120]. Decreasing the pH (near silk's isoelectric point (pI, ~4)) or adding a hydrophilic polymer [poly(ethylene oxide)] decreases protein–protein repulsion, and therefore causes faster gelation [120]. Silk solutions with calcium exhibited fastest gelation, two days, at pH values of 3–4, near the isoelectric point of silk ~4 [121]. Gels did not form at pH values below 1.5 or above 13 [121]. Other additivities such as poloxamers [119] and glycerol [122, 123] can also decrease gelation time with increasing concentrations. Furthermore, gelation was obtainable at pH 7 with a poloaxmer (not achievable without) [119] and increased the mechanical strength of the gels [124]. The addition of poloaxmer also enabled thermoreversible gelation [119]. Poloaxmer bonds with silk and interferes with silk intermolecular forces that are so strong without poloaxmer that gelation is irreversible. Pore sizes can also be affected by gelation parameters. Smaller pore sizes are achieved by increasing protein concentration or gelation temperature [120], and pore sizes can be increased by increasing the calcium concentration [120]. Mechanical properties are enhanced (likely due to gel ultrastructure) at higher protein concentrations or higher gelation temperatures [120].

5.3 *Gelation Strategies for Cell Studies*

The pH and time of gelation are two key issues for cell seeding and encapsulation with silk hydrogels. In one study, a silk hydrogel was formed in a citric acid solution at a pH less than the pI [125]. The resulting gel had a pH of 3.3, and so, in order to increase the pH for cell studies, the gel was extracted with saline solution. Although the pH only increased to 5.8, favorable cell and in vivo compatibility were still observed. The time of gelation (days) is obviously impractical and furthermore, eliminates the possibility of cell encapsulation. A number of techniques have been developed that dramatically decrease gelation time to hours or minutes, and some have led to viable cell encapsulation. The methods include vortex-induced hydrogelation [126], ultrasonication [127], alcohol-induced [128], and electrogelation [129–133]. In the case of electrogelation, a DC current was applied to an aqueous silk solution, and within seconds, a very soft (mucous-like) gel began to form at the positive electrode [130]. The gel remained in absence of the electric field; however, the process was reversible by switching the polarity. The formed gel also displayed thermoresponsive behavior. When the gel was heated above 60 °C, it turned into a solution, and when cooled back to room temperature, the solution reformed into a gel [130]. The mechanism by which these gels form is a local decrease in pH at the positive electrode, facilitating aggregation. Electric field strength, pH, and silk concentration affect gelation kinetics and gel mechanical properties [133]. Produced gels display a mostly random coil configuration. Chemical and physical cross-links may also be employed to thermally and mechanically stabilize the gels [131]. However, the percentage of amino acids with modifiable side chains is much more limited in silk than compared to mammalian extracellular matrix proteins [105]. Examples of cross-linking reactions include carbodiimide and glucose oxidase. Silk has also been 3D printed into defined, microporous structures, and with the recent advances in different gelation methods, more instances of microfabricated silk may be seen in the near future [134].

6 Other Proteins: Elastin, Keratin, and Tissue-specific ECM

6.1 *Elastin*

Elastin is the structural extracellular matrix protein largely responsible for a tissue's elasticity, specifically the skin, lungs, and blood vessels. It is particularly of interest for skin and vascular tissue engineering for its mechanical properties as well as natural cell binding and degradation sites [135, 136]. The monomer of elastin, tropoelastin, has both hydrophilic and hydrophobic domains. The hydrophilic domains are rich in amines, lysines, for cross-linking while the hydrophobic domains interact during self-assembly [135]. 75 % of elastin's amino acids are nonpolar amino acids, alanine, proline, valine, and glycine, producing one of the most hydrophobic proteins known [137]. At physiological temperature, salt concentration, and pH, tropoelastin monomers aggregate through hydrophobic domains and are subsequently heavily cross-linked with lysyl oxidase into insoluble elastin [137].

For biomaterials research, the insolubility of elastin has posed a great challenge for material processing [135]. However, recent strategies have emerged to facilitate its use. These strategies include solubilization techniques, utilization of the monomer tropoelastin, or peptide and recombinant proteins. Elastin-like peptides and recombinant techniques (including recombinant tropoelastin) are beyond the scope of this chapter, and thus are not discussed further. Although elastin-like polypeptides and recombinant tropoelastin dominate elastin-based biomaterials, there have been a few instances of elastin biomaterials, specifically alpha-elastin hydrogels derived from solubilization techniques. The two types of solubilization methods that have been developed are with oxalic acid to produce alpha-elastin and with potassium hydroxide to produce kappa-elastin [137]. The synthesis of elastin hydrogels occurs in a similar fashion as in vivo elastin formation from tropoelastin, and that is alpha-elastin assembles through hydrophobic interactions. The hydrogel is then cross-linked to stabilize and increase mechanical properties for cell culture [138–140]. High pressure CO_2 may be used during cross-linking to impart a porous structure for cell infiltration [138, 139]. The future of elastin-based hydrogel biomaterials, particularly for cell-encapsulating applications, will likely rely on peptide and recombinant strategies due to ease of use, mild conditions, and accessibility, although solubilized elastin hydrogels studies may continue, especially in the case of protein or polymer blends, as well as non-hydrogel biomaterials such as coatings and elastin tethering.

6.2 Keratin

Keratins are epithelia and epidermal cytoplasmic intermediate filament (IF) proteins that are responsible for the toughness of hair, hooves, nails, and horns. Uniquely, keratin extracted from human hair could be used as a cheap and renewable biomaterial; it is estimated that hundreds of thousands of tons of human hair are annually discarded worldwide [141]. Highly cross-linked hair fibers are composed of 50–60 % low-sulfur alpha keratins and 20–30 % high-sulfur matrix proteins [142]. The alpha keratins assemble into coiled-coil IFs of 10 nm length while the matrix proteins mostly serve as disulfide cross-linkers [143]. Keratin extraction and purification procedures have been developed. Extracted keratin has been processed into films, scaffolds (sponges), and fibers and studied for tissue engineering and regenerative medicine [144, 142]. Thus far, it has shown good cell and in vivo biocompatibility as well as biodegradability [144]. Recently, keratin hydrogels have been developed from soluble extracts [145–148]. The procedure of extracting alpha keratin from human hair consists of delipidization followed by solubilization using either oxidative or reductive reactions to break down disulfide bonds (Fig. 6) [144].

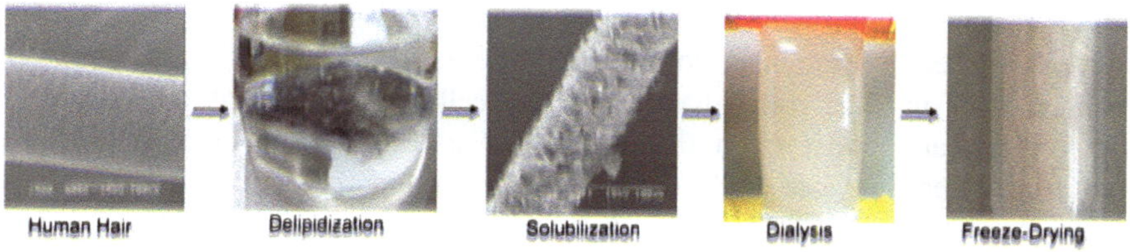

Fig. 6 Keratin extraction. Human hair is delipidized and is then solubilized by breaking down disulfide bonds. The extract is purified by dialysis, lyophilized, and then ground to form a powder that can then be reconstituted in saline solution. Reprinted from [144], Copyright 2014, with permission from Springer

The extract is purified by dialysis, concentrated, neutralized, lyophilized, and then ground to form a powder that can then be reconstituted in saline solution. Thereafter, keratin molecules self-assemble to form a gel. These have been investigated with promising results for nerve repair [145–147].

6.3 *Tissue-Specific Decellularized Extracellular Matrix (dECM)*

Decellularization is required for isolating specific structural proteins from tissue, like in the case of collagen, gelatin, fibrinogen, and elastin. More gentle methods have been developed in order to maintain the original architecture of animal or human cadaver tissues and organs (Fig. 7). For most researchers, the residual tissue matrix will then be recellularized by in vitro perfusion and by host infiltration after implantation to serve as an engineered repair or replacement tissue [149]. ECM is vital in many pathways that are not yet well understood and is a complex mixture of

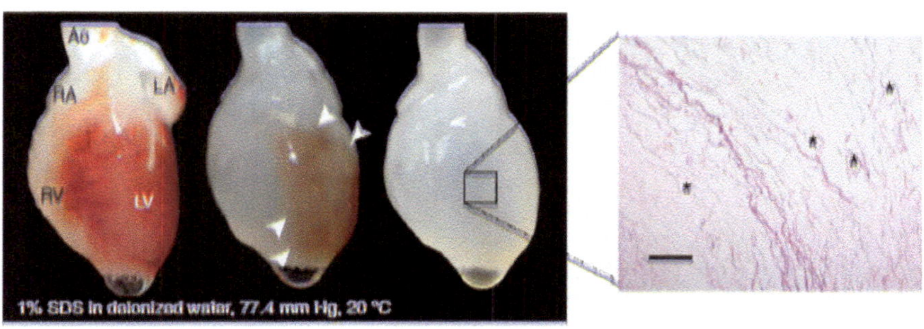

Fig. 7 The decellularization of a rat heart by perfusion with a detergent solution. The tissue becomes clear, leaving behind the extracellular matrix. The histological section (hematoxylin and eosin staining) shows that all cells have been removed. Reprinted from [159], Copyright 2008, with permission from Nature Publishing Group

proteins (structural, growth factors), glycoproteins, and glycosaminoglycans [150]. Because a specific protein is not isolated, all of these complex components are retained, and in fact, many have shown excellent retention of collagen, glycosaminoglycans, and growth factors [151]. Furthermore, the bioactivity of the matrix is evident by those who have shown stem differentiation into the appropriate tissue-specific lineage [152, 153]. Particularly for tissue engineering, this is a valuable approach as one does not need to add lost key biochemical signals back into the matrix. Hydrogels have recently been developed, including even one case of those 3D printed, from many tissue-specific ECMs: cardiac [154, 153, 155], liver [156, 157], adipose [152, 153], dermis [158], bladder [158], and cartilage [153]. After decellularization, the tissue is processed similar to how collagen fibrils are processed for solubilization. Lyophilized tissue is milled into a powder and solubilized by an acidic pepsin digestion (Fig. 8). After adjusting the salt concentration and pH of a cold dECM solution, the temperature is raised to induce self-assembly of the proteins to form a hydrogel. The ultrastructure of the gel can be modulated by changing pH, temperature, and ionic strength, similar to that which can be done with collagen gels. Particular challenges for the future study of dECM hydrogel biomaterials will be to identify the many components and how each impacts gel ultrastructure, mechanical properties, and bioactivity, as well as how processing conditions can impact material properties and bioactivity (i.e., decellularization method, length of pepsin digestion).

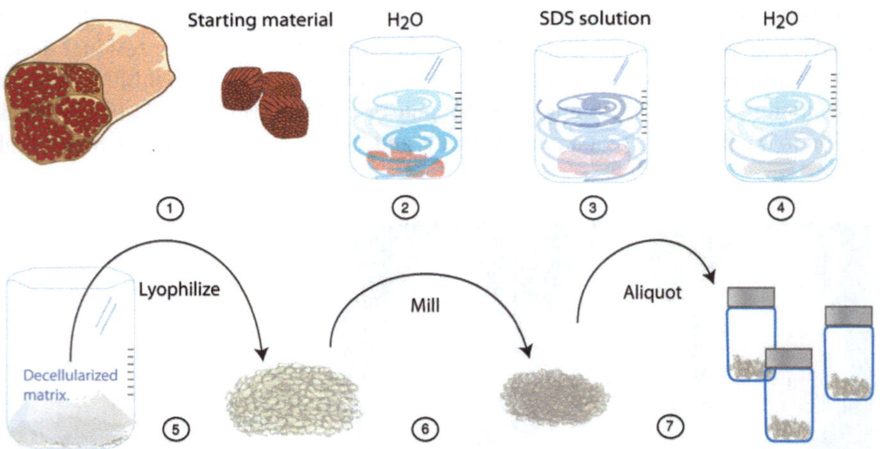

Fig. 8 The preparation of tissue-specific decellularized extracellular matrix (dECM) hydrogels. Tissue is minced, washed, and decellularized with detergent. The tissue is then lyophilized and milled into a fine powder, which is then suspended in an acidic pepsin solution to solubilize the components. Reprinted from [160]

7 Chemical Cross-linking

As mentioned above, with many of the discussed protein hydrogels, mechanical strength, thermal stability, and rapid degradation are typically concerns for end applications. Chemical cross-linking is performed in many cases to improve these properties. Since proteins are amino acid-based, the potential cross-linking reactions are almost universal for proteins, albeit conditions such as those for solubility are important considerations that may make some reactions not possible. The most abundant functional groups for reaction are amines and carboxylic acids. Amines are present at both the N terminus as well as in lysine and arginine residues. Likewise, carboxylic acids are found at the C terminus and glutamic and aspartic acid residues. Other functional groups exist, particularly hydroxyl and sulfhydryl groups, that may be utilized; although, these groups are typically not found in abundant enough quantities for substantial cross-linking. However, sulfhydryl and hydroxyl groups can be used for tethering, functionalization, or grafting purposes. The degree of cross-linking, and therefore, mechanical properties, degradation rate, and release rates of payloads, can be tuned by simply changing the concentration of the cross-linker.

Protein biomaterials are most commonly cross-linked by aldehydes (cross-links by a variety of mechanisms [161]) and amine–carboxylic acid carbodiimide coupling [51, 162, 163, 164]. Both of these cross-linkers, however, are cytotoxic and must be thoroughly removed prior to cell work, and thus, also eliminate the possibility of cell encapsulation. There have been instances of PEG cross-linkers (bi or multifunctional) with aldehyde or activated ester (intermediate in carboxylic–amine coupling succinimide) groups used for cross-linking [43, 165, 166, 167, 168]. Since PEG remains in the hydrogel, if used in large quantities, the properties of the gel will change (e.g., increase mechanical properties, increase swelling). Other small molecules such as isocyanate and genipin have also been used for cross-linking, although cytotoxicity is still cited as a concern with these. Genipin has been used to cross-link collagen [169, 170], gelatin [171], fibrin [172], and silk [173, 174]. Genipin is a natural small molecule derived from gardenia fruits that reacts with primary amines. Although these aforementioned cross-linkers carry a risk of toxicity, carbodiimide and genipin are regarded as safer than aldehyde cross-linking [175, 176]. Transglutaminase, the enzyme responsible for fibrin cross-linking in vivo, has also been used for cross-linking other proteins in vitro [177, 178]. For more details on these reactions (excluding enzymes) as well as other methods such as biomolecule attachment to proteins or PEGylation, readers are referred to the text "Bioconjugate Techniques" [179].

8 Functionalization and Cell-Friendly Encapsulation

To expand potential cross-linking reactions, proteins may be functionalized to present more utilizable functional groups. Namely, amines and carboxylic acids can be reacted with bifunctional small molecules that on the unreacted end, display the functional group of interest. Ethylenediamine has been used to change carboxylic acid groups to amines to catonize gelatin for controlled release of plasmid DNA [64, 180, 181]. The abundance of positive charges provides plasmid DNA complexation with its negatively charged phosphate backbone [182]. Functionalization with phenolic hydroxyl group had led to peroxidase-catalyzed cross-linked proteins [183–185]. More commonly, acrylation or methacrylation has been performed to create UV cross-linkable gels for cell (Fig. 2) [186–188] and growth factor [189] encapsulation and stabilization after additive manufacturing [190, 191]. Likewise, thiolation can also be performed to expand cross-linking to thiol-ene, thiol-maleimide, and thiol-vinyl sulfone reactions, all regarded as cell-friendly [192–194]. These thiol-based reactions are included in a group of biorthogonal reactions referred to as "click chemistry" [195]. Click reactions have been developed extensively in the past few years to include cell-friendly (metal-free and mild reaction conditions) chemistries [196]. In fact, PEG, hyaluronic acid, and gelatin hydrogels have been fabricated by click chemistry, and it is expected that more click hydrogels will be developed as this group of reactions becomes more widely used. In addition to cross-linking, functionalization of proteins is instrumental in tagging proteins with targeting motifs, drugs, or imaging modalities [197]. Functionalization of specific functional groups can enable site-specific modification, such as those performed with click chemistry. Covalent modification is especially important for hydrogels since small molecules passively diffuse through the material and larger or charged encapsulated molecules have quick release. It should be noted that of all proteins, gelatin is the most easily modified. Unlike collagen, the range of conditions at which gelatin is soluble as well as minimal risk of denaturing allow gelatin to be synthesized with other functional groups for chemical cross-linking and easily processed into other forms.

9 Conclusion

In summary, protein hydrogels are a valuable biomaterial for their inherent biocompatibility and biodegradability. Since most protein hydrogels undergo a mild sol–gel transition, this allows cells to be encapsulated in the solution and subsequently gelled. Cell-encapsulating protein hydrogels provide cell adhesion and cell degradation sequences and may be used for injectable cell therapies or tissue engineering. Protein hydrogels also provide a gentle means for encapsulation of biomolecule payloads for controlled release. In addition to drug delivery, tissue engineering and regenerative medicine, protein hydrogels may be useful for other applications, such as biosensors and bioelectronics.

Of all of the proteins discussed, collagen is by far the most widely studied. Due to the natural abundance of collagen in our own tissues, collagen will continue to be important for mimetic tissue engineering strategies. Derived from collagen, gelatin has also become an important biomaterial due to mild and easy solubilization. Gelatin still possesses the primary sequence of collagen and therefore, still has cell adhesion and degradation sites, yet with the loss of nanostructure. Gelatin can be processed with different IEPs to electrostatically bind and subsequently release charged biomolecules, and can also be easily modified with other functional groups to expand potential cross-linking strategies. Fibrin is the most bioactive of the discussed proteins, containing growth factors, growth factor-binding sequences, and ECM protein-binding sequences. Ideally, the soluble monomer (fibrinogen) can also be harvested from a patient to provide a patient-specific biomaterial. There is still much unknown about fibrin and its roles in tissue regeneration and disease, and therefore, as we know more, more biomaterial advances may ensue. Silk, the most widely used nonmammalian protein, has been established as biocompatible and is cited for its incredible mechanical strength; although with hydrogels, these high mechanical properties are not retained. Solubilization and gelation strategies for silk have been developed, including those for short time scales and conditions friendly for cell encapsulation. Although elastin is physiologically important for many of our compliant tissues, its solubilization has made utilization in biomaterials rather difficult. Some instances of elastin hydrogels have been produced, but recombinant elastin and elastin-like peptides dominate the field and will continue to do so. Emerging protein hydrogels include those containing keratin and dECM. Keratin is unique in the sense that it is isolated from discarded human hair and therefore provides a cheap and renewable biomaterial source. Tissue-specific dECM will likely prove to be instrumental in tissue engineering as many bioactive factors found in the native tissue are retained, and some researchers have proven enhanced stem cell differentiation to the specific lineages on these matrices.

As discussed above, parameters that influence protein monomer self-assembly have been uncovered, and thereby, have led to modulation of gel ultrastructure. More advances in this arena will continue, but far more promising technologies are emerging. Specifically, those that can create heterogeneous hydrogels, either on the nanoscale with specific chemistries or on the microscale with microfabrication technologies. Other emerging technologies include genetic engineering and biomanufacturing. Particularly recombinant proteins will prove to be easier for protein hydrogel processing, like in the case of elastin and spider silk. As well, bioactive peptides (e.g., cell adhesion, growth factor binding) and protein-like peptides (e.g., elastin-like, silk-like) will continue to be an important strategy for harnessing the strong bioactivity of protein hydrogels when the protein hydrogels themselves are not ideal for the end application.

References

1. Ricard-Blum S (2011) The Collagen Family. Cold Spring Harb Perspect Biol 3:1–19. doi:10.1101/cshperspect.a004978
2. Kadler KE, Holmes DF, Trotter JA, Chapman JA (1996) Collagen fibril formation. Biochem J 316 (Pt 1:1–11)

3. Antoine EE, Vlachos PP, Rylander MN (2014) Review of collagen I hydrogels for bioengineered tissue microenvironments: characterization of mechanics, structure, and transport. Tissue Eng Part B Rev 20:1–14. doi:10.1089/ten.TEB.2014.0086
4. Abou Neel EA, Bozec L, Knowles JC, Syed O, Mudera V, Day R, Hyun JK (2013) Collagen —emerging collagen based therapies hit the patient. Adv Drug Deliv Rev 65:429–456. doi:10.1016/j.addr.2012.08.010
5. Chattopadhyay S, Raines RT (2014) Review collagen-based biomaterials for wound healing. Biopolymers 101:821–833. doi:10.1002/bip.22486
6. Ferreira AM, Gentile P, Chiono V, Ciardelli G (2012) Collagen for bone tissue regeneration. Acta Biomater 8:3191–3200. doi:10.1016/j.actbio.2012.06.014
7. Kew SJ, Gwynne JH, Enea D, Abu-Rub M, Pandit A, Zeugolis D, Brooks RA, Rushton N, Best SM, Cameron RE (2011) Regeneration and repair of tendon and ligament tissue using collagen fibre biomaterials. Acta Biomater 7:3237–3247. doi:10.1016/j.actbio.2011.06.002
8. Wang X, Han C, Hu X, Sun H, You C, Gao C, Haiyang Y (2011) Applications of knitted mesh fabrication techniques to scaffolds for tissue engineering and regenerative medicine. J Mech Behav Biomed Mater 4:922–932. doi:10.1016/j.jmbbm.2011.04.009
9. Zheng W, Zhang W, Jiang X (2010) Biomimetic collagen nanofibrous materials for bone tissue engineering. Adv Eng Mater 12:B451–B466. doi:10.1002/adem.200980087
10. Walters BD, Stegemann JP (2014) Strategies for directing the structure and function of three-dimensional collagen biomaterials across length scales. Acta Biomater 10:1488–1501. doi:10.1016/j.actbio.2013.08.038
11. Carey SP, Kraning-Rush CM, Williams RM, Reinhart-King CA (2012) Biophysical control of invasive tumor cell behavior by extracellular matrix microarchitecture. Biomaterials 33:4157–4165. doi:10.1016/j.biomaterials.2012.02.029
12. Chrobak KM, Potter DR, Tien J (2006) Formation of perfused, functional microvascular tubes in vitro. Microvasc Res 71:185–196. doi:10.1016/j.mvr.2006.02.005
13. Yang YL, Motte S, Kaufman LJ (2010) Pore size variable type I collagen gels and their interaction with glioma cells. Biomaterials 31:5678–5688. doi:10.1016/j.biomaterials.2010.03.039
14. Gillette BM, Rossen NS, Das N, Leong D, Wang M, Dugar A, Sia SK (2011) Engineering extracellular matrix structure in 3D multiphase tissues. Biomaterials 32:8067–8076. doi:10.1016/j.biomaterials.2011.05.043
15. Wolf K, Alexander S, Schacht V, Coussens LM, von Andrian UH, van Rheenen J, Deryugina E, Friedl P (2009) Collagen-based cell migration models in vitro and in vivo. Semin Cell Dev Biol 20:931–941. doi:10.1016/j.semcdb.2009.08.005
16. Abou Neel EA, Cheema U, Knowles J, Brown R, Nazhat S (2006) Use of multiple unconfined compression for control of collagen gel scaffold density and mechanical properties. Soft Matter 2(11):986–992. doi:10.1039/b609784g
17. Brown RA, Wiseman M, Chuo CB, Cheema U, Nazhat SN (2005) Ultrarapid engineering of biomimetic materials and tissues: fabrication of nano- and microstructures by plastic compression. Adv Funct Mater 15:1762–1770. doi:10.1002/adfm.200500042
18. Feng Z, Tateishi Y, Nomura Y, Kitajima T, Nakamura T (2006) Construction of fibroblast-collagen gels with orientated fibrils induced by static or dynamic stress: toward the fabrication of small tendon grafts. J Artif Organs 9:220–225. doi:10.1007/s10047-006-0354-z
19. Gigante A, Cesari E, Busilacchi A, Manzotti S, Kyriakidou K, Greco F, Di Primio R, Mattioli-Belmonte M (2009) Collagen I membranes for tendon repair: Effect of collagen fiber orientation on cell behavior. J Orthop Res 27:826–832. doi:10.1002/jor.20812
20. Seliktar D, Black RA, Vito RP, Nerem RM (2000) Dynamic mechanical conditioning of collagen-gel blood vessel constructs induces remodeling in vitro. Ann Biomed Eng 28:351–362. doi:10.1114/1.275
21. Huang G, Wang L, Wang S, Han Y, Wu J, Zhang Q, Xu F, Lu TJ (2012) Engineering three-dimensional cell mechanical microenvironment with hydrogels. Biofabrication 4:042001. doi:10.1088/1758-5082/4/4/042001

22. Nguyen TD, Liang R, Woo SL-Y, Burton SD, Wu C, Almarza A, Sacks MS, Abramowitch S (2009) Effects of cell seeding and cyclic stretch on the fiber remodeling in an extracellular matrix-derived bioscaffold. Tissue Eng Part A 15:957–963. doi:10.1089/ten.tea.2007.0384
23. Pins GD, Christiansen DL, Patel R, Silver FH (1997) Self-assembly of collagen fibers. Influence of fibrillar alignment and decorin on mechanical properties. Biophys J 73:2164–2172. doi:10.1016/S0006-3495(97)78247-X
24. Ceballos D, Navarro X, Dubey N, Wendelschafer-Crabb G, Kennedy WR, Tranquillo RT (1999) Magnetically aligned collagen gel filling a collagen nerve guide improves peripheral nerve regeneration. Exp Neurol 158:290–300. doi:10.1006/exnr.1999.7111
25. Chen S, Hirota N, Okuda M, Takeguchi M, Kobayashi H, Hanagata N, Ikoma T (2011) Microstructures and rheological properties of tilapia fish-scale collagen hydrogels with aligned fibrils fabricated under magnetic fields. Acta Biomater 7:644–652. doi:10.1016/j.actbio.2010.09.014
26. Torbet J, Ronzière MC (1984) Magnetic alignment of collagen during self-assembly. Biochem J 219:1057–1059
27. Guo C, Kaufman LJ (2007) Flow and magnetic field induced collagen alignment. Biomaterials 28:1105–1114. doi:10.1016/j.biomaterials.2006.10.010
28. Cheng X, Gurkan UA, Dehen CJ, Tate MP, Hillhouse HW, Simpson GJ, Akkus O (2008) An electrochemical fabrication process for the assembly of anisotropically oriented collagen bundles. Biomaterials 29:3278–3288. doi:10.1016/j.biomaterials.2008.04.028
29. Saeidi N, Karmelek KP, Paten J a., Zareian R, DiMasi E, Ruberti JW (2012) Molecular crowding of collagen: A pathway to produce highly-organized collagenous structures. Biomaterials 33:7366–7374. doi:10.1016/j.biomaterials.2012.06.041
30. Hong S, Hsu H-J, Kaunas R, Kameoka J (2012) Collagen microsphere production on a chip. Lab Chip 12:3277. doi:10.1039/c2lc40558j
31. Chan BP, Hui TY, Yeung CW, Li J, Mo I, Chan GCF (2007) Self-assembled collagen-human mesenchymal stem cell microspheres for regenerative medicine. Biomaterials 28:4652–4666. doi:10.1016/j.biomaterials.2007.07.041
32. Cheng HW, Luk KDK, Cheung KMC, Chan BP (2011) In vitro generation of an osteochondral interface from mesenchymal stem cell-collagen microspheres. Biomaterials 32:1526–1535. doi:10.1016/j.biomaterials.2010.10.021
33. Matsunaga YT, Morimoto Y, Takeuchi S (2011) Molding cell beads for rapid construction of macroscopic 3D tissue architecture. Adv Mater 23:90–94. doi:10.1002/adma.201004375
34. Bian W, Liau B, Badie N, Bursac N (2009) Mesoscopic hydrogel molding to control the 3D geometry of bioartificial muscle tissues. Nat Protoc 4:1522–1534. doi:10.1038/nprot.2009.155
35. Tang MD, Golden AP, Tien J (2004) Fabrication of collagen gels that contain patterned, micrometer-scale cavities. Adv Mater 16:1345–1348. doi:10.1002/adma.200400766
36. Cross VL, Zheng Y, Won Choi N, Verbridge SS, Sutermaster BA, Bonassar LJ, Fischbach C, Stroock AD (2010) Dense type I collagen matrices that support cellular remodeling and microfabrication for studies of tumor angiogenesis and vasculogenesis in vitro. Biomaterials 31:8596–8607. doi:10.1016/j.biomaterials.2010.07.072
37. Golden AP, Tien J (2007) Fabrication of microfluidic hydrogels using molded gelatin as a sacrificial element. Lab Chip 7:720–725. doi:10.1039/b618409j
38. Nazhat SN, Abou Neel EA, Kidane A, Ahmed I, Hope C, Kershaw M, Lee PD, Stride E, Saffari N, Knowles JC, Brown RA (2007) Controlled microchannelling in dense collagen scaffolds by soluble phosphate glass fibers. Biomacromolecules 8:543–551. doi:10.1021/bm060715f
39. Stegemann JP, Kaszuba SN, Rowe SL (2007) Review: advances in vascular tissue engineering using protein-based biomaterials. Tissue Eng 13:2601–2613. doi:10.1089/ten.2007.0196
40. Lee W, Lee V, Polio S, Keegan P, Lee JH, Fischer K, Park JK, Yoo SS (2010) On-demand three-dimensional freeform fabrication of multi-layered hydrogel scaffold with fluidic channels. Biotechnol Bioeng 105:1178–1186. doi:10.1002/bit.22613

41. Smith CM, Stone AL, Parkhill RL, Stewart RL, Simpkins MW, Kachurin AM, Warren WL, Williams SK (2004) Three-dimensional bioassembly tool for generating viable tissue-engineered constructs. Tissue Eng 10:1566–1576. doi:10.1089/ten.2004.10.1566
42. Smith CM, Christian JJ, Warren WL, Williams SK (2007) Characterizing environmental factors that impact the viability of tissue-engineered constructs fabricated by a direct-write bioassembly tool. Tissue Eng 13:373–383. doi:10.1089/ten.2007.13.ft-338
43. Chan BK, Wippich CC, Wu CJ, Sivasankar PM, Schmidt G (2012) Robust and semi-interpenetrating hydrogels from poly(ethylene glycol) and collagen for elastomeric tissue scaffolds. Macromol Biosci 12:1490–1501. doi:10.1002/mabi.201200234
44. Piechocka IK, Van Oosten ASG, Breuls RGM, Koenderink GH (2011) Rheology of heterotypic collagen networks. Biomacromolecules 12:2797–2805. doi:10.1021/bm200553x
45. Saw MM, Chandler B, Ho KM (2012) Benefits and risks of using gelatin solution as a plasma expander for perioperative and critically ill patients: a meta-analysis. Anaesth Intensive Care 40:17–32
46. Rose JB, Pacelli S, El Haj AJ, Dua HS, Hopkinson A, White LJ, Rose FR a J (2014) Gelatin-based materials in ocular tissue engineering. Materials (Basel) 7:3106–3135. doi:10.3390/ma7043106
47. Baziwane D, He Q (2003) Gelatin: the paramount food additive. Food Rev Int 19:423–435. doi:10.1081/FRI-120025483
48. Van Den Bulcke AI, Bogdanov B, De Rooze N, Schacht EH, Cornelissen M, Berghmans H (2000) Structural and rheological properties of methacrylamide modified gelatin hydrogels. Biomacromolecules 1:31–38. doi:10.1021/bm990017d
49. Young S, Wong M, Tabata Y, Mikos AG (2005) Gelatin as a delivery vehicle for the controlled release of bioactive molecules. J Control Release 109:256–274. doi:10.1016/j.jconrel.2005.09.023
50. Lee J (2007) The Use of Gelatin Substrates for Traction Force Microscopy in Rapidly Moving Cells. Methods Cell Biol. doi: 10.1016/S0091-679X(07)83012-3
51. Kuijpers AJ, Engbers GH, Krijgsveld J, Zaat SA, Dankert J, Feijen J (2000) Cross-linking and characterisation of gelatin matrices for biomedical applications. J Biomater Sci Polym Ed 11:225–243. doi:10.1163/156856200743670
52. Yamamoto M, Ikada Y, Tabata Y (2001) Controlled release of growth factors based on biodegradation of gelatin hydrogel. J Biomater Sci Polym Ed 12:77–88
53. Kang HW, Tabata Y, Ikada Y (1999) Fabrication of porous gelatin scaffolds for tissue engineering. Biomaterials 20:1339–1344
54. Elzoghby AO (2013) Gelatin-based nanoparticles as drug and gene delivery systems: reviewing three decades of research. J Control Release 172:1075–1091. doi:10.1016/j.jconrel.2013.09.019
55. Iwanaga K, Yabuta T, Kakemi M, Morimoto K, Tabata Y, Ikada Y (2003) Usefulness of microspheres composed of gelatin with various cross-linking density. J Microencapsul 20:767–776. doi:10.1080/0265204031000l600523
56. Wang X, Yan Y, Pan Y, Xiong Z, Liu H, Cheng J, Liu F, Lin F, Wu R, Zhang R, Lu Q (2006) Generation of three-dimensional hepatocyte/gelatin structures with rapid prototyping system. Tissue Eng 12:83–90. doi:10.1089/ten.2006.12.ft-16
57. Thompson JA, Anderson KD, DiPietro JM, Zwiebel JA, Zametta M, Anderson WF, Maciag T (1988) Site-directed neovessel formation in vivo. Science 241:1349–1352. doi:10.1126/science.2457952
58. Yamada K, Tabata Y, Yamamoto K, Miyamoto S, Nagata I, Kikuchi H, Ikada Y (1997) Potential efficacy of basic fibroblast growth factor incorporated in biodegradable hydrogels for skull bone regeneration. J Neurosurg 86:871–875. doi:10.3171/jns.1997.86.5.0871
59. Yamamoto M, Tabata Y, Hong L, Miyamoto S, Hashimoto N, Ikada Y (2000) Bone regeneration by transforming growth factor β1 released from a biodegradable hydrogel. J Control Release 64:133–142. doi:10.1016/S0168-3659(99)00129-7

60. Yamamoto M, Takahashi Y, Tabata Y (2003) Controlled release by biodegradable hydrogels enhances the ectopic bone formation of bone morphogenetic protein. Biomaterials 24:4375–4383. doi:10.1016/S0142-9612(03)00337-5
61. Holland TA, Tabata Y, Mikos AG (2005) Dual growth factor delivery from degradable oligo (poly(ethylene glycol) fumarate) hydrogel scaffolds for cartilage tissue engineering. J Control Release 101:111–25. doi:10.1016/j.jconrel.2004.07.004
62. Kojima K, Ignotz RA, Kushibiki T, Tinsley KW, Tabata Y, Vacanti CA (2004) Tissue-engineered trachea from sheep marrow stromal cells with transforming growth factor β2 released from biodegradable microspheres in a nude rat recipient. J Thorac Cardiovasc Surg 128:147–153. doi:10.1016/j.jtcvs.2004.02.038
63. Nakahara T, Nakamura T, Kobayashi E, Inoue M, Shigeno K, Tabata Y, Eto K, Shimizu Y (2003) Novel approach to regeneration of periodontal tissues based on in situ tissue engineering: effects of controlled release of basic fibroblast growth factor from a sandwich membrane. Tissue Eng 9:153–162. doi:10.1089/107632703762687636
64. Fukunaka Y, Iwanaga K, Morimoto K, Kakemi M, Tabata Y (2002) Controlled release of plasmid DNA from cationized gelatin hydrogels based on hydrogel degradation. J Control Release 80:333–343
65. Ikada Y, Tabata Y (1998) Protein release from gelatin matrices. Adv Drug Deliv Rev 31:287–301
66. La Corte ALC, Philippou H, Arins RAS (2011) Role of fibrin structure in thrombosis and vascular disease, 1st edn. In: Advances in protein chemistry structural biology. doi:10.1016/B978-0-12-381262-9.00003-3
67. Janmey PA, Winer JP, Weisel JW (2009) Fibrin gels and their clinical and bioengineering applications. J R Soc Interface 6:1–10. doi:10.1098/rsif.2008.0327
68. Spotnitz WD (2010) Fibrin sealant: past, present, and future: a brief review. World J Surg 34:632–634. doi:10.1007/s00268-009-0252-7
69. Buchta C, Hedrich HC, Macher M, Höcker P, Redl H (2005) Biochemical characterization of autologous fibrin sealants produced by CryoSeal® and Vivostat® in comparison to the homologous fibrin sealant product Tissucol/Tisseel®. Biomaterials 26:6233–6241. doi:10.1016/j.biomaterials.2005.04.014
70. Clark RAF (2003) Fibrin is a many splendored thing. J Invest Dermatol 121:1. doi:10.1046/j.1523-1747.2003.12575.x
71. Carless PA, Anthony DM, Henry DA (2002) Systematic review of the use of fibrin sealant to minimize perioperative allogeneic blood transfusion. Br J Surg 89:695–703. doi:10.1046/j.1365-2168.2002.02098.x
72. Ahmed TAE, Dare EV, Hincke M (2008) Fibrin: a versatile scaffold for tissue engineering applications. Tissue Eng Part B Rev 14:199–215. doi:10.1089/ten.teb.2007.0435
73. Anitua E, Prado R, Orive G (2013) Endogenous morphogens and fibrin bioscaffolds for stem cell therapeutics. Trends Biotechnol 31:364–374. doi:10.1016/j.tibtech.2013.04.003
74. Kollman JM, Pandi L, Sawaya MR, Riley M, Doolittle RF (2009) Crystal structure of human fibrinogen. Biochemistry 48:3877–3886. doi:10.1021/bi802205g
75. Ryan EA, Mockros LF, Weisel JW, Lorand L (1999) Structural origins of fibrin clot rheology. Biophys J 77:2813–2826. doi:10.1016/S0006-3495(99)77113-4
76. Blombiick B, Carlsson K, Hessel B, Liljeborg A, Procyk R, Aslund N (1989) Native fibrin gel networks observed by 3D microscopy, permeation and turbidity. Biochim Biophys Acta 997:96–110
77. Soon ASC, Lee CS, Barker TH (2011) Modulation of fibrin matrix properties via knob:hole affinity interactions using peptide-PEG conjugates. Biomaterials 32:4406–4414. doi:10.1016/j.biomaterials.2011.02.050
78. Stabenfeldt SE, Gourley M, Krishnan L, Hoying JB, Barker TH (2012) Engineering fibrin polymers through engagement of alternative polymerization mechanisms. Biomaterials 33:535–544. doi:10.1016/j.biomaterials.2011.09.079

79. Dubey N, Letourneau PC, Tranquillo RT (2001) Neuronal contact guidance in magnetically aligned fibrin gels: effect of variation in gel mechano-structural properties. Biomaterials 22:1065–1075. doi:10.1016/S0142-9612(00)00341-0
80. Le Guehennec L, Goyenvalle E, Aguado E, Pilet P, Spaethe R, Daculsi G (2007) Influence of calcium chloride and aprotinin in the in vivo biological performance of a composite combining biphasic calcium phosphate granules and fibrin sealant. J Mater Sci Mater Med 18:1489–1495. doi:10.1007/s10856-006-0086-x
81. Davis HE, Miller SL, Case EM, Leach JK (2011) Supplementation of fibrin gels with sodium chloride enhances physical properties and ensuing osteogenic response. Acta Biomater 7:691–699. doi:10.1016/j.actbio.2010.09.007
82. Wang M-C, Pins GD, Silver FH (1995) Preparation of fibrin glue: the effects of calcium chloride and sodium chloride. Mater Sci Eng C 3:131–135. doi:10.1016/0928-4931(95)00116-6
83. Eyrich D, Brandl F, Appel B, Wiese H, Maier G, Wenzel M, Staudenmaier R, Goepferich A, Blunk T (2007) Long-term stable fibrin gels for cartilage engineering. Biomaterials 28:55–65. doi:10.1016/j.biomaterials.2006.08.027
84. Brown AEX, Litvinov RI, Discher DE, Purohit PK, Weisel JW (2009) Multiscale mechanics of fibrin polymer: gel stretching with protein unfolding and loss of water. Science 325:741–744. doi:10.1126/science.1172484
85. Kang H, Wen Q, Janmey PA, Tang JX, Conti E, MacKintosh FC (2009) Nonlinear elasticity of stiff filament networks: strain stiffening, negative normal stress, and filament alignment in fibrin gels. J Phys Chem B 113:3799–3805. doi:10.1021/jp807749f
86. Matsumoto T, Sasaki JI, Alsberg E, Egusa H, Yatani H, Sohmura T (2007) Three-dimensional cell and tissue patterning in a strained fibrin gel system. PLoS ONE 2:1–6. doi:10.1371/journal.pone.0001211
87. Alsberg E, Feinstein E, Joy MP, Prentiss M, Ingber DE (2006) Magnetically-guided self-assembly of fibrin matrices with ordered nano-scale structure for tissue engineering. Tissue Eng 12(11):3247–3256
88. Cui X, Boland T (2009) Human microvasculature fabrication using thermal inkjet printing technology. Biomaterials 30:6221–6227. doi:10.1016/j.biomaterials.2009.07.056
89. Xu T, Gregory CA, Molnar P, Cui X, Jalota S, Bhaduri SB, Boland T (2006) Viability and electrophysiology of neural cell structures generated by the inkjet printing method. Biomaterials 27:3580–3588. doi:10.1016/j.biomaterials.2006.01.048
90. Breen A, O'Brien T, Pandit A (2009) Fibrin as a delivery system for therapeutic drugs and biomolecules. Tissue Eng Part B Rev 15:201–214. doi:10.1089/ten.teb.2008.0527
91. Whelan D, Caplice NM, Clover AJP (2014) Fibrin as a delivery system in wound healing tissue engineering applications. J Control Release 196:1–8. doi:10.1016/j.jconrel.2014.09.023
92. Joo JY, Amin ML, Rajangam T, An SS a. (2015) Fibrinogen as a promising material for various biomedical applications. Mol Cell Toxicol 11:1–9. doi:10.1007/s13273-015-0001-y
93. Falvo MR, Gorkun OV, Lord ST (2010) The molecular origins of the mechanical properties of fibrin. Biophys Chem 152:15–20. doi:10.1016/j.bpc.2010.08.009
94. De la Puente P, Ludeña D (2014) Cell culture in autologous fibrin scaffolds for applications in tissue engineering. Exp Cell Res 322:1–11. doi:10.1016/j.yexcr.2013.12.017
95. Ceccarelli J, Putnam AJ (2014) Sculpting the blank slate: how fibrin's support of vascularization can inspire biomaterial design. Acta Biomater 10:1515–1523. doi:10.1016/j.actbio.2013.07.043
96. Morin KT, Tranquillo RT (2013) In vitro models of angiogenesis and vasculogenesis in fibrin gel. Exp Cell Res 319:2409–2417. doi:10.1016/j.yexcr.2013.06.006
97. Bjork JW, Johnson SL, Tranquillo RT (2011) Ruthenium-catalyzed photo cross-linking of fibrin-based engineered tissue. Biomaterials 32:2479–2488. doi:10.1016/j.biomaterials.2010.12.010
98. Syedain ZH, Bjork J, Sando L, Tranquillo RT (2009) Controlled compaction with ruthenium-catalyzed photochemical cross-linking of fibrin-based engineered connective tissue. Biomaterials 30:6695–6701. doi:10.1016/j.biomaterials.2009.08.039

99. Schense JC, Hubbell JA (1999) Cross-linking exogenous bifunctional peptides into fibrin gels with factor XIIIa. Bioconjug Chem 10:75–81. doi:10.1021/bc9800769
100. Schense JC, Bloch J, Aebischer P, Hubbell JA (2000) Enzymatic incorporation of bioactive peptides into fibrin matrices enhances neurite extension. Nat Biotechnol 18:415–419. doi:10.1038/74473
101. Martino MM, Tortelli F, Mochizuki M, Traub S, Ben-David D, Kuhn GA, Müller R, Livne E, Eming SA, Hubbell JA (2011) Engineering the growth factor microenvironment with fibronectin domains to promote wound and bone tissue healing. Sci Transl Med 3:100ra89. doi:10.1126/scitranslmed.3002614
102. Sacchi V, Mittermayr R, Hartinger J, Martino MM, Lorentz KM, Wolbank S, Hofmann A, Largo RA, Marschall JS, Groppa E, Gianni-Barrera R, Ehrbar M, Hubbell J a, Redl H, Banfi A (2014) Long-lasting fibrin matrices ensure stable and functional angiogenesis by highly tunable, sustained delivery of recombinant VEGF164. Proc Natl Acad Sci USA 111:6952–6957. doi:10.1073/pnas.1404605111
103. Sakiyama-Elbert SE (2001) Development of growth factor fusion proteins for cell-triggered drug delivery. FASEB J 2:363–374. doi:10.1096/fj.00-0564fje
104. Zisch AH, Schenk U, Schense JC, Sakiyama-Elbert SE, Hubbell JA (2001) Covalently conjugated VEGF-fibrin matrices for endothelialization. J Control Release 72:101–113. doi:10.1016/S0168-3659(01)00266-8
105. Vepari C, Kaplan DL (2007) Silk as a biomaterial. Prog Polym Sci 32:991–1007. doi:10.1016/j.progpolymsci.2007.05.013
106. Fu C, Shao Z, Fritz V (2009) Animal silks: their structures, properties and artificial production. Chem Commun (Camb) 6515–6529. doi:10.1039/b911049f
107. Heim M, Keerl D, Scheibel T (2009) Spider silk: from soluble protein to extraordinary fiber. Angew Chemie Int Ed 48:3584–3596. doi:10.1002/anie.200803341
108. Kluge JA, Rabotyagova O, Leisk GG, Kaplan DL (2008) Spider silks and their applications. Trends Biotechnol 26:244–251. doi:10.1016/j.tibtech.2008.02.006
109. Tokareva O, Jacobsen M, Buehler M, Wong J, Kaplan DL (2014) Structure-function-property-design interplay in biopolymers: spider silk. Acta Biomater 10:1612–1626. doi:10.1016/j.actbio.2013.08.020
110. Yucel T, Lovett ML, Kaplan DL (2014) Silk-based biomaterials for sustained drug delivery. J Control Release 190:381–397. doi:10.1016/j.jconrel.2014.05.059
111. Kundu B, Rajkhowa R, Kundu SC, Wang X (2013) Silk fibroin biomaterials for tissue regenerations. Adv Drug Deliv Rev 65:457–470. doi:10.1016/j.addr.2012.09.043
112. Wang Y, Kim HJ, Vunjak-Novakovic G, Kaplan DL (2006) Stem cell-based tissue engineering with silk biomaterials. Biomaterials 27:6064–6082. doi:10.1016/j.biomaterials.2006.07.008
113. Meinel L, Kaplan DL (2012) Silk constructs for delivery of musculoskeletal therapeutics. Adv Drug Deliv Rev 64:1111–1122. doi:10.1016/j.addr.2012.03.016
114. Seib FP, Kaplan DL (2013) Silk for drug delivery applications: opportunities and challenges. Isr J Chem 53:756–766. doi:10.1002/ijch.201300083
115. Wenk E, Merkle HP, Meinel L (2011) Silk fibroin as a vehicle for drug delivery applications. J Control Release 150:128–141. doi:10.1016/j.jconrel.2010.11.007
116. Hota MK, Bera MK, Kundu B, Kundu SC, Maiti CK (2012) A natural silk fibroin protein-based transparent bio-memristor. Adv Funct Mater 22:4493–4499. doi:10.1002/adfm.201200073
117. Omenetto FG, Kaplan DL (2008) A new route for silk. Nat Photonics 2:641–643. doi:10.1038/nphoton.2008.207
118. Tao H, Kaplan DL, Omenetto FG (2012) Silk materials—a road to sustainable high technology. Adv Mater 24:2824–2837. doi:10.1002/adma.201104477
119. Kang G, Nahm J, Park J, Moon J, Cho C, Yeo J (2000) Effects of poloxamer on the gelation of silk fibroin. Macromol Rapid Commun 21:788–791. doi:10.1002/1521-3927(20000701)21:11<788::AID-MARC788>3.0.CO;2-X

120. Kim UJ, Park J, Li C, Jin HJ, Valluzzi R, Kaplan DL (2004) Structure and properties of silk hydrogels. Biomacromolecules 5:786–792. doi:10.1021/bm0345460
121. Ayub Haider Z, Arai M, Hirabayashi K (1993) Mechanism of the gelation of fibroin solution. Biosci Biotechnol Biochem 57:1910–1912. doi:10.1271/bbb.57.1910
122. Hanawa T, Watanabe A, Tsuchiya T, Ikoma R, Hidaka M, Sugihara M (1995) New oral dosage form for elderly patients: preparation and characterization of silk fibroin gel. Chem Pharm Bull (Tokyo) 43:284–288. doi:10.1248/cpb.37.3229
123. Motta A, Migliaresi C, Faccioni F, Torricelli P, Fini M, Giardino R (2004) Fibroin hydrogels for biomedical applications: preparation, characterization and in vitro cell culture studies. J Biomater Sci Polym Ed 15:851–864. doi:10.1163/1568562041271075
124. Yoo MK, Kweon HY, Lee KG, Lee HC, Cho CS (2004) Preparation of semi-interpenetrating polymer networks composed of silk fibroin and poloxamer macromer. Int J Biol Macromol 34:263–270. doi:10.1016/j.ijbiomac.2004.06.002
125. Fini M, Motta A, Torricelli P, Giavaresi G, Nicoli Aldini N, Tschon M, Giardino R, Migliaresi C (2005) The healing of confined critical size cancellous defects in the presence of silk fibroin hydrogel. Biomaterials 26:3527–3536. doi:10.1016/j.biomaterials.2004.09.040
126. Yucel T, Cebe P, Kaplan DL (2009) Vortex-induced injectable silk fibroin hydrogels. Biophys J 97:2044–2050. doi:10.1016/j.bpj.2009.07.028
127. Wang X, Kluge JA, Leisk GG, Kaplan DL (2008) Sonication-induced gelation of silk fibroin for cell encapsulation. Biomaterials 29:1054–1064. doi:10.1016/j.biomaterials.2007.11.003
128. Numata K, Katashima T, Sakai T (2011) State of water, molecular structure, and cytotoxicity of silk hydrogels. Biomacromolecules 12:2137–2144. doi:10.1021/bm200221u
129. Kojic N, Panzer MJ, Leisk GG, Raja WK, Kojic M, Kaplan DL (2012) Ion electrodiffusion governs silk electrogelation. Soft Matter 8:6897. doi:10.1039/c2sm25783a
130. Leisk GG, Lo TJ, Yucel T, Lu Q, Kaplan DL (2010) Electrogelation for protein adhesives. Adv Mater 22:711–715. doi:10.1002/adma.200902643
131. Lin Y, Xia X, Shang K, Elia R, Huang W, Cebe P, Leisk G, Omenetto F, Kaplan DL (2013) Tuning chemical and physical cross-links in silk electrogels for morphological analysis and mechanical reinforcement. Biomacromolecules 14:2629–2635. doi:10.1021/bm4004892
132. Lu Q, Huang Y, Li M, Zuo B, Lu S, Wang J, Zhu H, Kaplan DL (2011) Silk fibroin electrogelation mechanisms. Acta Biomater 7:2394–2400. doi:10.1016/j.actbio.2011.02.032
133. Yucel T, Kojic N, Leisk GG, Lo TJ, Kaplan DL (2010) Non-equilibrium silk fibroin adhesives. J Struct Biol 170:406–412. doi:10.1016/j.jsb.2009.12.012
134. Sun L, Parker ST, Syoji D, Wang X, Lewis J a, Kaplan DL (2012) Direct-write assembly of 3D silk/hydroxyapatite scaffolds for bone co-cultures. Adv Healthc Mater 1:729–35. doi:10.1002/adhm.201200057
135. Almine JF, Bax DV, Mithieux SM, Nivison-Smith L, Rnjak J, Waterhouse A, Wise SG, Weiss AS (2010) Elastin-based materials. Chem Soc Rev 39:3371–3379. doi:10.1039/b919452p
136. Daamen WF, Veerkamp JH, van Hest JCM, van Kuppevelt TH (2007) Elastin as a biomaterial for tissue engineering. Biomaterials 28:4378–4398. doi:10.1016/j.biomaterials.2007.06.025
137. Wise SG, Mithieux SM, Weiss AS (2009) Engineered tropoelastin and elastin-based biomaterials. Adv Protein Chem Struct Biol 78:1–24. doi:10.1016/S1876-1623(08)78001-5
138. Annabi N, Mithieux SM, Boughton EA, Ruys AJ, Weiss AS, Dehghani F (2009) Synthesis of highly porous crosslinked elastin hydrogels and their interaction with fibroblasts in vitro. Biomaterials 30:4550–4557. doi:10.1016/j.biomaterials.2009.05.014
139. Jiankang H, Dichen L, Yaxiong L, Bo Y, Hanxiang Z, Qin L, Bingheng L, Yi L (2009) Preparation of chitosan-gelatin hybrid scaffolds with well-organized microstructures for hepatic tissue engineering. Acta Biomater 5:453–461. doi:10.1016/j.actbio.2008.07.002
140. Leach JB, Wolinsky JB, Stone PJ, Wong JY (2005) Crosslinked alpha-elastin biomaterials: Towards a processable elastin mimetic scaffold. Acta Biomater 1:155–164. doi:10.1016/j.actbio.2004.12.001

141. Reichl S (2009) Films based on human hair keratin as substrates for cell culture and tissue engineering. Biomaterials 30:6854–6866. doi:10.1016/j.biomaterials.2009.08.051
142. Rouse JG, Van Dyke ME (2010) A review of keratin-based biomaterials for biomedical applications. Materials (Basel) 3:999–1014. doi:10.3390/ma3020999
143. Schweizer J, Bowden PE, Coulombe PA, Langbein L, Lane EB, Magin TM, Maltais L, Omary MB, Parry DAD, Rogers MA, Wright MW (2006) New consensus nomenclature for mammalian keratins. J Cell Biol 174:169–174. doi:10.1083/jcb.200603161
144. Lee H, Noh K, Lee SC, Kwon I, Han D, Lee I, Hwang Y (2014) Human hair keratin and its-based biomaterials for biomedical applications. Tissue Eng Regen Med 11:255–265. doi:10.1007/s13770-014-0029-4
145. Hill PS, Apel PJ, Barnwell J, Smith T, Koman LA, Atala A, Van Dyke M (2011) Repair of peripheral nerve defects in rabbits using keratin hydrogel scaffolds. Tissue Eng Part A 17:1499–1505. doi:10.1089/ten.tea.2010.0184
146. Pace LA, Plate JF, Smith TL, Van Dyke ME (2013) The effect of human hair keratin hydrogel on early cellular response to sciatic nerve injury in a rat model. Biomaterials 34:5907–5914. doi:10.1016/j.biomaterials.2013.04.024
147. Sierpinski P, Garrett J, Ma J, Apel P, Klorig D, Smith T, Koman LA, Atala A, Van Dyke M (2008) The use of keratin biomaterials derived from human hair for the promotion of rapid regeneration of peripheral nerves. Biomaterials 29:118–128. doi:10.1016/j.biomaterials.2007.08.023
148. Wang S, Taraballi F, Tan LP, Ng KW (2012) Human keratin hydrogels support fibroblast attachment and proliferation in vitro. Cell Tissue Res 347:795–802. doi:10.1007/s00441-011-1295-2
149. Badylak SF, Taylor D, Uygun K (2010) Whole organ tissue engineering: decellularization and recellularization of three-dimensional matrix scaffolds. Annu Rev Biomed Eng 13:110301095218061. doi:10.1146/annurev-bioeng-071910-124743
150. Hynes RO (2009) The extracellular matrix: not just pretty fibrils. Science 326:1216–1219. doi:10.1126/science.1176009
151. Soto-Gutierrez A, Zhang L, Medberry C, Fukumitsu K, Faulk D, Jiang H, Reing J, Gramignoli R, Komori J, Ross M, Nagaya M, Lagasse E, Stolz D, Strom SC, Fox IJ, Badylak SF (2011) A whole-organ regenerative medicine approach for liver replacement. Tissue Eng Part C Methods 17:677–686. doi:10.1089/ten.tec.2010.0698
152. Choi JS, Yang HJ, Kim BS, Kim JD, Kim JY, Yoo B, Park K, Lee HY, Cho YW (2009) Human extracellular matrix (ECM) powders for injectable cell delivery and adipose tissue engineering. J Control Release 139:2–7. doi:10.1016/j.jconrel.2009.05.034
153. Pati F, Jang J, Ha D, Won Kim S, Rhie J, Shim J, Kim D, Cho D (2014) Printing three-dimensional tissue analogues with decellularized extracellular matrix bioink. Nat Commun 5:3935. doi:10.1038/ncomms4935
154. Johnson TD, Lin SY, Christman KL (2011) Tailoring material properties of a nanofibrous extracellular matrix derived hydrogel. Nanotechnology 22:494015. doi:10.1088/0957-4484/22/49/494015
155. Singelyn JM, DeQuach JA, Seif-Naraghi SB, Littlefield RB, Schup-Magoffin PJ, Christman KL (2009) Naturally derived myocardial matrix as an injectable scaffold for cardiac tissue engineering. Biomaterials 30:5409–5416. doi:10.1016/j.biomaterials.2009.06.045
156. Enat R, Jefferson DM, Ruiz-Opazo N, Gatmaitan Z, Leinwand LA, Reid LM (1984) Hepatocyte proliferation in vitro: its dependence on the use of serum-free hormonally defined medium and substrata of extracellular matrix. Proc Natl Acad Sci USA 81:1411–1415. doi:10.1073/pnas.81.5.1411
157. Skardal A, Smith L, Bharadwaj S, Atala A, Soker S, Zhang Y (2012) Tissue specific synthetic ECM hydrogels for 3-D in vitro maintenance of hepatocyte function. Biomaterials 33:4565–4575. doi:10.1016/j.biomaterials.2012.03.034
158. Wolf MT, Daly KA, Brennan-Pierce EP, Johnson SA, Carruthers CA, D'Amore A, Nagarkar SP, Velankar SS, Badylak SF (2012) A hydrogel derived from decellularized

dermal extracellular matrix. Biomaterials 33:7028–7038. doi:10.1016/j.biomaterials.2012.06.051

159. Ott HC, Matthiesen TS, Goh S-K, Black LD, Kren SM, Netoff TI, Taylor DA (2008) Perfusion-decellularized matrix: using nature's platform to engineer a bioartificial heart. Nat Med 14:213–21. doi:10.1038/nm1684
160. DeQuach JA, Mezzano V, Miglani A, Lange S, Keller GM, Sheikh F, Christman KL (2010) Simple and high yielding method for preparing tissue specific extracellular matrix coatings for cell culture. PLoS One 5:e13039. doi:10.1371/journal.pone.0013039
161. Migneault I, Dartiguenave C, Bertrand MJ, Waldron KC (2004) Glutaraldehyde: behavior in aqueous solution, reaction with proteins, and application to enzyme crosslinking. Biotechniques 37:790–802. doi:10.2144/3705A0790
162. Lai JY, Li YT (2010) Functional assessment of cross-linked porous gelatin hydrogels for bioengineered cell sheet carriers. Biomacromolecules 11:1387–1397. doi:10.1021/bm100213f
163. Liu Y, Griffith M, Watsky MA, Forrester JV, Kuffová L, Grant D, Merrett K, Carlsson DJ (2006) Properties of porcine and recombinant human collagen matrices for optically clear tissue engineering applications. Biomacromolecules 7:1819–1828. doi:10.1021/bm060160o
164. Liu W, Merrett K, Griffith M, Fagerholm P, Dravida S, Heyne B, Scaiano JC, Watsky MA, Shinozaki N, Lagali N, Munger R, Li F (2008) Recombinant human collagen for tissue engineered corneal substitutes. Biomaterials 29:1147–1158. doi:10.1016/j.biomaterials.2007.11.011
165. Liang Y, Jeong J, DeVolder RJ, Cha C, Wang F, Tong YW, Kong H (2011) A cell-instructive hydrogel to regulate malignancy of 3D tumor spheroids with matrix rigidity. Biomaterials 32:9308–9315. doi:10.1016/j.biomaterials.2011.08.045
166. Rafat M, Li F, Fagerholm P, Lagali NS, Watsky MA, Munger R, Matsuura T, Griffith M (2008) PEG-stabilized carbodiimide crosslinked collagen-chitosan hydrogels for corneal tissue engineering. Biomaterials 29:3960–3972. doi:10.1016/j.biomaterials.2008.06.017
167. Rutz AL, Hyland KE, Jakus AE, Burghardt WR, Shah RN (2015) A multimaterial bioink method for 3D printing tunable, cell-compatible hydrogels. Adv Mater 27:1607–1614. doi:10.1002/adma.201405076
168. Singh RK, Seliktar D, Putnam AJ (2013) Capillary morphogenesis in PEG-collagen hydrogels. Biomaterials 34:9331–9340. doi:10.1016/j.biomaterials.2013.08.016
169. Hwang YJ, Larsen J, Krasieva TB, Lyubovitsky JG (2011) Effect of genipin crosslinking on the optical spectral properties and structures of collagen hydrogels. ACS Appl Mater Interfaces 3:2579–2584. doi:10.1021/am200416h
170. MacAya D, Ng KK, Spector M (2011) Injectable collagen-genipin gel for the treatment of spinal cord injury: In vitro studies. Adv Funct Mater 21:4788–4797. doi:10.1002/adfm.201101720
171. Kirchmajer DM, Watson CA, Ranson M, Panhuis M in Het (2013) Gelapin, a degradable genipin cross-linked gelatin hydrogel. RSC Adv 3:1073. doi:10.1039/c2ra22859a
172. Schek RM, Michalek AJ, Iatridis JC (2011) Genipin-crosslinked fibrin hydrogels as a potential adhesive to augment intervertebral disc annulus repair. Eur Cells Mater 21:373–383. doi:10.1016/j.biotechadv.2011.08.021.Secreted
173. Silva SS, Motta A, Rodrigues MT, Pinheiro AFM, Gomes ME, Mano JF, Reis RL, Migliaresi C (2008) Novel genipin-cross-linked chitosan/silk fibroin sponges for cartilage engineering strategies. Biomacromolecules 9:2764–2774. doi:10.1021/bm800874q
174. Xiao W, Liu W, Sun J, Dan X, Wei D, Fan H (2012) Ultrasonication and genipin cross-linking to prepare novel silk fibroin-gelatin composite hydrogel. J Bioact Compat Polym 27:327–341. doi:10.1177/0883911512448692
175. Liang HC, Chang WH, Liang HF, Lee MH, Sung HW (2004) Crosslinking structures of gelatin hydrogels crosslinked with genipin or a water-soluble carbodiimide. J Appl Polym Sci 91:4017–4026. doi:10.1002/app.13563
176. Sung HW, Huang DM, Chang WH, Huang RN, Hsu JC (1999) Evaluation of gelatin hydrogel crosslinked with various crosslinking agents as bioadhesives: in vitro study.

J Biomed Mater Res 46:520–530. doi:10.1002/(SICI)1097-4636(19990915)46:4<520::AID-JBM10>3.0.CO;2-9
177. McDermott MK, Chen T, Williams CM, Markley KM, Payne GF (2004) Mechanical properties of biomimetic tissue adhesive based on the microbial transglutaminase-catalyzed crosslinking of gelatin. Biomacromolecules 5:1270–1279. doi:10.1021/bm034529a
178. Yung CW, Wu LQ, Tullman JA, Payne GF, Bentley WE, Barbari TA (2007) Transglutaminase crosslinked gelatin as a tissue engineering scaffold. J Biomed Mater Res —Part A 83:1039–1046. doi:10.1002/jbm.a.31431
179. Hermanson GT (2013). Bioconjugate techniques, 3rd edn. Academic press, Elsevier. ISBN: 978-0-12-382239-0
180. Kushibiki T, Tomoshige R, Fukunaka Y, Kakemi M, Tabata Y (2003) In vivo release and gene expression of plasmid DNA by hydrogels of gelatin with different cationization extents. J Control Release 90:207–216. doi:10.1016/S0168-3659(03)00197-4
181. Santoro M, Tatara AM, Mikos AG (2014) Gelatin carriers for drug and cell delivery in tissue engineering. J Control Release 190:210–218. doi:10.1016/j.jconrel.2014.04.014
182. Nezhadi SH, Choong PFM, Lotfipour F, Dass CR (2009) Gelatin-based delivery systems for cancer gene therapy. J Drug Target 17:731–738. doi:10.3109/10611860903096540
183. Sakai S, Hirose K, Taguchi K, Ogushi Y, Kawakami K (2009) An injectable, in situ enzymatically gellable, gelatin derivative for drug delivery and tissue engineering. Biomaterials 30:3371–3377. doi:10.1016/j.biomaterials.2009.03.030
184. Wang LS, Boulaire J, Chan PPY, Chung JE, Kurisawa M (2010) The role of stiffness of gelatin-hydroxyphenylpropionic acid hydrogels formed by enzyme-mediated crosslinking on the differentiation of human mesenchymal stem cell. Biomaterials 31:8608–8616. doi:10.1016/j.biomaterials.2010.07.075
185. Wang LS, Chung JE, Pui-Yik Chan P, Kurisawa M (2010) Injectable biodegradable hydrogels with tunable mechanical properties for the stimulation of neurogenesic differentiation of human mesenchymal stem cells in 3D culture. Biomaterials 31:1148–1157. doi:10.1016/j.biomaterials.2009.10.042
186. Chen YC, Lin RZ, Qi H, Yang Y, Bae H, Melero-Martin JM, Khademhosseini A (2012) Functional human vascular network generated in photocrosslinkable gelatin methacrylate hydrogels. Adv Funct Mater 22:2027–2039. doi:10.1002/adfm.201101662
187. Nichol JW, Koshy ST, Bae H, Hwang CM, Yamanlar S, Khademhosseini A (2010) Cell-laden microengineered gelatin methacrylate hydrogels. Biomaterials 31:5536–5544. doi:10.1016/j.biomaterials.2010.03.064
188. Nikkhah M, Eshak N, Zorlutuna P, Annabi N, Castello M, Kim K, Dolatshahi-Pirouz A, Edalat F, Bae H, Yang Y, Khademhosseini A (2012) Directed endothelial cell morphogenesis in micropatterned gelatin methacrylate hydrogels. Biomaterials 33:9009–9018. doi:10.1016/j.biomaterials.2012.08.068
189. Hu X, Ma L, Wang C, Gao C (2009) Gelatin hydrogel prepared by photo-initiated polymerization and loaded with TGF-beta1 for cartilage tissue engineering. Macromol Biosci 9:1194–1201. doi:10.1002/mabi.200900275
190. Billiet T, Gevaert E, De Schryver T, Cornelissen M, Dubruel P (2014) The 3D printing of gelatin methacrylamide cell-laden tissue-engineered constructs with high cell viability. Biomaterials 35:49–62. doi:10.1016/j.biomaterials.2013.09.078
191. Kolesky DB, Truby RL, Gladman AS, Busbee TA, Homan KA, Lewis JA (2014) 3D bioprinting of vascularized, heterogeneous cell-laden tissue constructs. Adv Mater 26:3124–3130. doi:10.1002/adma.201305506
192. Fu Y, Xu K, Zheng X, Giacomin AJ, Mix AW, Kao WJ (2012) 3D cell entrapment in crosslinked thiolated gelatin-poly(ethylene glycol) diacrylate hydrogels. Biomaterials 33:48–58. doi:10.1016/j.biomaterials.2011.09.031
193. Phelps EA, Enemchukwu NO, Fiore VF, Sy JC, Murthy N, Sulchek TA, Barker TH, García AJ (2012) Maleimide cross-linked bioactive PEG hydrogel exhibits improved reaction kinetics and cross-linking for cell encapsulation and in situ delivery. Adv Mater 24:64–70. doi:10.1002/adma.201103574

194. Vanderhooft JL, Alcoutlabi M, Magda JJ, Prestwich GD (2009) Rheological properties of cross-linked hyaluronan-gelatin hydrogels for tissue engineering. Macromol Biosci 9:20–28. doi:10.1002/mabi.200800141
195. Nimmo CM, Shoichet MS (2011) Regenerative biomaterials that "click": Simple, aqueous-based protocols for hydrogel synthesis, surface immobilization, and 3D patterning. Bioconjug Chem 22:2199–2209. doi:10.1021/bc200281k
196. Jiang Y, Chen J, Deng C, Suuronen EJ, Zhong Z (2014) Click hydrogels, microgels and nanogels: emerging platforms for drug delivery and tissue engineering. Biomaterials 35:4969–4985. doi:10.1016/j.biomaterials.2014.03.001
197. Zheng M, Zheng L, Zhang P, Li J, Zhang Y (2015) Development of bioorthogonal reactions and their applications in bioconjugation. Molecules 20:3190–3205. doi:10.3390/molecules20023190

Sterculia Gum-Based Hydrogels for Drug Delivery Applications

Amit Kumar Nayak and Dilipkumar Pal

Abstract Sterculia gum is one of the medicinally important plant-derived water soluble polysaccharides obtained from the exudate of the tree, *Sterculia urens* (Family: sterculiaceae). It is recognized as a promising biodegradable material in the development of various biomedical applications including drug delivery applications, wound dressing applications, etc. Sterculia gum is also employed as excipient in the designing of various pharmaceutical applications. In recent years, several attempts for the modification of sterculia gum have been undertaken to develop sterculia gum-based hydrogels for controlling the rate of hydration and swelling, and also tailoring the release profile of various types of drugs. In the development of these sterculia gum-based hydrogels, modifications of sterculia gum through polymer blending, cross-linking, interpenetrated polymer network (IPN) formation, polymer grafting, etc., were investigated for improved drug delivery applications. Most of these already reported sterculia gum-based hydrogels were found effective for gastroretentive deliveries as wound dressings for sustained release of various drugs. The current chapter deals with a comprehensive and useful discussion on already investigated sterculia gum-based hydrogels for the use in drug delivery applications, where the first portion of the chapter contains source, composition, and properties of sterculia gum and the latter portion contains discussion on the formulations of various sterculia gum-based hydrogel systems used for various types of drug delivery applications.

Keywords Sterculia gum · Hydrogels · Cross-linking · Polymer blending · Drug delivery

A.K. Nayak (✉)
Department of Pharmaceutics, Seemanta Institute of Pharmaceutical Sciences, Mayurbhanj, Odisha 757086, India
e-mail: amitkrnayak@yahoo.co.in

D. Pal
Department of Pharmaceutical Sciences, Guru Ghasidas Vishwavidyalaya, Koni, Bilaspur, Chhattisgarh 495009, India

S. Kalia (ed.), *Polymeric Hydrogels as Smart Biomaterials*,
Springer Series on Polymer and Composite Materials,
DOI 10.1007/978-3-319-25322-0_5

1 Introduction

Naturally derived materials are gaining importance day by day [1–5]. Currently, natural polysaccharides are the choice of potential biomaterials used in various biomedical and pharmaceutical applications due to their biocompatibility and biodegradability [6–8]. These are also available from natural renewable sources. These are also capable of high swelling ability, aqueous solubility, and stable in various pH environments [9, 10]. Besides these, natural polysaccharides have a wide variety of compositions and properties which allow these for appropriately tailoring chemical as well as physical modifications [11–14]. In recent years, various modified (either chemical or physical) natural polysaccharides are extensively used to develop hydrogels for various biomedical applications including drug delivery, wound dressing, tissue engineering, etc. [11, 15–19]. Modifications of natural polysaccharides include polymer blending [20–24], chemical cross-linking [25–29], polymer grafting [30–34], carboxymethylation [35–38], carbomoylethylation [39, 40], thiolation [41–43], esterification [44–46], interpenetrating polymer networks (IPNs) [47–50], polyelectrolyte complexes [51–53], etc. These modifications are able to overcome some important limitations of natural polysaccharides such as uncontrolled hydration rate, fall in viscosity on storage, chemical instability, low mechanical properties, proneness to microbial contamination, etc. [12]. Recently, various types of hydrogels are formulated using various modified natural polysaccharides for the use in biomedical and pharmaceutical applications.

Hydrogels are cross-linked three-dimensional networks of hydrophilic polymeric systems which are capable of holding large amount of water or biological fluids without losing their structure [54]. In swollen state, hydrogels are soft and rubbery [55]. However, hydrogels do not dissolve quickly in water, aqueous solutions, or biological fluids [56]. These resemble with living tissues exhibiting excellent mechanical ability [55]. Hydrogels are of special interest in controlled release applications in which drugs are dispersed throughout the matrix and are capable of delivering drugs at a constant rate over an extended period of time [56–58]. In the light of the above discussion, the present chapter deals with a comprehensive and useful discussion on already investigated hydrogel systems made of sterculia gum, plant-derived nontoxic natural polysaccharides and its derivatives for the use in drug delivery applications. The first portion of the chapter contains source, composition, and properties of sterculia gum. Then, the latter portion contains discussion on the formulations of various sterculia gum-based hydrogel systems and their applications in drug delivery.

2 Sterculia Gum: Source, Chemical Composition, Properties, and Uses

2.1 *Source*

Sterculia gum is one of the medicinally important plant-derived water soluble, high molecular weight polysaccharides commonly known as karaya gum [59]. It is obtained from a low-cost source, the exudate of the tree, *Sterculia urens* (Family: sterculiaceae) [59, 60]. This tree is a small to medium-sized tree with a pale-colored trunk or by making deep gashes at the base of the trunk with an axe. It is found in India and Burma. The crude gum is conventionally tapped as exudate through cutting or peeling back the bark of the tree.

2.2 *Chemical Composition*

Sterculia gum is a partially acetylated polysaccharide [61]. It is composed of three different polysaccharide chains. The first chain (50 % of the total polysaccharide) contains repeating units of four galacturonic acid residues containing β-D-galactose branches and L-rhamnose residues at the reducing end of the unit. The second chain (17 % of the total polysaccharide) contains an oligorhamnan having D-galacturonic acid branch residues and interrupted occasionally by D-galactose residues. The third chain (33 % of the total polysaccharide) contains D-glucuronic acid residues which constitutes about 13–26 % galactose, 15–30 % rhamnose, and approximately 40 % uronic acid residues [60, 62].

2.3 *Properties and Uses*

Sterculia gum exhibits some unique properties like greater acidic stability, good viscosity, high swelling, and water retention ability [63, 64]. Sterculia gum is graded as 'Generally Recognized as Safe' ('GRAS') in USA [65]. It is reported as nontoxic, nonallergic, nonmutagenic, and nonteratogenic [63, 64, 66]. Sterculia gum has antimicrobial property [64]. It has the ability to reduce cholesterol to improve glucose metabolism without hampering mineral balances [67].

Sterculia gum has been studied for its use in treatment of ulcers [68], diarrhea [69], chronic colonic disorders [70], and irritable bowel syndrome [71]. It is also reported as balk laxative [72]. Although the intake of sterculia gum without adequate amount of water is harmful to our body, no side effect of it has been reported till date [66]. Sterculia gum is used as emulsifier, stabilizer, and thickener in foods [73]. It is also used as adulterate gum tragacanth due to their similar physical characteristics [74].

The gelation of sterculia gum by the influence of polyvalent metal cations like 'Egg-box model' of alginate has been reported in the literature [75]. It could be due to the formation of intermolecular junction zones involving ionotropic interaction between homogalacturonic segments present in sterculia gum chains and the polyvalent metal cations. Also, the ionotropic interaction of sterculia gum could occur between two or more galacturonic acid residues on different main chains or glucuronic in branched or nonregular chain segments of the sterculia gum [76].

During last few decades', sterculia gum is recognized as a promising biodegradable polysaccharide excipient used in the development of various drug delivery dosage forms [77–80]. However, several attempts for the modification of sterculia gum to develop sterculia gum-based hydrogels have been undertaken by various research groups to control the rate of hydration and swelling, and also to tailor the release profile of drugs. The modifications of sterculia gum like polymer blending [75, 81], cross-linking [82, 83], IPN formation [62, 84], polyelectrolyte complex formation [85], grafting [66, 86], etc., were already investigated to develop various sterculia gum-based hydrogels for improved drug delivery applications.

3 Sterculia Gum-Alginate Beads for Use in Antiulcer Drug Delivery

During last few decades', polymer blends of sterculia gum and a widely used natural polysaccharide, sodium alginate has been extensively studied to develop various potential hydrogel systems for the use in drug delivery [75, 81, 87]. Sodium alginate is a biodegradable, biocompatible anionic natural heteropolysaccharide, which is widely used in various drug delivery dosage forms [88–91]. It is able to form hydrogel microparticles/beads through ionotropic gelation in the presence of divalent and trivalent metal cations like Ca^{2+}, Ba^{2+}, Zn^{2+}, Al^{3+}, etc. [92]. The ionotropically gelled alginate-based hydrogel beads have been employed in encapsulation of a variety of drug molecules and other therapeutic agents [93–97]. However, these alginate-based hydrogel beads have some drawbacks like low encapsulation efficiency and premature release of encapsulated small molecular drugs [20, 92]. To overcome these drawbacks, polymer blends of alginate with second natural biocompatible polymers have been researched to develop new and effective controlled release drug delivery matrices [23, 98–100].

Few hydrogel beads of sterculia gum-alginate blends have been investigated for the use in oral drug delivery [62, 75, 81, 87]. Most of the sterculia gum-alginate blends hydrogel beads were prepared using ionotropic cross-linkers like Ca^{2+}, Ba^{2+}, etc. [61, 75, 81, 87]. Actually, sterculia gum and sodium alginate contain anionic groups (i.e., –COOH groups) in their structures. Both gums exhibit the characteristics of electrostatic interaction by the influence of divalent metal cations. When the blends of sterculia gum and sodium alginate come in contact with the divalent metal ions, an ionotropic interaction occurs between the positively charged metal

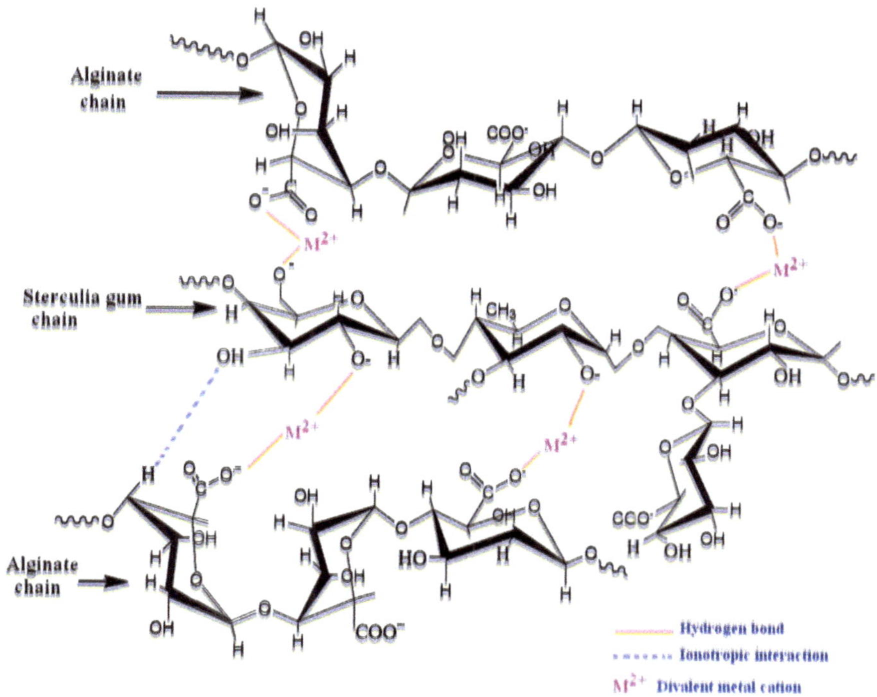

Scheme 1 Schematic presentation of proposed interactions between sterculia gum, sodium alginate, and divalent cations (M^{2+})

cations and –COOH groups of both sterculia gum and sodium alginate chains. The divalent metal cations compete with the monovalent Na^+ ions of the anionic sites of the sterculia gum–sodium alginate polymer blends and replace it, thus bringing the two polymeric chains together. Divalent metal ions get accommodated in the interstices of two polyuronate chains having a close ion pair interaction with –COO^- anions and sufficient coordination by other electronegative oxygen atoms [75, 101]. Besides the ionotropic interaction, the hydrogen bonding occurs between two polysaccharide chains present in the sterculia gum–sodium alginate polymer blends [75]. The schematic presentation of proposed interactions between sterculia gum, sodium alginate, and divalent cations (M^{2+}) is presented in Scheme 1.

In view of pharmacological importance related with antiulcer properties of sterculia gum and alginate, Singh et al. [75] have formulated sterculia gum-alginate beads and floating sterculia gum-alginate beads containing an antiulcer drug, pentoprazole using calcium chloride ($CaCl_2$) as ionotropic cross-linker. After preliminary evaluation, optimized formula of Ca^{2+} ion cross-linked sterculia gum-alginate beads of pentoprazole was considered as polymer blends of 2 % w/v sodium alginate, 1.25 % w/v sterculia gum, and 0.1 M $CaCl_2$. To prepare the floating sterculia gum-alginate beads of pentoprazole in this work, the effervescent technique was employed. These Ca^{2+} ion cross-linked sterculia gum-alginate floating beads of pentoprazole were prepared using 2 % w/v calcium carbonate ($CaCO_3$) in the sterculia gum–sodium alginate polymer blend solutions (2 % w/v sodium alginate and 1.25 % w/v sterculia gum). The sterculia gum–sodium alginate polymer blend solutions were added dropwise to the 0.1 M $CaCl_2$ containing 10 % acetic acid (v/v). In case of the Ca^{2+} ion cross-linked sterculia gum-alginate floating beads of pentoprazole, both external and internal gelations occurred. Upon contact with an acidic medium, ionotropic gelation through Ca^{2+} ions took place to facilitate a gel barrier at the surface of the formulated beads. The $CaCO_3$ effervesced with releasing carbon dioxide (CO_2) and Ca^{2+} ions. The released CO_2 was entrapped into the gel network to provide buoyancy to those formulated beads. Then, the Ca^{2+} ions interacted with the –COOH groups of these two anionic polysaccharides (i.e., sterculia gum and sodium alginate) to produce Ca^{2+} ion cross-linked three-dimensional gel network that restricted further diffusion of CO_2.

The shape of these Ca^{2+} ion cross-linked sterculia gum-alginate beads of pentoprazole was found spherical as the sodium alginate concentration increased in the polymer blend solutions. This can be attributed that the mean diameter of these beads increased due to increment in microviscosity of the polymer blend with the increment in sodium alginate concentrations in the polymer blend solutions. However, sterculia gum concentration in the polymer blend solutions and the cross-linker concentration did not influence the bead shape and diameter. The mean diameter of the optimized Ca^{2+} ion cross-linked sterculia gum-alginate floating beads of pentoprazole was found as 1.35 ± 0.30 mm. The mean diameter increment with the incorporation of $CaCO_3$ in the formula might be due to the cause that when $CaCO_3$ reacted with acetic acid present in the cross-linking medium, CO_2 was formed and escaped from the bead matrix. The scanning electron microscope (SEM) image of optimized Ca^{2+} ion cross-linked sterculia gum-alginate non-floating beads of pentoprazole possessed rough surface (Fig. 1). On the other hand, SEM image of optimized Ca^{2+} ion cross-linked sterculia gum-alginate floating beads of pentoprazole prepared through the incorporation of $CaCO_3$ confirmed comparative smooth surface (Fig. 2). The Ca^{2+} ions from $CaCO_3$ might have contributed to the homogeneous sterculia gum-alginate beads formation, which may be responsible for smooth surface of the beads. This might be due to the bursting effect of larger amount of CO_2 evolved before the walls get sufficiently hardened.

The optimized Ca^{2+} ion cross-linked sterculia gum-alginate non-floating and floating beads of pentoprazole were characterized by electron dispersion X-ray (EDX) analysis and the EDX results demonstrated the presence of carbon, oxygen, and hydrogen, which are the key constituents of polysaccharides. Additionally, the elemental peak for calcium in optimized Ca^{2+} ion cross-linked sterculia gum-alginate non-floating and floating beads of pentoprazole supported the fact of Ca^{2+} ion-induced cross-linking of these beads. Fourier transform infrared (FTIR) analyses supported that both the Ca^{2+} ion cross-linked sterculia gum-alginate non-floating and floating beads of pentoprazole had significant characteristic peaks of sterculia gum and alginate.

In swelling study of these beads, the effect of alginate amount, sterculia gum amount, $CaCl_2$ (ionotropic cross-linker) concentration, and pH were evaluated.

(a)
(b)
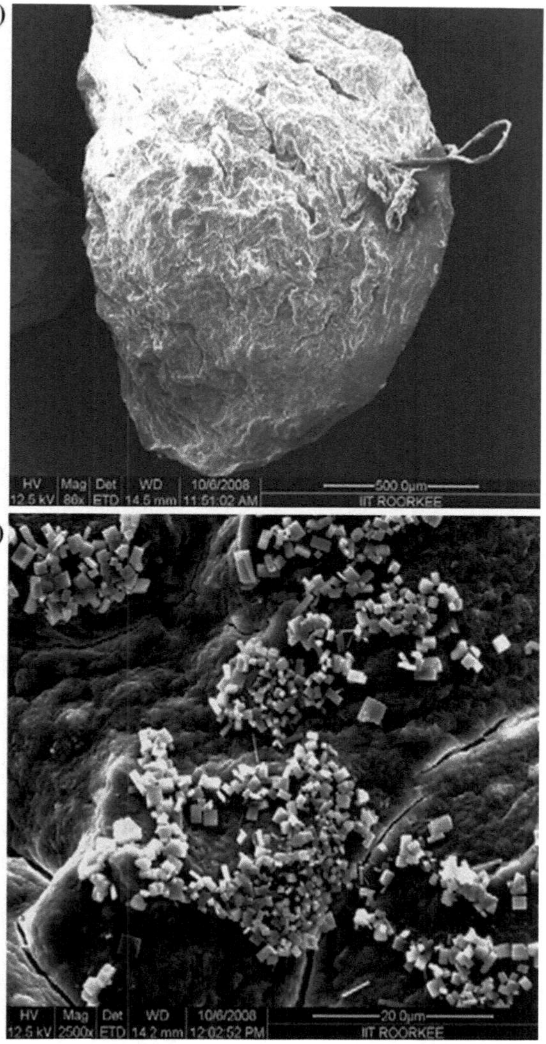

Fig. 1 SEM image of optimized Ca^{2+} ion cross-linked sterculia gum-alginate non-floating beads of pentoprazole at different magnifications **a** ×86 and **b** ×2500 [75]. Copyright © 2010. The Institution of Chemical Engineers, with permission from Elsevier B.V.

The effect of alginate amount varying from 0.5 to 2.5 % w/v on the swelling of Ca^{2+} ion cross-linked sterculia gum-alginate beads of pentoprazole in distilled water displayed an irregular trend of swelling up to 5 h. However, after attaining the swelling equilibrium after 24 h, a decrease in swelling was observed. Though the swelling of beads prepared using 0.5 % w/v sodium alginate was found more, the sodium alginate concentration for the further preparation of optimized Ca^{2+} ion cross-linked sterculia gum-alginate non-floating and floating beads of pentoprazole was taken as 2.0 % w/v. when the sterculia gum amount in these beads varied from 0.25 to 1.25 % w/v, it was observed that the swelling of these beads increased with the increment of the sterculia gum content. The maximum swelling of these beads

(a)
(b)
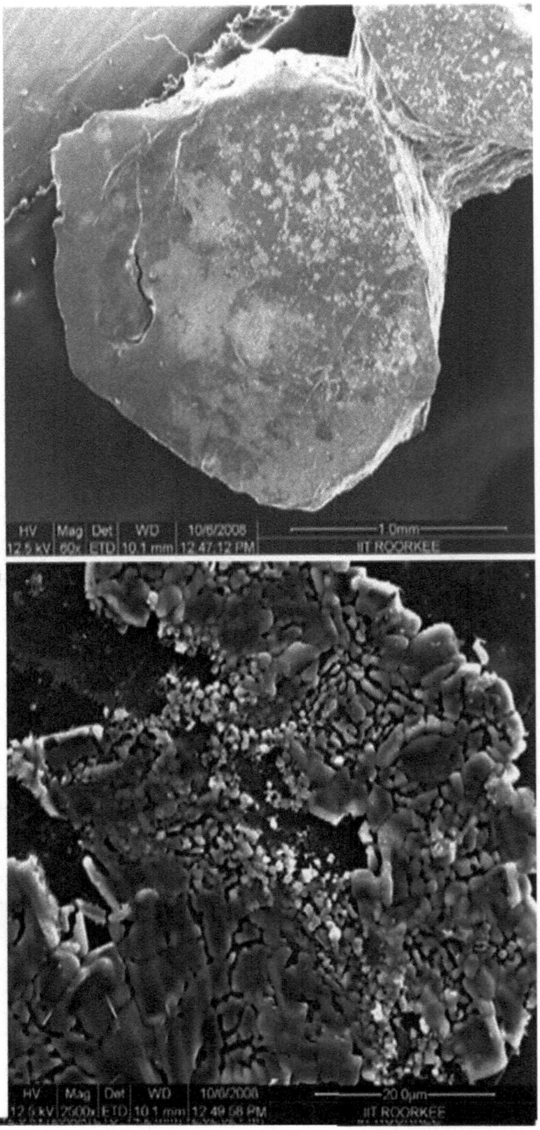

Fig. 2 SEM image of optimized Ca^{2+} ion cross-linked sterculia gum-alginate floating beads of pentoprazole at different magnifications **a** ×60 and **b** ×2500 [75]. Copyright © 2010. The Institution of Chemical Engineers, with permission from Elsevier B.V.

was seen in case of beads prepared with 1.25 % w/v sterculia gum. The effect of $CaCl_2$ concentration as cross-linker on the swelling of these beads showed a decrease in swelling with the increment of the $CaCl_2$ concentration and maximum swelling of these beads was experienced in case of the beads prepared by 0.1 M $CaCl_2$. The effect of pH on the optimized Ca^{2+} ion cross-linked sterculia gum-alginate non-floating beads of pentoprazole was evaluated. In this study, it was experienced that the swelling in pH 7.4 was found increased up to 1 h and thereafter, dissolution of these beads took place. In pH 2.2, the swelling of these beads was found higher as compared with that in distilled water. Similar trends of swelling of optimized Ca^{2+} ion cross-linked sterculia gum-alginate floating beads of pentoprazole in various pHs (distilled water, pH 7.4 and 2.2) were observed. Almost all these floating beads exhibited buoyant behavior for a longer period. The drug loadings in these optimized Ca^{2+} ion cross-linked sterculia gum-alginate non-floating and floating beads of pentoprazole were measured as 83.90 and 66.10 %, respectively. All these non-floating and floating beads of pentoprazole showed sustained drug release pattern over 24 h in distilled water and in pH 2.2 (Figs. 3 and 4). The drug release mechanism from these beads was found to follow the Fickian diffusion mechanism.

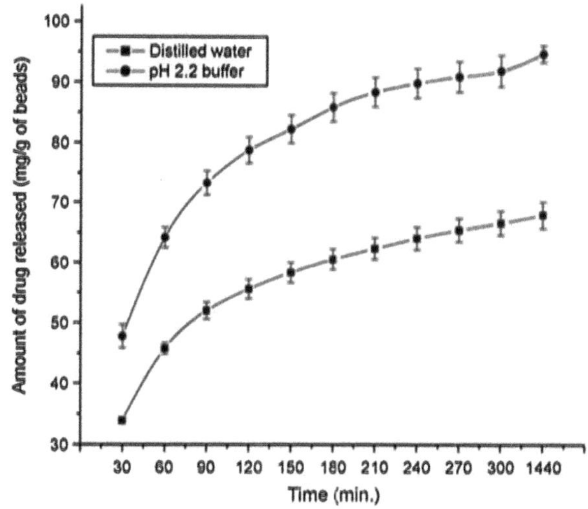

Fig. 3 Release profile of pantoprazole from optimized Ca^{2+} ion cross-linked sterculia gum-alginate non-floating and floating beads of pentoprazole in different media at 37 °C [75]. Copyright © 2010. The Institution of Chemical Engineers, with permission from Elsevier B.V.

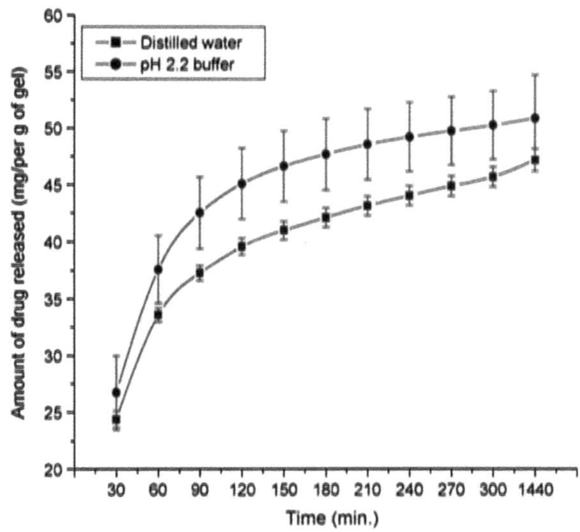

Fig. 4 Release profile of pantoprazole from optimized Ca^{2+} ion cross-linked sterculia gum-alginate floating and floating beads of pentoprazole in different media at 37 °C [75]. Copyright © 2010. The Institution of Chemical Engineers, with permission from Elsevier B.V.

In another report, the same research group has investigated same types of sterculia gum-alginate floating beads of pentoprazole using $BaCl_2$ as cross-linker through ionotropic gelation [87]. These Ba^{2+} ion cross-linked sterculia gum-alginate beads of pentoprazole were prepared through a similar method as mentioned in the preparation of Ca^{2+} ion cross-linked sterculia gum-alginate beads of pentoprazole [75]. But, 0.1 M $BaCl_2$ was used as cross-linked instead of $CaCl_2$; while other ingredients were remained the same as before. The drug loadings in these Ba^{2+} ion cross-linked sterculia gum-alginate non-floating and floating beads of pentoprazole were measured as 67.90 and 61.60 %, respectively. Therefore, the drug loadings in Ba^{2+} ion cross-linked sterculia gum-alginate beads of petoprazole were lesser than that of Ca^{2+} ion cross-linked sterculia gum-alginate beads of petoprazole prepared in the previous study by the same research group. The swelling and drug release of these Ba^{2+} ion cross-linked sterculia gum-alginate floating beads of petoprazole were studied in distilled water, pH 7.4 and 2.2. The swelling of these floating beads was found much higher in pH 7.4 than pH 2.2 and distilled water. In case of the Ba^{2+} ion cross-linked sterculia gum-alginate floating beads of petoprazole, less swelling was observed as compared to the Ba^{2+} ion cross-linked sterculia gum-alginate non-floating beads of petoprazole. At the same time, the Ba^{2+} ion cross-linked beads showed more stability in pH 7.4 than that of Ca^{2+} ion cross-linked beads. This phenomenon can be explained on the basis of the size of Ba^{2+} ions used in cross-linking in this study. Since both Ba^{2+} and Ca^{2+} ions are divalent, the bondings to alginate by Ba^{2+} and Ca^{2+} ions are expected to occur in a planner two-dimensional manner inside the beads. However, the Ba^{2+} ions possess the larger radius (1.74 Å) than that of Ca^{2+} ions (1.14 Å). Therefore, it is supposed to fill a larger space between the cross-linking chains, this produces a tight arrangement with a smaller voids. So, the exchange of the larger Ba^{2+} ions in these beads with the monovalent Na^+ ions and also their removal in the form of insoluble barium phosphate is hindered, which might result in the lowest water uptake and higher stability. The release of drug from these floating beads was found to be sustained over 24 h in all tested mediums. Also, the drug release from these beads occurred through the Fickian diffusion mechanism.

4 Sterculia Gum-Alginate IPN Microparticles for Sustained Release of Antidiabetic Drug

IPN hydrogels are polymeric systems composed of two or more hydrophilic polymers, which are obtained when at least one polymer network is synthesized and/or cross-linked independently in immediate presence of others [47, 48]. According to IUPAC, an IPN is defined as a polymer comprising two or more networks, which are at least partially interlocked on a molecular scale; but cannot be separated unless chemical bonds are broken. A mixture of two or more preformed polymer networks is not an IPN [102]. An IPN can be distinguished from a polymer blend in the way that an IPN swells; but does not dissolve in solvents, creep and flow are suppressed [103]. IPNs can be distinguished from other multiple polymeric systems through their bicontinuous structure ideally formed by cross-linking of two polymers that are in intimate contact but without any chemical contact and yields a material with improved properties depending on their composition and degree of cross-linking [104]. Sterculia gum–sodium alginate blends have also been investigated to prepare IPN hydrogel systems for prolonged drug release applications [62].

In an investigation by Kulkarni et al. [62], IPN microparticles of sterculia gum–sodium alginate blends were prepared to load an antidiabetic drug (repaglinide) by ionotropic gelation and emulsion cross-linking method. In the preparation of these IPN microparticles containing repaglinide, $CaCl_2$, $BaCl_2$, and $AlCl_3$ solutions were employed as cross-linking solutions. The researchers have explained that the exchange of Na^+ ions occurred with Ca^{2+}, Ba^{2+}, and Al^{3+} ions at the carboxylate site and a second strand of sodium alginate might be connected to the same through the formation of a link in which, the counter ions (Ca^{2+}, Ba^{2+}, and Al^{3+} ions) could be attached to two or three sodium alginate strands jointly. The further treatment of these ionotropically cross-linked sterculia gum-alginate microparticles with glutaraldehyde, an acetal structure had been formed between the –CHO groups of glutaraldehyde and –OH groups of sodium alginate and sterculia gum to form an IPN structure. The schematic presentation of proposed interactions to form an IPN structure between sterculia gum and sodium alginate is presented in Scheme 2.

The drug encapsulation efficiency in these IPN microparticles of sterculia gum–sodium alginate blends containing repaglinide was measured as, 81.10–91.70 % and it was found to decrease with the decrease of sodium alginate concentration. The drug encapsulation efficiency of the microparticles prepared by Al^{3+} ion-induced cross-linking was found to be higher than the microparticles prepared by Ba^{2+} ion-induced cross-linking, which in turn was comparatively higher than the microparticles prepared by Ca^{2+} ion-induced cross-linking. In case of the Ca^{2+} ion cross-linked microparticles, the polymeric network might loose with the larger pores that may result in leakage of entrapped drug molecules into the external medium from the polymeric matrix during the preparation of microparticles, which may lead to lower drug encapsulation efficiency. In case of the Al^{3+} ion cross-linked microparticles, the polymeric network should be rigid due to faster cross-linking rate and the presence of extra positive charge compared to other divalent metal ions (each Al^{3+} ion is able to bind extra caroxylate ions of the alginate backbone than other divalent metal ions) [96]. These phenomenons might lead to higher drug encapsulation efficiency of the Al^{3+} ion cross-linked microparticles. The drug encapsulation efficiency of these sterculia gum-alginate microparticles containing repaglinide prepared by dual cross-linking (i.e., sequential ionotropic and covalent cross-linking) was found higher than the microparticles prepared by ionotropic cross-linking. This event might be explained by the formation of stiffer matrix that could reduce the leakage of the entrapped drug molecules from the sterculia gum-alginate IPN matrix investigated in this work. The average size of these microparticles was measured as 19.75–61.52 μm. The size of these microparticles

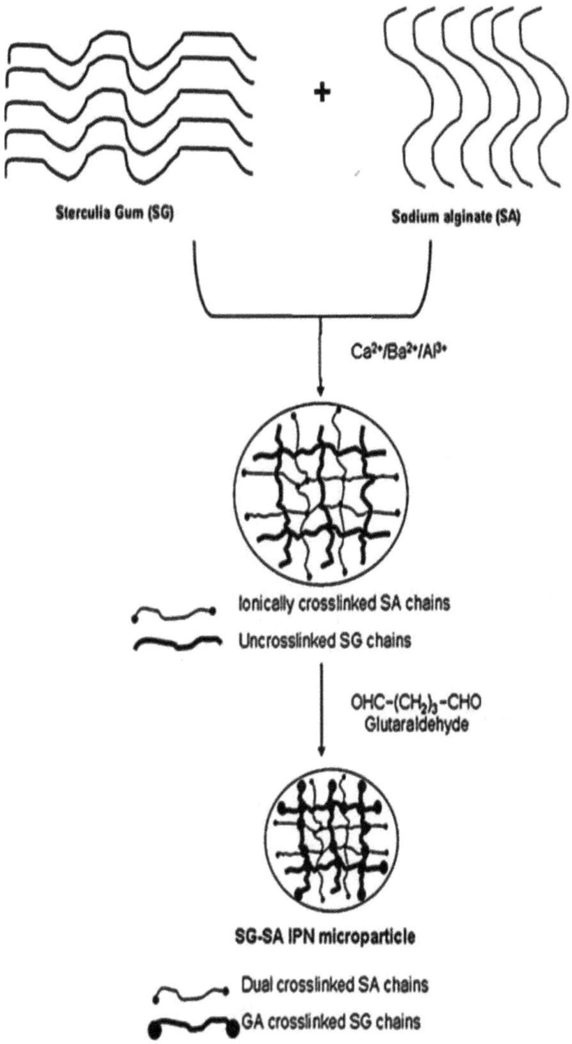

Scheme 2 Schematic presentation of proposed interactions to form an IPN structure between sterculia gum and sodium alginate [62]. Copyright © 2014 with permission from Elsevier B.V.

was found to be increased as the sodium alginate concentration was increased. The microparticle size was also found dependent upon the type of the ionotropic cross-linking agents. $AlCl_3$ produced smaller microparticles than $BaCl_2$, which in turn produced smaller microparticles than $CaCl_2$. At the time of cross-linking of sterculia gum-alginate polymer blends, the polymeric network might have undergone the syneresis, which might form smaller microparticles at higher cross-linking densities. When ionotropically cross-linked and dual cross-linked sterculia gum-alginate microparticles were compared, comparative smaller microparticles of sterculia gum-alginate blends containing repaglinide was experienced in case of dual cross-linking. This might be due to rapid shrinking of the polymeric matrix through the formation of covalent cross-links between the sterculia gum and alginate chains resulting in rigid IPN network.

SEM analyses of the sterculia gum-alginate IPN micoparticles containing repaglinide revealed smooth surface. FTIR spectroscopy and thermogravimetric (TG) analyses confirmed the formation of IPN structure between sterculia gum and alginate chains in these IPN micoparticles containing repaglinide. Differential scanning calorimetry (DSC) and X-ray diffraction (XRD) suggested the occurrence of uniform and amorphous dispersion of drug in the sterculia gum-alginate IPN matrix.

The in vitro drug release from these IPN microparticles was evaluated in 0.1 N HCl (pH 1.2) for 2 h and in phosphate buffer (pH 1.2) for the next hours. The results indicated that the drug release from these IPN microparticles was sustained over 24 h (Fig. 5). The drug release was found slower from Al^{3+} ion cross-linked microparticles as compared to Ba^{2+} ion cross-linked microparticles, which in turn was slower than that of Ca^{2+} ion cross-linked microparticles. The glutaraldehyde treatment of ionotropic cross-linked sterculia gum-alginate blends microparticles produced further slower profile of drug release. The drug release was found slower as the sodium alginate concentration was increased in these microparticles. The ionotropically cross-linked microparticles discharged encapsulated drug quickly; while the dual cross-linked IPN microparticles showed an extended in vitro drug

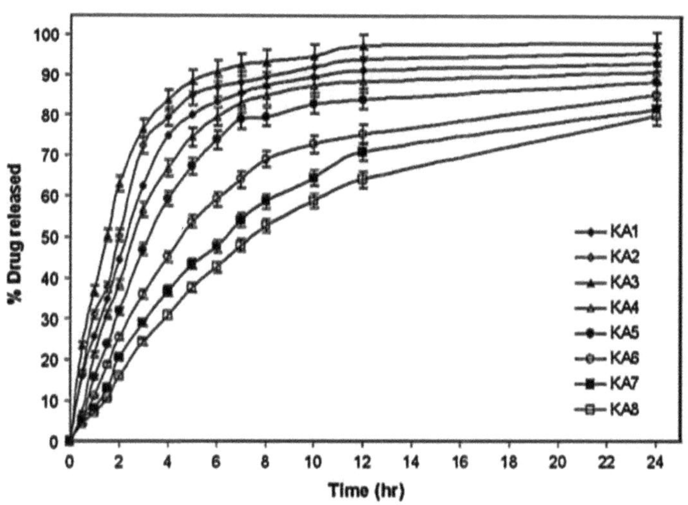

Fig. 5 In vitro drug release from sterculia gum-alginate IPN microparticles evaluated in 0.1 N HCl (pH 1.2) for 2 h and in phosphate buffer (pH 1.2) for next hours [Key: KA1, KA2, and KA3 were Ca^{2+} ion cross-linked microparticles; KA4 was Ba^{2+} ion cross-linked microparticle; KA5 was Al^{3+} ion cross-linked microparticle; KA6 was dual cross-linked IPN microparticle (sequential ionotropic cross-linking by Ca^{2+} ions and covalent cross-linking); KA7 was dual cross-linked IPN microparticle (sequential ionotropic cross-linking by Ba^{2+} ions and covalent cross-linking); KA8 was dual cross-linked IPN microparticle (sequential ionotropic cross-linking by Al^{3+} ions and covalent cross-linking)] [62]. Copyright © 2014 with permission from Elsevier B.V.

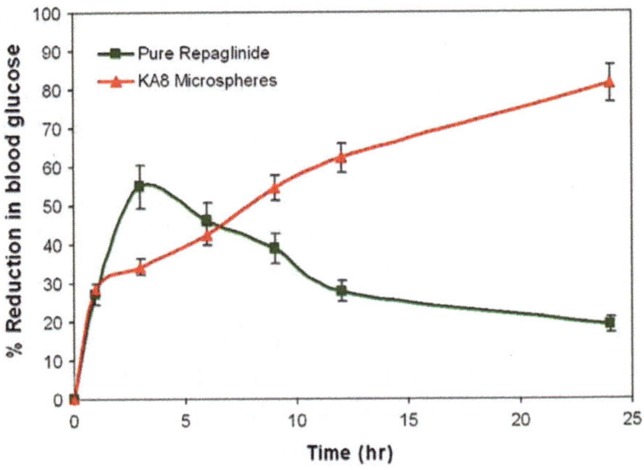

Fig. 6 Percentage reduction of blood glucose in streptozotocin-induced diabetic rats treated with pristine repaglinide and repaglinide-loaded KA8 IPN microparticles (dual cross-linked IPN microparticles by sequential ionotropic cross-linking by Al^{3+} ions and covalent cross-linking) [62]. Copyright © 2014 with permission from Elsevier B.V.

release for prolonged period. The in vitro drug release from sterculia gum-alginate IPN microparticles containing repaglinide was found to follow the non-Fickian transport mechanism.

The in vivo antidiabetic activity of sterculia gum-alginate IPN microparticles containing repaglinide was performed in streptozotocin-induced diabetic rats. The results showed that the pristine repaglinide produced a sudden reduction in blood glucose level in streptozotocin-induced diabetic rats up to 3 h and afterward, the blood glucose level was recovered. In case of the diabetic rats treated with sterculia gum-alginate IPN microparticles containing repaglinide, the percentage reduction of glucose level was slower as compared to pristine repaglinide within 3 h, but, it was increased gradually to 81.27 % up to 24 h. This indicated the slower drug release from these sterculia gum-alginate IPN microparticles containing repaglinide over a longer period (Fig. 6). Therefore, the developed sterculia gum-alginate IPN microparticles containing repaglinide was found useful to release repaglinide in a sustained manner for the effective treatment of diabetes mellitus.

5 Oil-Entrapped Sterculia Gum-Alginate Buoyant Systems for Gastroretentive Drug Delivery

Guru et al. [81] have developed and optimized oil-entrapped buoyant beads made of sterculia gum–sodium alginate blends for gastroretentive delivery of a NSAID (nonsteroidal anti-inflammatory drug), aceclofenac by ionotropic emulsion–gelation technique using $CaCl_2$ as ionotropic cross-linker. A 3^2-factorial design-based computer-aided optimization technique was employed for the formulation optimization of oil-entrapped buoyant beads containing aceclofenac. Through this 3^2-factorial design, the effects of process variables like polymer-to-drug ratio by weight and sodium alginate-to-sterculia gum ratio by weight on the drug entrapment and drug release of these oil-entrapped buoyant beads containing aceclofenac were analyzed. From the 3^2-factorial design-based optimization process, optimized formula of these oil-entrapped buoyant beads was obtained through numerical optimization, where polymer-to-drug ratio by weight and sodium alginate-to-sterculia gum ratio by weight were considered as 4.99 and 2.17, respectively. The drug entrapment efficiency of these oil-entrapped beads ranged from 63.28 ± 0.55 to 90.92 ± 2.34 %, while the optimized oil-entrapped beads showed the drug entrapment efficiency of 83.73 ± 0.81 %. The drug entrapment efficiency of these beads was found increased with decreasing sodium alginate-to-sterculia gum ratio and increasing polymer-to-drug ratio. Actually, the decrease in sodium alginate-to-sterculia gum ratio in the preparation of these oil-entrapped beads might produce an increase in viscosity of the polymeric solution by increasing addition of sterculia gum amount in the polymer blend solutions, so that this might have prevented leaching of the drug molecules to the cross-linking solutions, which may increase the drug entrapment efficiency in these beads. Increase in polymer-to-drug ratio in these oil-entrapped buoyant beads might produce entanglements of higher drug amounts inside the intricate cross-linked sterculia gum-alginate gel network, which may facilitate the increase of drug entrapment efficiency. The average bead sizes of these beads ranged from 1.32 ± 0.04 to 1.72 ± 0.12 % mm. The average size of the optimized beads was measured as 1.62 ± 0.08 mm. The densities of these oil-entrapped beads were measured less than the density of simulated gastric fluid (pH 1.2), imparting buoyancy. These oil-entrapped beads were also found to float within 6 min after being placed in simulated gastric fluid (pH 1.2) and exhibited buoyancy over 7 h, in vitro. The entrapment of liquid paraffin as low-density oil by these sterculia gum-alginate beads was responsible for imparting low density as well as buoyancy of these oil-entrapped beads containing aceclofenac.

SEM image of the optimized oil-entrapped sterculia gum-alginate buoyant beads containing aceclofenac showed a rough surface with small pores or channels (Fig. 7). The surface had an 'orange peel' appearance with corrugations. FTIR spectroscopy analyses confirmed the compatibility of the aceclofenac with sodium alginate and sterculia gum, which were used to prepare optimized oil-entrapped sterculia gum-alginate buoyant beads containing aceclofenac. XRD results suggested the crystalline nature of aceclofenac in the pure drug sample and the crystalline nature of aceclofenac was found to decrease significantly in the optimized oil-entrapped sterculia gum-alginate beads containing aceclofenac. This could be due to the effect of polymers used or the formulation process employed.

In vitro drug release from these oil-entrapped sterculia gum-alginate buoyant beads containing aceclofenac in simulated gastric fluid (pH 1.2) was studied and the results showed prolonged sustained release of aceclofenac over 7 h (Fig. 8). An increase in aceclofenac release from these beads was observed with the increase of polymer-to-drug ratio. The slower drug release was also observed with the increment of liquid paraffin entrapment in these beads and this can be explained as the

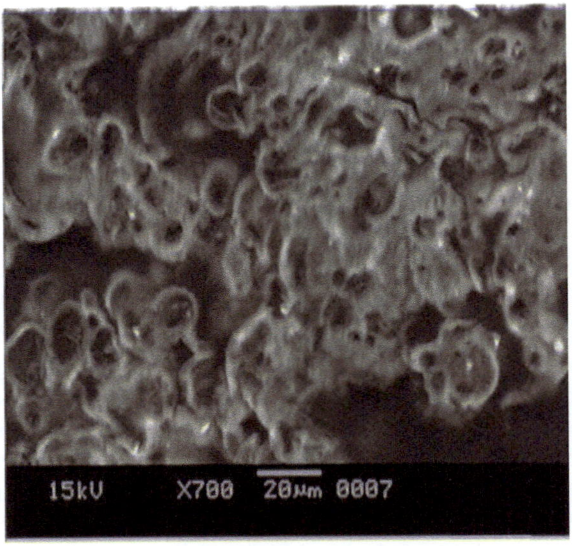

Fig. 7 SEM image of optimized oil-entrapped sterculia gum-alginate buoyant beads containing aceclofenac at magnification of ×700 [81]. Copyright © 2012 with permission from Elsevier B.V.

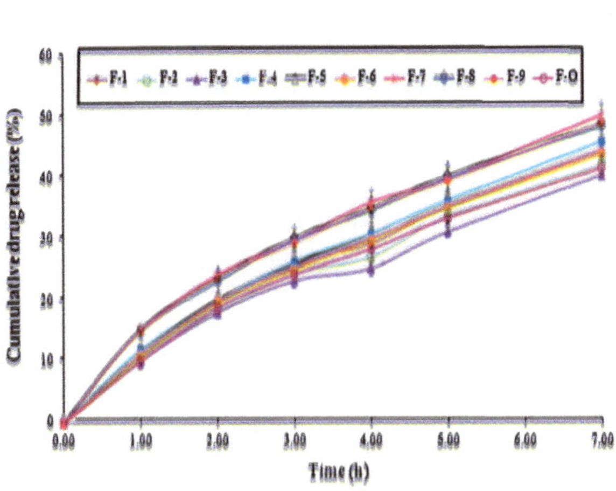

Fig. 8 In vitro drug release profile of oil-entrapped sterculia gum-alginate buoyant beads containing aceclofenac in simulated gastric fluid (pH 1.2) [81]. Copyright © 2012 with permission from Elsevier B.V.

most of the drugs remained saturated and dispersed in the oil pockets of the beads to form a drug–oil dispersed matrix. Actually, two steps of drug dissolution from these oil-entrapped beads may occur. First, the drug may diffuse out of the oil pockets into calcium alginate matrix and second, it may diffuse out of the calcium alginate matrix into the dissolution medium. The in vitro drug release from these oil-entrapped beads followed the Koresmeyer–Peppas model over a period of 7 h and the anolomous (non-Fickian) drug release mechanism, indicating both diffusion-controlled and swelling-controlled drug release. Therefore, the oil-entrapped sterculia gum-alginate buoyant beads containing aceclofenac were combined of excellent drug entrapment efficiency, good buoyant ability with a minimum lag time and suitable drug release pattern over a longer period, which should be effective for gastroretentive drug delivery. In addition, the antiulcer activity of sterculia gum and alginate might be beneficial for such a formulation loaded with a NSAID like aceclofenac, which often can produce ulceration as a result of prolonged use.

6 Alginate-Sterculia Gum Gel-Coated Oil-Entrapped Alginate Buoyant Systems for Gastroretentive Drug Delivery

In an investigation, Bera et al. [61] have employed Ca^{2+} ion cross-linked alginate-sterculia gum gel coating onto olive oil-entrapped alginate beads containing risperidone to improve gastroretention through a combination mechanism of floatation and mucoadhesion. In this work, researcher have placed optimized oil-entrapped alginate beads containing risperidone in 1.0 % w/w alginate-sterculia gum (1:1) aqueous dispersion and subsequently introduced into 5 % w/v $CaCl_2$ solution for 10 min to harden. Then, these coated beads were washed with distilled water and dried overnight at room temperature. These alginate-sterculia gum gel-coated oil-entrapped alginate beads containing risperidone possessed drug entrapment efficiency of 81.63 ± 1.54 %, mean diameter of 2.49 ± 0.12 mm, and density of 0.66 ± 0.15 g/cm^3.

SEM analyses of the alginate-sterculia gum gel-coated oil-entrapped alginate beads containing risperidone revealed spherical shape with smooth surface relative to optimized uncoated oil-entrapped alginate beads containing risperidone (Fig. 9a, b). The coated membrane could reduce the number of cracks and pores on the outer surface of the beads. The SEM view of the cross section of the coated beads showed a sponge-like structure, in which oil was entrapped (Fig. 9c, d). In addition, it showed that a thin-film membrane deposited onto the surface of the oil-entrapped beads as coating.

The alginate-sterculia gum gel-coated oil-entrapped alginate beads containing risperidone exhibited sustained in vitro drug release in simulated gastric fluid (pH 1.2) over 8 h of the in vitro dissolution, which was found much slower than the uncoated oil-entrapped alginate beads containing risperidone. The coating membrane might act as a barrier to retard risperidone release from the coated beads. Moreover, the coating could block the surface pores of the uncoated beads, which could lead to reduced water penetration and slower drug release rate. In addition, these coated beads followed the Koresmeyer–Peppas model kinetics and the Fickian diffusion mechanism during in vitro drug release. These alginate-sterculia

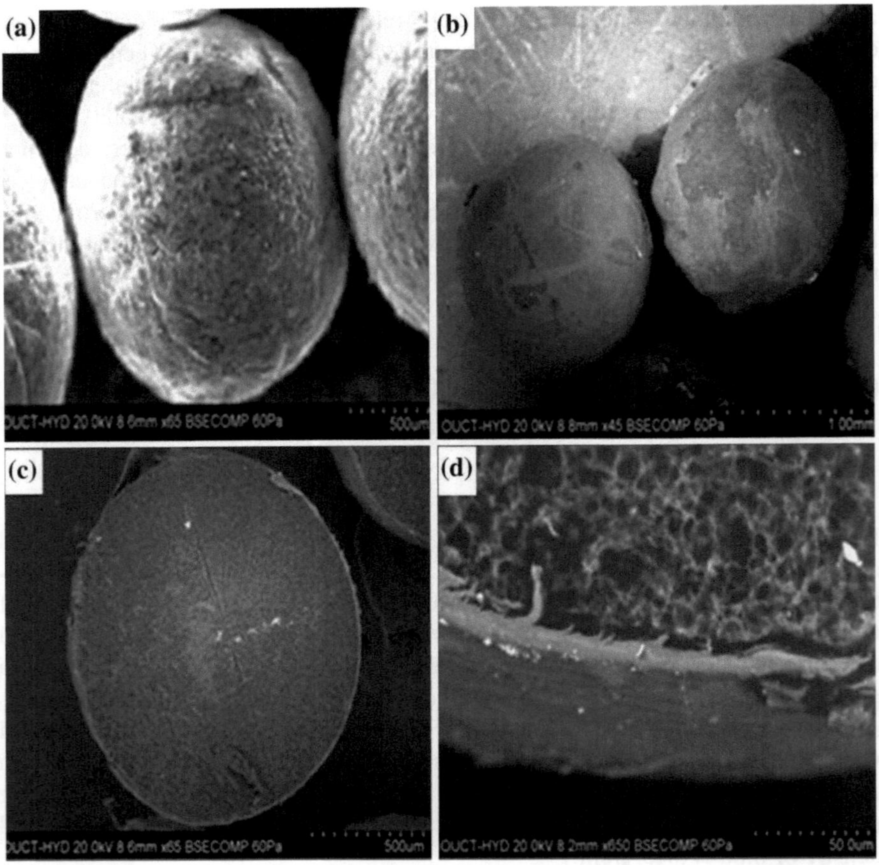

Fig. 9 SEM images of the surface of the uncoated oil-entrapped alginate beads containing risperidone (**a**), alginate-sterculia gum gel-coated oil-entrapped alginate beads containing risperidone (**b**), cross-sectional view of the coated beads with less magnification of ×65 (**c**) and high magnification of ×650 (**d**) [61]. Copyright © 2014 with permission from Elsevier Ltd.

gum gel-coated oil-entrapped alginate beads showed improved swelling, buoyancy with minimum lag time, and good mucoadhesion as compared to that of uncoated beads containing risperidone.

7 Pectinate–Sterculia Gum Interpenetrating Polymer Network (IPN) Beads for Gastroretentive Drug Delivery

In recent years, extensive research efforts have been paid to develop pectinate-based hydrogel beads for oral drug delivery due to its high biodegradability, excellent biocompatibility and acid stability [105–107]. Pectin is a plant-derived polysaccharide, which consists of linearly connected α-(1-4)-D-galacturonic acid regularly interspersed with α-(1-2) linked α-L-rhamnopyranose residues [108, 109]. Low methoxy pectin is able to cross-link inotropically with various divalent metal cations like Ca^{2+}, Zn^{2+}, etc., to produce rigid hydrogel beads, which has been intensively exploited as drug delivery carriers [22, 110]. The high aqueous solubility and swelling of these ionotropically cross-linked pectinate hydrogel beads led to poor encapsulation efficiency of low molecular drugs and also premature release of the entrapped drug molecules [10, 22]. Previous researches have already revealed that the blending of low methoxy pectin with other second biocompatible polymers is an effective approach to diminish the above-discussed drawbacks of ionotropically cross-linked pectinate gel beads as oral drug delivery carriers [10, 22, 109–111]. Based on this approach, novel dual cross-linked IPN beads made of pectinate–sterculia gum were developed for intragastric delivery of an atypical antipsychotic drug, ziprasidone HCl [84].

The IPN beads made of low methoxy pectin–sterculia gum blends containing ziprasidone HCl were accomplished by simultaneous ionotropic gelation by zinc acetate and covalent cross-linking by glutaraldehyde. In this work, the effects of low methoxy pectin and sterculia gum amounts on the drug encapsulation and drug release were studied to optimize the formulation of the floating mucoadhesive IPN beads containing ziprasidone HCl using 3^2-factorial design. In fact, Zn^{2+} ions (having a lower coordination number) could induce strong noncovalent associations (viz., hydrogen bonding and hydrophobic interactions) between the carbohydrate chains of the pectinate gel network [105]. Also, the ionotropic interactions between the homogalacturonic segments, galacturonic acid, or glucurunic acid residues of sterculia gum molecules and divalent ions could form intermolecular zones like 'Egg-box model'. Thus, sterculia gum could be associated with the ionotropic gelation process [75]. The hydrogen bonding and the electrostatic interactions between both low methoxy pectin and sterculia gum chains might also play a crucial role in the formation of hydrogel beads. Moreover, the –CHO groups of glutaraldehyde reacted with the –OH groups of the polysaccharides to create an acetal linkage leading to formation of IPN structure between low methoxy pectin and sterculia gum. The optimized IPN beads made of low methoxy pectinate–sterculia gum blends containing ziprasidone HCl displayed good drug encapsulation efficiency of 87.98 ± 1.15 % and mean diameter of 2.17 ± 0.13 mm.

SEM analyses of these optimized IPN beads of ziprasodone HCl revealed a spherical-shaped structure with characteristic large wrinkles and cracks on the bead surface (Fig. 10). The rough surface of the bead contained channels and small pores, which could be possibly formed as a result of the migration of the water molecules during the drying process and subsequent shrinkage of the polymer networks. The SEM image of surface of the IPN beads did not reveal the presence of drug crystals on the bead surface, indicating that the drug was in finely dispersed state in the polymeric matrix after formulation. In addition, some polymeric debris was seen on the bead surface, which could be due to the method of preparation (i.e., simultaneous gel bead formation and the formation of the polymeric matrix) [23]. FTIR analyses confirmed the absence of drug–excipients incompatibility and

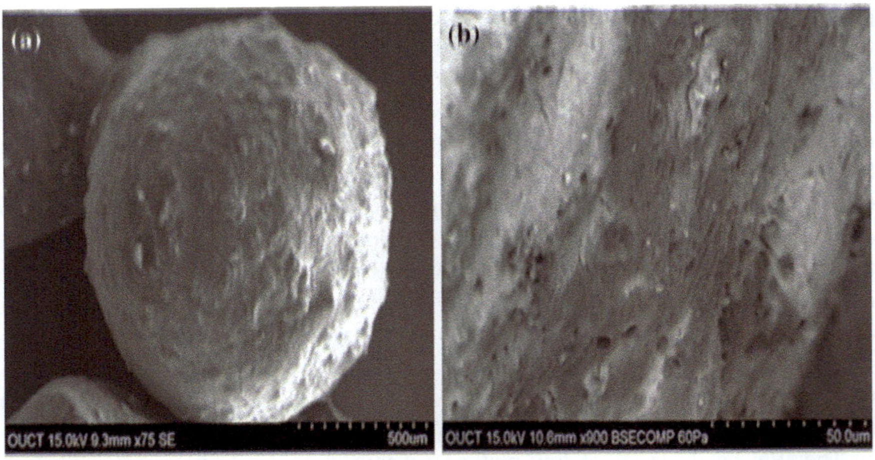

Fig. 10 SEM images of optimized pectinate–sterculia gum IPN beads containing ziprasidone HCl (F-O) showing rough surface magnification of ×75 (**a**) and presence of pores and channels at high magnification of ×900 (**b**) [84]. Copyright © 2015 with permission from Elsevier Ltd.

formation of IPN structure between low methoxy pectin and sterculia gum in the optimized IPN beads. DSC and XRD analyses revealed the physical state of the encapsulated drug (i.e., ziprasidone HCl) within the IPN matrix.

All these pectinate–sterculia gum IPN beads exhibited in vitro sustained ziprasidone HCl release over 8 h in simulated gastric fluid (pH 1.2) (Fig. 11). The delayed release of drug could probably due to reduced free volume spaces of

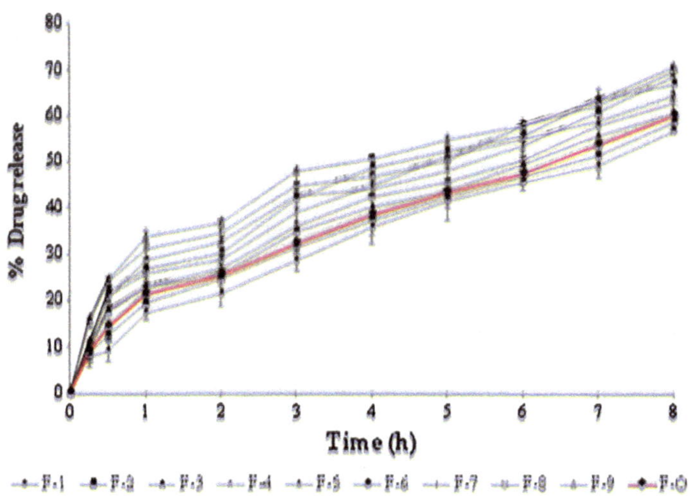

Fig. 11 In vitro drug release profiles of pectinate–sterculia gum IPN beads containing ziprasidone HCl in simulated gastric fluid (pH 1.2) [84]. Copyright © 2015 with permission from Elsevier Ltd.

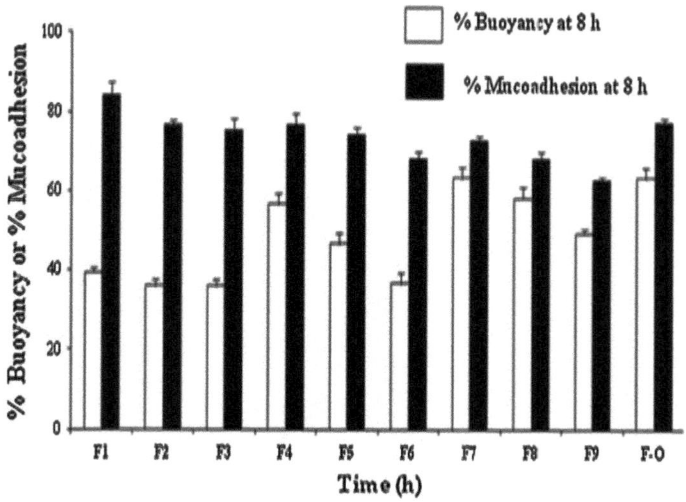

Fig. 12 The percentage buoyancy (**a**) and percentage mucoadhesion (**b**) of the pectinate–sterculia gum IPN beads containing ziprasidone HCl at 8 h in simulated gastric fluid (pH 1.2) [84]. Copyright © 2015 with permission from Elsevier Ltd.

glutaraldehyde-treated IPN matrices, which could restrict the drug diffusion from the IPN network [62]. In most of the cases, the drug release behavior obeyed the Higuchi kinetics with the anolomous transport or non-Fickian release mechanism. The swelling profiles of all the IPN beads in simulated gastric fluid (pH 1.2) demonstrated that the beads swelled to a lesser extent with gradual increments up to 6 h. The optimized IPN beads also exhibited excellent buoyancy with floating lag time of <2 min and percentage buoyancy at 8 h of 63 % (Fig. 12). It also showed good mucoadhesivity with the goat gastric mucosa (Fig. 12).

8 Sterculia Gum Cross-Linked Polyacrylamide [Poly (AAm)] Hydrogel for Use in Antiulcer Drug Delivery

Sigh and Sharma [112] have synthesized cross-linked hydrogels of sterculia gum and acrlamide (AAm), sterculia gum-*cl*-poly(AAm) hydrogels for an antiulcer drug (ranitidine HCl) delivery using *N*,*N*′-methylene bisacrylamide (*N*,*N*′-MBAAm) as cross-linker and ammonium persulfate (APS) as initiator. The optimum reaction parameters were evaluated for the synthesis of sterculia gum-*cl*-poly(AAm) hydrogels through performing the preliminary trials by varying AAm concentration, APS concentration, *N*,*N*′-MBAAm concentration, and sterculia gum amount on the basis of swelling of the synthesized hydrogels in distilled water after 24 h of swelling, shape and structural integrity were maintained by the hydrogels during swelling. After performing the preliminary trials, the optimum reaction condition

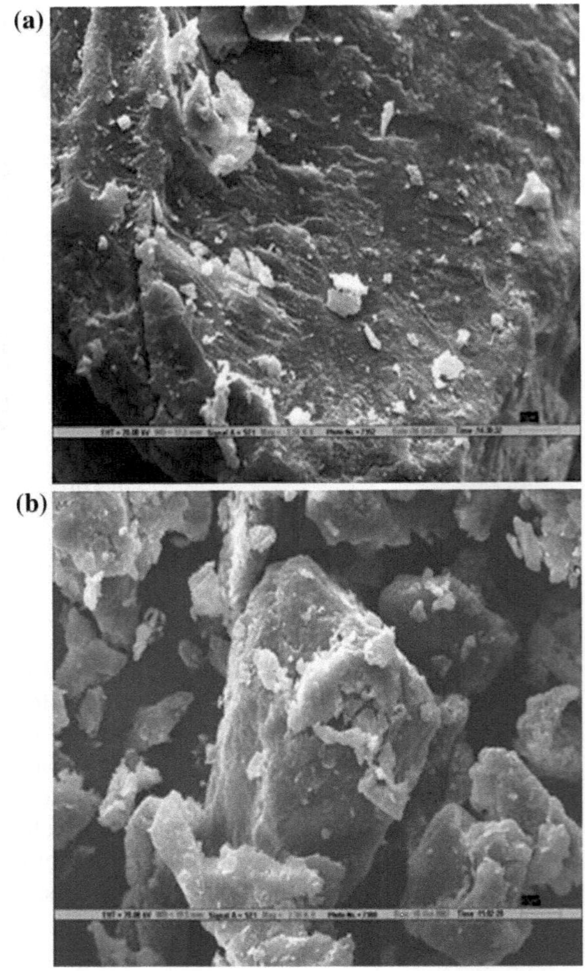

Fig. 13 SEM images showing the surface morphologies of sterculia gum (**a**) and sterculia gum-*cl*-poly(AAm) hydrogel (**b**) [112]. Copyright © 2008 with permission from Elsevier Ltd.

for the synthesis of sterculia gum-*cl*-poly(AAm) was considered as AAm concentration = 1.125 mol/l, APS concentration = 13.158 mmol/l, sterculia gum amount = 0.8 g, and *N,N′*-MBAAm concentration = 6.486 mmol/l.

SEM analyses indicated smooth and homogeneous morphology of sterculia gum; whereas structural heterogeneity was evidenced in case of the synthesized sterculia gum-cl-poly(AAm) hydrogel (Fig. 13). FTIR analyses suggested the cross-linking between sterculia gum and AAm to form sterculia gum-*cl*-poly(AAm) hydrogel.

The swelling of these synthesized sterculia gum-*cl*-poly(AAm) hydrogels was studied as a function of AAm concentration, APS concentration, *N,N′*-MBAAm concentration, and sterculia gum amount, pH and [NaCl] of the swelling mediums. The swelling of these sterculia gum-*cl*-poly(AAm) hydrogels decreased with increment of monomer (i.e., AAm) concentration in the polymeric hydrogel matrix. At the AAm concentration of 1.125 mol/l, maximum water uptake was seen after 24 h of swelling and the formed cross-linked hydrogels maintained a cylindrical shape. It was also seen that the hydrogels prepared with different monomer concentrations followed the Fickian diffusion mechanism. The swelling of these synthesized sterculia gum-*cl*-poly(AAm) hydrogels was decreased with the increase of APS (initiator) concentration used during synthesis. This could be due to the fact that higher concentration of initiator has started the formation of the number of the polymeric chains, which decreased the chain length and affected the network formation. It was observed that these cross-linked hydrogels prepared by different concentrations of initiators followed the non-Fickian diffusion mechanism. The swelling of the synthesized sterculia gum-*cl*-poly(AAm) hydrogels was found to be increased with the increment of sterculia gum amount in the composition of hydrogel matrix. This could be higher degree of hydration of the hydrophilic sterculia gum with its increasing amount during synthesis of sterculia gum-*cl*-poly (AAm) hydrogels. The increasing amount of sterculia gum in these hydrogels might increase the number of intimate contacts between sterculia gum particles and water leading to higher degree of swelling. The diffusion of water from these cross-linked hydrogels prepared with different amounts of sterculia gum followed the Fickian diffusion mechanism. The swelling of synthesized cross-linked sterculia gum-based hydrogels was found decreased with the increment of the *N*,*N*′-MBAAm (cross-linker) concentration during synthesis. This could be due to the reason that with the increment of cross-linker during synthesis, cross-linking density increases and this may lead to decrease in the size of the pores in the cross-linked hydrogel matrix and thereafter, to decrease in the swelling of the cross-linked matrix. The diffusion of the water from these sterculia gum-based hydrogels prepared with different concentrations of cross-linkers followed the non-Fickian diffusion mechanism. The swelling of the optimized cross-linked sterculia gum-*cl*-poly(AAm) hydrogels in different pH mediums indicated increase of swelling with the increment of the pH of swelling mediums. At the lower pH, the $–CONH_2$ groups are unable to ionize and maintain the hydrogel network at its collapse state. At the higher pH, the $–CONH_2$ groups get partially ionized and the charged $–COO^-$ groups repel each other and cause more swelling of the hydrogel. Moreover, the swelling of these optimized cross-linked sterculia gum-based hydrogels in 0.9 % NaCl was observed lesser than that in distilled water. The Fickian diffusion mechanism was observed in case of the all swelling mediums investigated in this study.

The loading of an antiulcer drug onto the optimized cross-linked sterculia gum-*cl*-poly(AAm) hydrogel was carried out by swelling equilibrium method. The in vitro release profile of ranitidine HCl from these hydrogel was studied in distilled water, pH 2.2 and pH 7.4 buffers. The amount of drug released from per gram of the cross-linked hydrogel was higher in pH 2.2 buffer as compared to drug release in pH 7.4 buffer and distilled water (Figs. 14 and 15). The drug release trends were not corresponding to the swelling pattern of the newly synthesized hydrogel, where swelling was observed higher in distilled water and pH 7.4 buffer. This may be

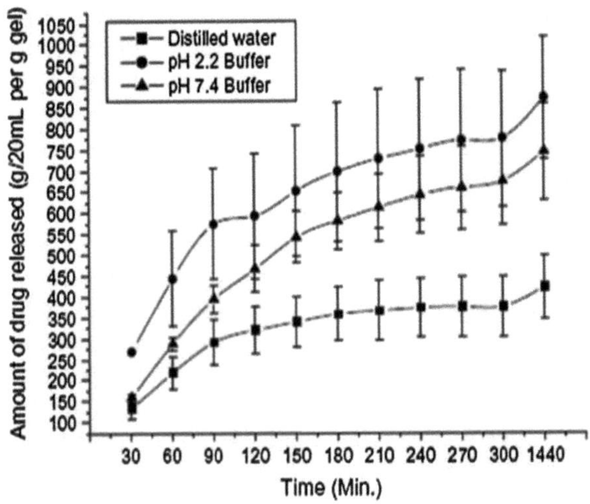

Fig. 14 Release profile of drug (ranitidine HCl) release from these drug-loaded sterculia gum-*cl*-poly(AAm) hydrogel in different mediums at 37 °C [112]. Copyright © 2008 with permission from Elsevier Ltd.

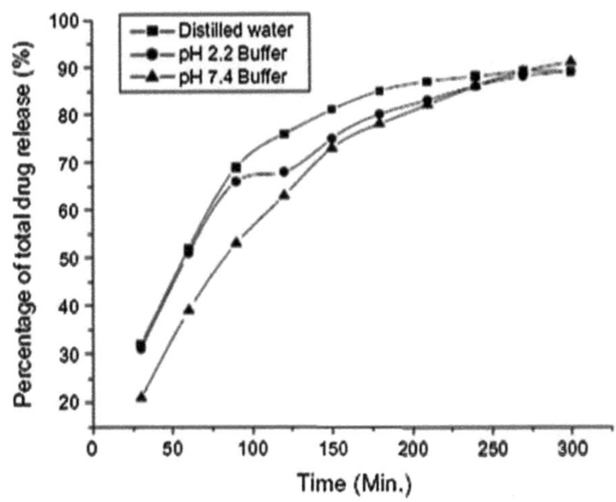

Fig. 15 Percentage of the total release of drug (ranitidine HCl) release from these drug-loaded sterculia gum-*cl*-poly(AAm) hydrogel in different mediums at 37 °C [112]. Copyright © 2008 with permission from Elsevier Ltd.

because of more solubility of ranitidine HCl in pH 2.2 buffer. The drug (ranitidine HCl) release from these drug-loaded optimized cross-linked sterculia gum-*cl*-poly (AAm) hydrogel followed the Fickian diffusion mechanism in distilled water and pH 2.2 buffer, and the non-Fickian diffusion mechanism in pH 7.4 buffer. Thus, the synthesized sterculia gum-*cl*-poly(AAm) hydrogel exhibited suitability as carrier for sustained delivery of antiulcer drug.

9 Sterculia Gum Cross-Linked Polymethacrylate [Poly (MAAc)] Hydrogel for Use in Antiulcer Drug Delivery

In an investigation, modification of sterculia gum with methacrylic acid (MAAc) was carried out to synthesize sterculia gum-*cl*-poly(MAAc) hydrogels using *N,N′*-MBAAm as cross-linker and APS as initiator for the use in drug delivery [86]. After performing the preliminary trials, the optimum reaction condition for the synthesis of sterculia gum-*cl*-poly(MAAc) hydrogels were selected as MAAc concentration = 1.179 mol/l, APS concentration = 21.930 mmol/l, sterculia gum amount = 0.6 g, and *N,N′*-MBAAm concentration = 6.486 mmol/l on the basis of swelling after 24 h in distilled water, shape and structural integrity were maintained by the synthesized hydrogels after swelling.

The SEM photograph of the morphology of sterculia gum-*cl*-poly(MAAc) hydrogel showed structural heterogeneity (Fig. 16). Similar type of morphology was seen in case of another cross-linked sterculia gum-based hydrogel, sterculia gum-*cl*-poly(AAm) hydrogel. FTIR analyses supported the cross-linking between sterculia gum and MAAc within the synthesized sterculia gum-*cl*-poly(MAAc) hydrogel.

The swelling of these synthesized sterculia gum-*cl*-poly(MAAc) hydrogels was evaluated as a function of MAAc concentration, APS concentration, *N,N′*-MBAAm concentration, and sterculia gum amount, pH and [NaCl] of the swelling mediums. With the increment of monomer (MAAc) concentration or during the synthesis, a decrease in the swelling of the sterculia gum-*cl*-poly(MAAc) hydrogels was experienced. The swelling of these hydrogels with different concentrations of

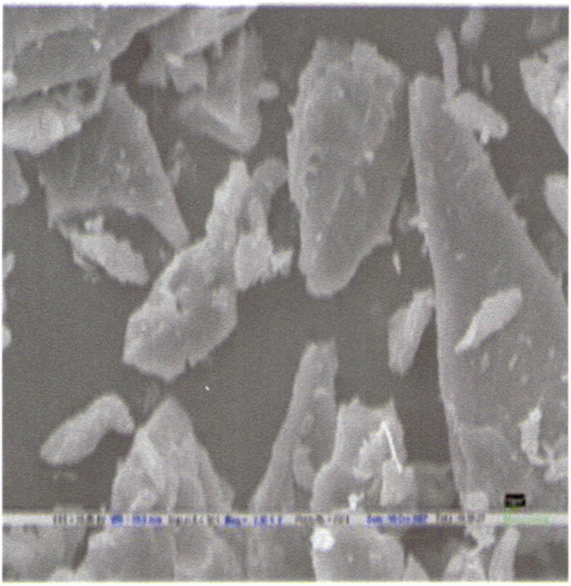

Fig. 16 SEM image showing the surface morphology of sterculia gum-*cl*-poly(MAAc) hydrogel [86]. Copyright © 2008 with permission from Elsevier B.V.

MAAc was followed by the Fickian diffusion mechanism. The swelling of these hydrogels increased with the increase in the concentration of the initiator (APS) used in the hydrogel synthesis. The diffusion of the water molecules from these hydrogels prepared with different concentrations of APS as initiator occurred through the non-Fickian diffusion mechanism. The swelling of the hydrogels was also found to increase with time and with the increase in sterculia gum amounts in the composition of the hydrogel matrix up to 0.6 g of sterculia gum amount and after that, the swelling of these hydrogels was found to be increased. This might occur due to the fact of higher degree of sterculia gum hydration with increased number of intimate contacts between the sterculia gum particles and water. This might lead to higher rate of swelling of the polymeric matrix. The diffusion of water from the hydrogel matrix prepared with different amounts of sterculia gum was found to follow the non-Fickian diffusion mechanism. The swelling of the synthesized hydrogels was decreased with the increase in concentration of the cross-linker (*N*,*N*′-MBAAm) used in the synthesis of hydrogels. This might occur due to the fact that with the increase in the concentration of the cross-linker, cross-linking density increases, which decreases the pore size in the cross-linked matrix and thereafter decreases the swelling of the polymeric matrix. The non-Fickian diffusion mechanism was followed for the diffusion of water from these hydrogels. The hydrogels synthesized using optimum reaction conditions were studied for the effect of swelling mediums on the swelling of these hydrogels. The swelling of these hydrogels were also evaluated in distilled water, pH 2.2 buffer, and pH 7.4 buffer for 24 h. It was experienced from this study that the swelling of these hydrogels was increased with the increase in the pH of the swelling mediums after 24 h. A similar result was also seen in case of the sterculia gum-*cl*-Poly (AAm) hydrogels [112]. The non-Fickian diffusion mechanism was found to follow for the diffusion of the water from these hydrogels, in distilled water, pH 2.2 and pH 7.4 buffers. To study the effect of salt concentration, the swelling of the sterculia gum-*cl*-poly(MAAc) hydrogels was evaluated in 0.9 % NaCl solution at 37 °C and the results of it exhibited lesser swelling as compared to the distilled water. In 0.9 % NaCl solution, the Fickian diffusion mechanism was observed for the diffusion of water from these hydrogel matrices.

For the loading of an antiulcer drug, ranitidine HCl within these optimized cross-linked sterculia gum-*cl*-poly(MAAc) hydrogel was carried out by swelling equilibrium method. The ranitidine HCl-loaded hydrogel was evaluated for in vitro drug release dynamics in distilled water, pH 2.2 and pH 7.4 buffers. The results of the in vitro drug release from these drug-loaded hydrogel exhibited higher amount of drug release in pH 7.4 buffer and distilled water, which was not corresponding to the results of swelling of these hydrogel (Figs. 17 and 18). Similar type of occurrence was seen in case of sterculia gum-*cl*-poly(AAm) hydrogel [112], and this can be explained on the basis of high solubility of ranitidine HCl in pH 2.2 buffer. The drug release from these hydrogels was found to follow non-Fickian diffusion mechanism in all the release mediums (distilled water, pH 2.2 and pH 7.4

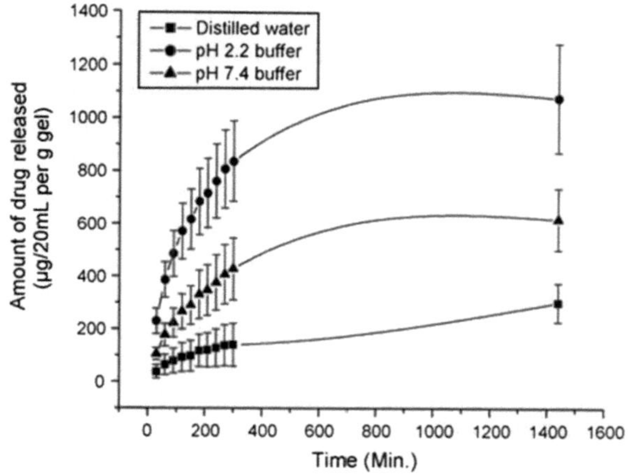

Fig. 17 Release profile of drug (ranitidine HCl) release from these drug-loaded sterculia gum-*cl*-poly(MAAc) hydrogel in different mediums at 37 °C [86]. Copyright © 2008 with permission from Elsevier B.V.

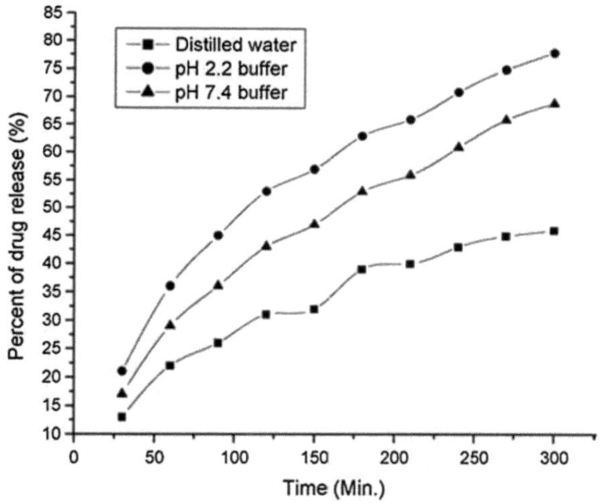

Fig. 18 Percentage of the total release of drug (ranitidine HCl) release from these drug-loaded optimized cross-linked sterculia gum-*cl*-poly(MAAc) hydrogel in different mediums at 37 °C [86]. Copyright © 2008 with permission from Elsevier B.V.

buffers) used in the study of drug release dynamics of drug-loaded sterculia gum-*cl*-poly(AAm) hydrogel. Hence, synthesized sterculia gum-*cl*-poly(MAAc) hydrogel exhibited efficacy as carrier for the effective delivery of antiulcer drug.

10 Sterculia Gum Cross-Linked Poly (N-Vinylpyrrolidone) [Poly(NVP)] Hydrogel for Use in Antidiarrhoeal Drug Delivery

In view of the antidiarrhoeal property of the sterculia gum and the antimicrobial property of NVP, a novel sterculia gum-*cl*-poly(NVP) hydrogel was developed through the functionalization of sterculia gum with NVP for the use in the treatment of diarrhea [63]. These sterculia gum-*cl*-poly(NVP) hydrogels were synthesized by free radical graft copolymerization of NVP onto sterculia gum using *N,N'*-MBAAm as cross-linker and APS as reaction initiator. The optimum reaction parameters for the synthesis of sterculia gum-*cl*-poly(NVP) hydrogels were searched by varying amounts of sterculia gum, concentrations of NVP, APS, and *N,N'*-MBAAm on the basis of swelling of these novel hydrogels in distilled water, shape, and structural integrity after 24 h of swelling. The optimization reaction parameters for the synthesis of optimized cross-linked sterculia gum-*cl*-poly(NVP) hydrogel were obtained from the preliminary trial as 0.375 mol/l of NVP, 21.930 mmol/l of APS, 1.0 g of sterculia gum, and 19.459 mmol/lof *N,N'*-MBAAm.

SEM image of optimized cross-linked sterculia gum-*cl*-poly(NVP) hydrogel exhibited the structural homogeneity. FTIR analyses and TG analyses suggested the cross-linking between sterculia gum and NVP in this novel hydrogel.

The swelling of synthesized sterculia gum-*cl*-poly(NVP) hydrogels first increased with the increasing of NVP concentration up to 0.563 mol/l and then decreased. These trends can be explained on the reason that after forming the pore of optimum size, further increase in the monomer concentration led to decrease in pore size and increase in the network density of the hydrogels. Swelling of these hydrogels prepared with different concentrations of NVP occurred through the non-Fickian diffusion mechanism. The swelling of these hydrogels was increased as the concentration of APS increased up to 13.158 mmol/l and after that swelling decreased. The swelling of these hydrogels prepared with varying concentrations of APS also occurred through the non-Fickian diffusion mechanism. Swelling was also found to increase with the increment of sterculia gum amount as composition of these hydrogels during synthesis. The non-Fickian diffusion mechanism occurred for the swelling of these hydrogels prepared with varying amounts of sterculia gum. Decrease in swelling of these hydrogels was observed with the increase of cross-linker concentrations during synthesis. This might be due to decrease in pore size and polymer chain flexibility. The swelling of these hydrogels occurred through the non-Fickian diffusion mechanism, when the cross-linker concentrations were varied during the synthesis. The swelling of the optimized hydrogel was much higher in distilled water than that in pH 2.2 and pH 7.4 buffers. This might be due to the nonionic nature of NVP. Swelling of the optimized hydrogel occurred through the Fickian diffusion mechanism in pH 2.2 and pH 7.4 buffers; while it exhibited the non-Fickian diffusion mechanism in distilled water. Swelling of the optimized hydrogel in 0.9 % NaCl solution was observed lesser than that in distilled water and it followed the Fickian diffusion mechanism.

The optimized sterculia gum-*cl*-poly(NVP) hydrogel was loaded with an antidiarrhoeal drug, ornidazole and the drug loading into the optimized cross-linked hydrogel was carried out by swelling equilibrium method. The in vitro drug release from the drug-loaded hydrogel showed higher amount of drug release in pH 2.2 buffer than that in distilled water and pH 7.4 buffer (Fig. 19). The diffusion of the drug release from these hydrogel followed the Fickian diffusion mechanism in all cases. Thus, the optimized sterculia gum-*cl*-poly(NVP) hydrogel was found effective for the sustained release of antidiarrhoeal drug. In addition with the

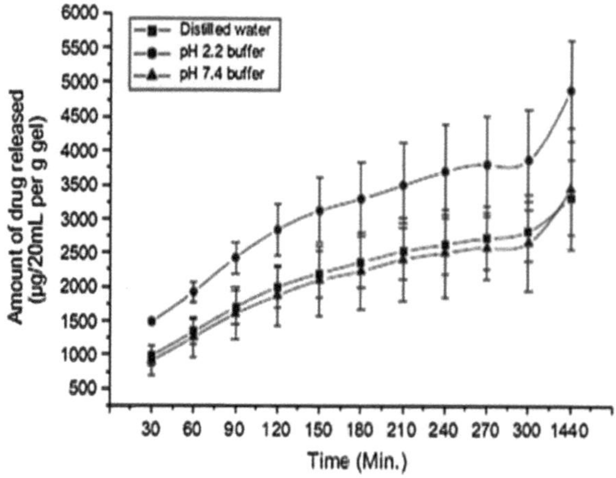

Fig. 19 Release profile of antidiarrhoeal drug (ornidazole) release from these drug-loaded sterculia gum-*cl*-poly(NVP) hydrogel in different mediums at 37 °C [63]. Copyright © 2010 with permission from Elsevier B.V.

antidiarrhoeal property, sterculia gum has bulk laxative action [72]. This may produce a decrease in the fluidity of the existing bowel contents and increase in fecal bulk. Sterculia gum also has the capacity to absorb water and it is transferred to a gelatinous state at a higher level. Thus, it is transported more rapidly through the intestinal tract and also, it increases the volume of the stool. This also softens the stool and stimulates the intestinal motility. Therefore, sterculia gum-*cl*-poly (NVP) hydrogel can be used for double potential antidiarrhoeal drug delivery, which is due to therapeutic importance of sterculia gum as an antidiarrhoeal therapeutic importance of sterculia gum and the loaded antidiarrhoeal drug (ornidazole).

11 Sterculia Gum Cross-Linked Polyvinyl Alcohol (PVA) Hydrogel Membrane as Wound Dressings for Delivery of Antimicrobial Drug

In an investigation, modification of sterculia gum with PVA to develop novel wound dressings of sterculia gum-*cl*-PVA hydrogel for the delivery of an antimicrobial agent (tetracycline HCl) was investigated in view of the gel-forming nature and antimicrobial activity of sterculia gum [113]. The optimum reaction parameters for the synthesis of sterculia gum-*cl*-PVA hydrogel were searched on the basis of swelling by varying the content of glutaraldehyde as cross-linker and di-*n*-butylphthalate (DBP) as plasticizer, and concentration of sterculia gum solution and PVA solution. After performing the preliminary trial, the optimum reaction parameters were chosen for the synthesis of optimized cross-linked sterculia gum-*cl*-PVA hydrogel as 3 ml of glutaraldehyde, 3 ml of 2.5 % sterculia gum solution, 0.1 ml of DBP, 5 ml of 5 % PVA solution. FTIR analyses supported the formation of cross-linking between sterculia gum and PVA within the sterculia gum-*cl*-PVA hydrogel matrix.

The swelling of these sterculia gum-*cl*-PVA hydrogel was found increased with the increase of glutaraldehyde content as cross-linker during synthesis. This might be due to the fact that increasing in glutaraldehyde content in the formula of hydrogel membrane can increase the cross-linking between the polymeric chains. This increases the retractive force in the polymeric hydrogel network. The swelling of these hydrogel matrices prepared by varying amounts of glutaraldehyde in the synthesis followed the non-Fickian diffusion mechanism. The amount of DBP used in the synthesis affected the swelling of these hydrogel matrices and the trends showed that the swelling of the hydrogel matrices decreased with the increase in the DBP contents as plasticizer in the sterculia gum-*cl*-PVA hydrogel formula. This could be due to the fact that plasticization occurs when plasticizer molecules interact with the polar groups of the polymers and replace the polymer/polymer interactions with the plasticizer/polymer interactions. This occurs the shielding of the polymeric chains from interacting with each other and with water molecules. On the other hand, the nonpolar groups present in the plasticizers also may contribute in the reduction of same interactions in the polymeric chains [114]. The non-Fickian diffusion mechanism was evidenced in case of the swelling of all these hydrogel matrices prepared by varying amounts of DBP as plasticizer. The swelling of the sterculia gum-*cl*-PVA hydrogel matrices increased with the increment of PVA amount in the hydrogel matrices and this might be due to the hydrophilic property of PVA. So, the increase in PVA amounts in these hydrogel matrices increased the hydrophilicity, which increased the water uptake to swell. The swelling of these hydrogel matrices followed non-Fickian diffusion mechanism when the varying amount of PVA was used in the synthesis of hydrogel matrices. The swelling of the sterculia gum-*cl*-PVA hydrogel matrices also found to be increased with the increase in the sterculia gum amounts in the hydrogel formula. This is probably due to the increase in hydration with the increment of hydrophilic gum in these hydrogel matrices. The non-Fickian diffusion mechanism was observed in case of these hydrogels prepared by varying amounts of sterculia gum. In addition, the swelling of the sterculia gum-*cl*-PVA hydrogel matrix prepared using optimum reaction parameter was observed to increase with the increment of time, when it was evaluated in distilled water, pH 2.2 and pH 7.4 buffers. The swelling of this optimized hydrogel matrix was found comparatively higher in distilled water than that in pH 2.2 and pH 7.4 buffers. The non-Fickian diffusion mechanism was observed for the swelling of the optimized sterculia gum-*cl*-PVA hydrogel matrix in the distilled water and the case II transport mechanism in both pH 2.2 and pH 7.4 buffers. The swelling of the optimized sterculia gum-*cl*-PVA hydrogel matrix in simulated wound fluid was evaluated to assess the viability of the hydrogel for the use as wound dressings and the result indicated less swelling in simulated wound fluid than that in distilled water. The case II transport mechanism was occurred for the swelling of the optimized hydrogel matrices in simulated wound fluid.

An antimicrobial drug, tetracycline HCl was loaded to the optimized sterculia gum-*cl*-PVA hydrogel matrix membrane and evaluated for in vitro drug release dynamics in distilled water, pH 2.2 and pH 7.4 buffers. Tetracycline HCl released from the tetracycline HCl-loaded optimized sterculia gum-*cl*-PVA hydrogel matrix membrane showed comparative higher amount of drug release in distilled water

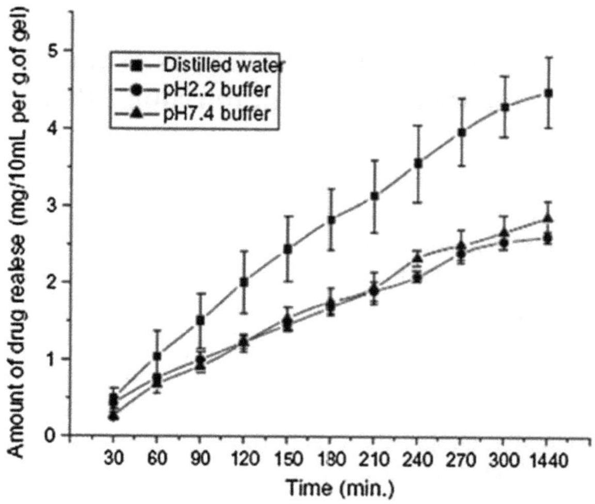

Fig. 20 Release profile of antimicrobial drug (tetracycline HCl) release from tetracycline HCl-loaded sterculia gum-*cl*-PVA hydrogel matrix membrane as wound dressings in different mediums at 37 °C [113]. Copyright © 2008 with permission from Elsevier Ltd.

than that in pH 2.2 and pH 7.4 buffers (Fig. 20). The drug release trends corresponding to the swelling pattern of these hydrogel matrix membrane and the drug release mechanism in all three mediums were found to follow the non-Fickian diffusion mechanism. When tetracycline HCl release from the tetracycline HCl-loaded optimized sterculia gum-*cl*-PVA hydrogel matrix membrane was evaluated in simulated wound fluid, the drug release was found slower than that in distilled water, pH 2.2 and pH 7.4 buffers though the non-Fickian diffusion mechanism was evidenced (Fig. 21). Therefore, tetracycline HCl-loaded optimized sterculia gum-*cl*-PVA hydrogel matrix membrane as wound dressings can be effective for double potential action; first, due to the inherent antimicrobial property of sterculia gum and second, due to sustain release of loaded antimicrobial drug from the tetracycline HCl.

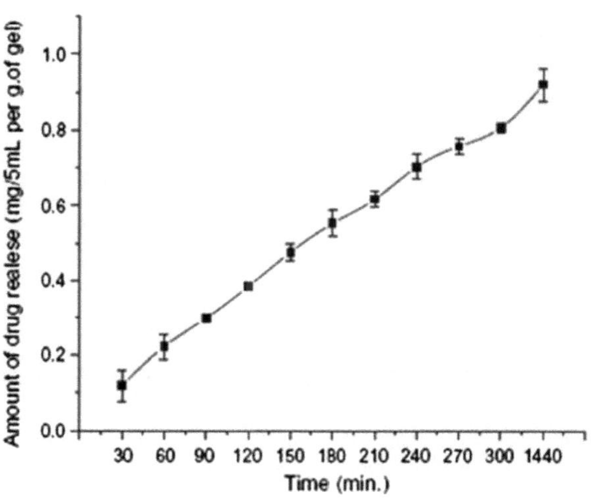

Fig. 21 Release profile of antimicrobial drug (tetracycline HCl) release from tetracycline HCl-loaded sterculia gum-*cl*-PVA hydrogel matrix membrane as wound dressings in simulated wound fluid at 37 °C [113]. Copyright © 2008 with permission from Elsevier Ltd.

12 Sterculia Gum Cross-Linked PVA and PVA-Poly (AAm) Hydrogel Films as Wound Dressings for Releases of Antimicrobial Drugs

In an investigation, Singh and Pal [82] have synthesized sterculia gum-*cl*-PVA and sterculia gum-*cl*-poly(VA-*co*-AAm) hydrogels using N,N'-MBAAm as cross-linker and APS as reaction initiator. The reaction conditions for the synthesis of sterculia gum-*cl*-PVA hydrogel were 0.5 g of sterculia gum, 0.5 g of PVA, 0.065 mmol of N,N'-MBAAm, 0.132 mmol of APS, and 15 ml of water. The reaction conditions for the synthesis of sterculia gum-*cl*-poly(VA-*co*-AAm) hydrogel were 0.4 g of sterculia gum, 7.03 mmol of AAm, 0.5 g of PVA, 0.065 mmol of N,N'-MBAAm, 0.132 mmol of APS, and 20 ml of water. At these above reaction conditions, sterculia gum-*cl*-PVA and sterculia gum-*cl*-poly(VA-co-AAm) hydrogels were synthesized. Hydrogel films were prepared from these two synthesized hydrogels by solution casting method and these hydrogel films were evaluated for their suitability as wound dressings for sustained release of antimicrobial drugs, tetracycline HCl, and gentamicin sulfate, respectively.

SEM images obtained by SEM analyses of the sterculia gum-*cl*-PVA and sterculia gum-*cl*-poly(VA-*co*-AAm) hydrogel films revealed smooth sheet-like networks in the cross-linked structure (Figs. 22 and 23). EDAX spectrum of these films showed the presence of carbon, oxygen, nitrogen, and hydrogen, which are the main constituents of polysaccharide. The incorporation of nitrogen in these polymeric hydrogel films was due to the presence of cross-linker. FTIR and thermogravimetric analyses suggested the cross-linking of the polymers present within these synthesized cross-linked hydrogel films.

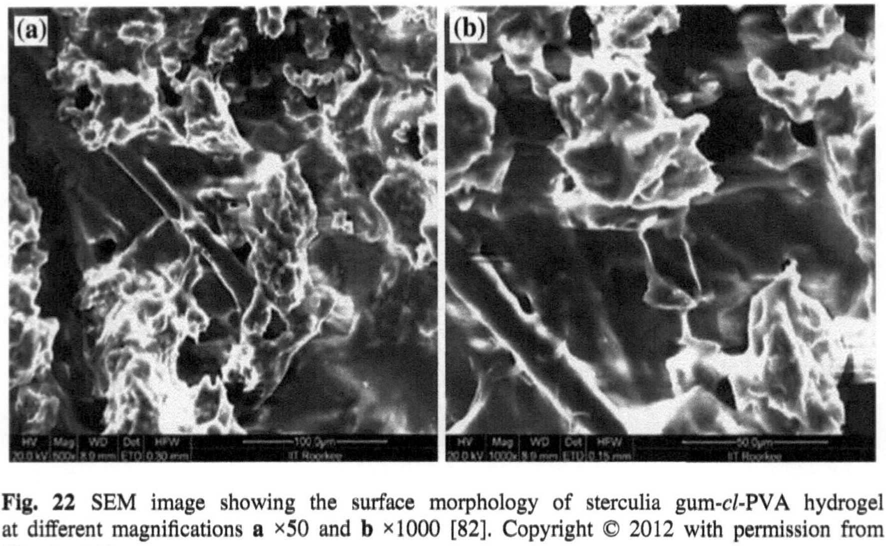

Fig. 22 SEM image showing the surface morphology of sterculia gum-*cl*-PVA hydrogel at different magnifications **a** ×50 and **b** ×1000 [82]. Copyright © 2012 with permission from Elsevier Ltd.

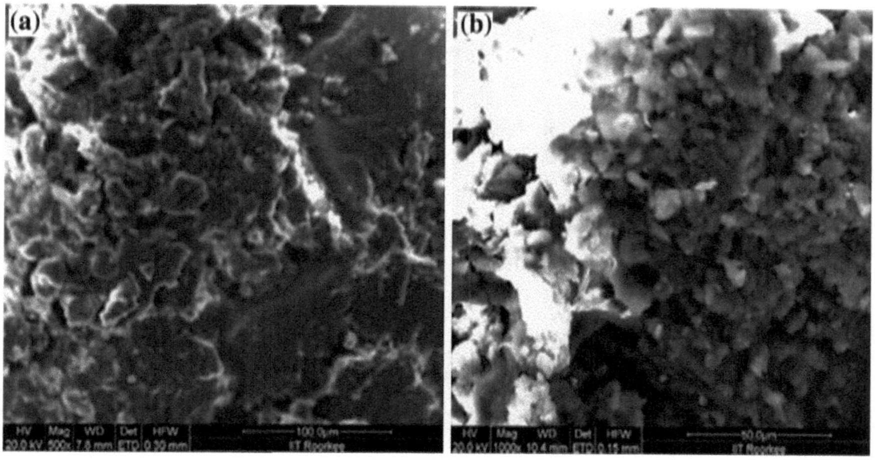

Fig. 23 SEM image showing the surface morphology of sterculia gum-*cl*-poly(VA-*co*-AAm) hydrogel at different magnifications **a** ×50 and **b** ×1000 [82]. Copyright © 2012 with permission from Elsevier Ltd.

The polymeric films of sterculia gum-*cl*-PVA and sterculia gum-*cl*-poly(VA-*co*-AAm) hydrogels were evaluated for swelling analyses in distilled water, pH 2.2 buffer, pH 7.4 buffer, and simulated wound fluid. Almost equal amounts of swellings were observed in distilled water, pH 2.2 and pH 7.4 buffers. However, swellings of these two hydrogel films in simulated wound fluid were found lesser than that in other swelling mediums investigated. The lesser swelling of these hydrogel films in simulated wound fluid could be due to very high ionic strength of the simulated wound fluid [115]. The increase in ionic strength reduces the swelling due to difference in concentration of mobile ions between the gel and solutions causing a reduction in osmotic swelling pressure of mobile ions inside in any gel [116]. From the swelling results, it was observed that the sterculia gum-*cl*-poly(VA-*co*-AAm) hydrogel films captured more amount of water as compared to that of sterculia gum-*cl*-PVA hydrogel film. These films showed case II transport mechanism. Therefore, the results of the swelling study indicated that these hydrogel films were found as capable of absorbing a high volume of fluid in relation to their physical dimensions and suitable for moderate to high exuding wounds.

In vitro release dynamics of the loaded antimicrobial drugs from both the hydrogel films were evaluated in various releasing mediums like distilled water, pH 2.2 buffer, pH 7.4 buffer, and simulated wound fluid. The release of tetracycline HCl from the sterculia gum-*cl*-PVA hydrogel film and gentamicin sulfate release from the sterculia gum-*cl*-poly(VA-*co*-AAm) hydrogel films were found to be sustained over 24 h of release (Figs. 24 and 25). The antimicrobial drug releases in simulated wound fluid occurred through the case II transport diffusion mechanism in case of tetracycline HCl-loaded sterculia gum-*cl*-PVA hydrogel film and the non-Fickian diffusion mechanism in case of gentamicin sulfate-loaded sterculia gum-*cl*-poly(VA-*co*-AAm) hydrogel film.

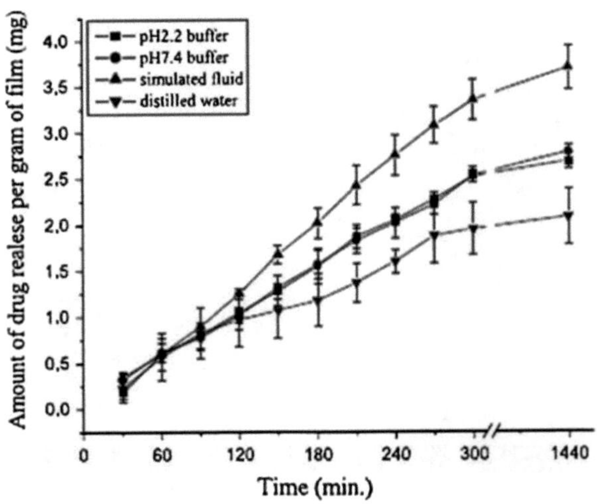

Fig. 24 Release profile of tetracycline HCl from drug-loaded sterculia gum-*cl*-PVA hydrogel film [82]. Copyright © 2012 with permission from Elsevier Ltd.

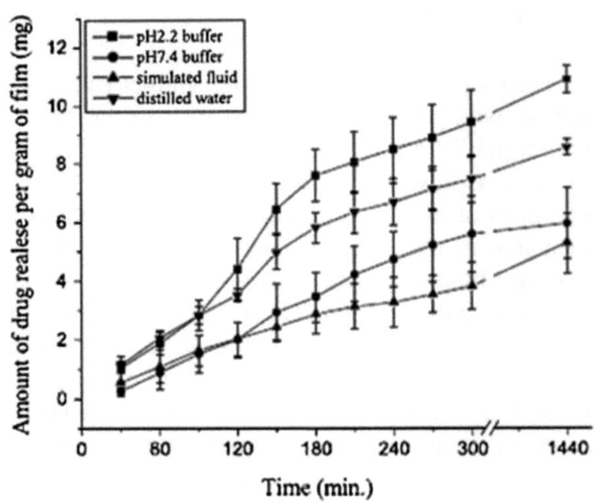

Fig. 25 Release profile of gentamicin sulfate from drug-loaded sterculia gum-*cl*-poly(VA-*co*-AAm) hydrogel film [82]. Copyright © 2012 with permission from Elsevier Ltd.

As the evaluation of blood compatibility is an important property of any material employed for wound dressing applications, thrombogenicity and haemolytic potential of these two hydrogel films were assessed. In the thrombogenicity evaluation, it was observed that clot formation was less in case of both hydrogel films than in the control, indicating the nonthrombogenic character of these synthesized hydrogel matrices. In addition, it was also observed that the thrombus formation in case of sterculia gum-*cl*-PVA hydrogel film was higher than that of sterculia gum-*cl*-poly(VA-*co*-AAm) hydrogel film. This could be due to more hydrophilicity of sterculia gum-*cl*-PVA hydrogel. When placed in contact with a hydrophobic surface, proteins adsorb to it in a strong and irreversible way, while at hydrophilic surfaces, proteins adsorb weakly and reversibly [117]. The haemolytic percentage of sterculia gum-*cl*-PVA and sterculia gum-*cl*-poly(VA-*co*-AAm) hydrogel films were measured as 2.64 and 5.58 %, respectively. From this result, it can be said that sterculia gum-*cl*-PVA hydrogel film was partially haemolytic. Hence, it can be used for the preparation of wound dressings. On the other hand, sterculia gum-*cl*-poly (VA-*co*-AAm) hydrogel film showed haemolytic percentage above 5 % and hence, it was haemolytic in nature. Therefore, sterculia gum-*cl*-poly(VA-*co*-AAm) hydrogel film may be used in other biomedical applications after proper processing. It is reported that material for which, the envisaged final application is on the biomedical applications, should not promote haemolysis and by definition, a blood compatible material should be nonhaemolytic in nature [118, 119].

When mucoadhesion of sterculia gum-*cl*-PVA and sterculia gum-*cl*-poly(VA-*co*-AAm) hydrogel films were evaluated, maximum detachment force and work of adhesion were found in case of the sterculia gum-*cl*-PVA hydrogel film than the sterculia gum-*cl*-poly(VA-*co*-AAm) hydrogel film. Thus, sterculia gum-*cl*-poly (VA-*co*-AAm) hydrogel film indicated good adhesion with the mucous membrane in simulated wound fluid indicating the ability to adhere to the wound site and also to protect the wound from the external milieu, which is a primary requisite for a material to act as wound dressing. Both the hydrogel films showed sufficient mechanical strength, which is also considered as an important parameter for wound dressing materials. Sterculia gum-*cl*-poly(VA-*co*-AAm) hydrogel film exhibited higher oxygen permeability than sterculia gum-*cl*-PVA hydrogel film, while both the hydrogel films confirmed their high permeability of water vapor and impermeability to the microorganisms. All these results indicated that these synthesized sterculia gum-*cl*-PVA hydrogel film and sterculia gum-*cl*-poly(VA-*co*-AAm) hydrogel film as wound dressings for sustained drug release.

13 Radiation-Induced Graft Copolymerization of Sterculia Gum Cross-Linked PVA Hydrogel for Delivery of Antimicrobial Drug

Singh et al. [66] have made an attempt to develop novel hydrogel of sterculia gum and PVA through radiation-induced graft copolymerization method for the use in drug delivery. In this radiation-induced method, solutions of sterculia gum and PVA were prepared in a beaker and stirred for half an hour to attain homogenous mixture. The reaction mixture was irradiated with the gamma rays for a definite time in ^{60}Co gamma chamber. The optimum reaction parameters were searched for the synthesis of this sterculia gum–PVA-based radiation-induced hydrogels by varying various reaction parameters like sterculia gum content, PVA content, and total radiation dose on the basis of the swelling of these hydrogels and the surface consistency maintained by the hydrogels after 24 h of swelling. After preliminary

Fig. 26 SEM image showing the surface morphology of sterculia gum–PVA-based radiation-induced hydrogel [66]. Copyright © 2011 with permission from Elsevier Ltd.

evaluation, the researcher have selected the optimum parameters as 0.125 g sterculia gum, 0.250 g PVA, 8.42 kGy total radiation dose and 30 ml solvent for the synthesis of optimized sterculia gum–PVA-based hydrogel through gamma ray-induced radiation method.

SEM analyses of the radiation-induced sterculia gum–PVA-based hydrogel demonstrated a sheet-like structure with pressed surface and the hydrogel networks indicating the change in morphology (Fig. 26). The SEM image of the synthesized hydrogel showed the structural homogeneity. During the alteration of surface morphology the hydrogel was found clearly and also, cross-linked network was also evidenced in the SEM image. Similar type of structure was already reported in the previous literature [120]. FTIR spectroscopy and various therogravimetric analyses suggested the cross-linking through radiation-induced graft copolymerization of sterculia gum and PVA.

The swelling of these synthesized sterculia gum–PVA-based radiation-induced hydrogels was increased with the increment of sterculia gum amount in the composition of the hydrogels copolymerization. This might be due to the increase in polymers to cross-linker ratio with the increase in sterculia gum amount, which decreased the cross-linking density. Another way, the increase in swelling of hydrogels may also be explained on the basis of higher degree of hydration of polymeric hydrogel matrices, with the increment in sterculia gum amounts during synthesis. This might increase the number of interactions with water, which could lead to higher swelling of the polymeric hydrogel matrix. The swelling of these hydrogels prepared with varying amounts of sterculia gum followed the non-Fickian diffusion mechanism. These hydrogels were not formed when PVA content was less than 0.150 g in the reaction system during synthesis. The swelling decreased as the PVA contents in the reaction mixture increased from 0.150 to 0.200 g and after that increase in PVA content in the reaction mixture did not influence significantly on the swelling of these hydrogels. The swelling of these hydrogels occurred through non-Fickian diffusion mechanism when PVA content was varied. The influence of the water content on the networks of these hydrogels was studied by varying water 9–13 ml during hydrogel synthesis. An increase in water content did not exert any significant effect on the swelling of these hydrogels prepared through radiation-induced method. This might be due to the effect of dilution, which was very less. The swelling of these hydrogels prepared by varying water content occurred through the non-Fickian diffusion mechanism. The effect of total radiation dose on the swelling of the hydrogels was evaluated by varying total radiation dose 8.42–50 kGy during synthesis and a decrease in swelling was observed with the increment of total radiation dose. This might be due to the increment of cross-linking density of the hydrogel with the increase of total radiation dose employed during the synthesis of these sterculia gum–PVA-based radiation-induced hydrogels. Similar results of radiation dose on the swelling of radiation-induced polymeric hydrogels were also reported by Zhai et al. [121] and Maziad [122]. Actually, cross-linking is the most important effect of polymer irradiation and the increase in total radiation dose lessens the number of small chains. Therefore, hydrogels prepared through higher amount of total radiation dose should produce higher cross-linking density than the hydrogels prepared through lower amount of radiation dose. This supports that higher amount of radiation dose can produce a decrease in the number of average molar mass between the cross-links, while a lower amount of radiation dose increases the number of average molar mass between the cross-links [123]. The swelling of these hydrogels prepared by varying radiation doses was followed by the non-Fickian diffusion mechanism. The swelling of these radiation-induced hydrogel prepared using optimizing reaction parameters was investigated in distilled water, pH 2.2 and pH 7.4 buffers. The swelling of optimized hydrogel was found higher in pH 7.4 buffer as compared to other medium investigated and the swelling occurred through non-Fickian diffusion mechanism in all these three swelling mediums.

In the sterculia gum–PVA-based radiation-induced hydrogel, tetracycline HCl (an antimicrobial drug) was loaded by swelling equilibrium method to study the release dynamics of these hydrogel in different release mediums like distilled water, pH 2.2 and pH 7.4 buffers. The amount of the drug release from the drug-loaded optimized cross-linked sterculia gum–PVA-based radiation-induced hydrogel was found higher in distilled water than that in pH 2.2 and pH 7.4 buffers (Fig. 27). The drug release from the drug-loaded hydrogel occurred through non-Fickian diffusion mechanism. Therefore, the synthesized sterculia gum–PVA-based radiation-induced hydrogel could be effective for the use in sustained drug release therapy.

Fig. 27 Release profile of antimicrobial drug (tetracycline HCl) release from the drug-loaded sterculia gum–PVA-based radiation-induced hydrogel in different mediums at 37 °C [66]. Copyright © 2011 with permission from Elsevier Ltd.

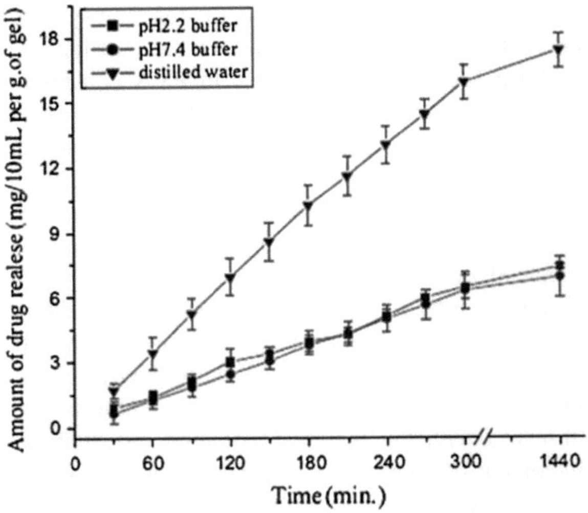

14 Radiation-Induced Cross-Linked Sterculia Gum–PVA–NVP-Based Hydrogel as Wound Dressing for Antimicrobial Drug Release

Singh and Pal [124] made an attempt to synthesize a novel hydrogel [sterculia gum-*cl*-poly(VA-*co*-NVP)] made of sterculia gum, PVA, and NVP as wound dressing for slower release of an antimicrobial drug (doxycycline hyclate) through radiation-induced cross-linking method. In this work, solutions of sterculia gum, PVA, and NVP were prepared and mixed to get a homogeneous polymeric mixture. The polymeric mixture was irradiated with the gamma rays in ^{60}Co gamma chamber for a definite time. The optimum reaction conditions were searched for the synthesis of sterculia gum-*cl*-poly(VA-*co*-NVP) hydrogels by varying amounts of sterculia gum, PVA, and NVP, radiation dose, and water on the basis of swelling of the synthesized hydrogels and surface consistency after 24 h of swelling. After preliminary evaluation, the optimized reaction conditions were selected as 0.125 g sterculia gum, 0.250 g PVA, 4.69 mmol NVP, 25.17 kGy radiation dose, and 13 ml water.

SEM analyses indicated the change in surface morphology of the cross-linked sterculia gum-*cl*-poly(VA-*co*-NVP) hydrogel (Fig. 28). FTIR analyses and thermogravimetric analyses of the sterculia gum-*cl*-poly(VA-*co*-NVP) hydrogel supported the cross-linking between these polymers (i.e., sterculia gum, PVA, and NVP) used as composition of the synthesized hydrogel.

The effect of various reaction parameters investigated in this work on the structure of radiation-induced sterculia gum-*cl*-poly(VA-*co*-NVP) hydrogel polymer network was determined by studying the swelling behavior of the hydrogels. Swelling of the sterculia gum-*cl*-poly(VA-*co*-NVP) hydrogels was increased with

Fig. 28 SEM image showing the surface morphology of radiation-induced sterculia gum-*cl*-poly(VA-*co*-NVP) hydrogel [124]. Copyright © 2011 with permission from Elsevier B.V.

increase in NVP concentration during the synthesis and this might be due to the reduction in the crystallinity of NVA segments with increase in NVP contents in the hydrogel polymer network. The swelling of the sterculia gum-*cl*-poly(VA-*co*-NVP) hydrogels prepared with different NVP concentration followed a non-Fickian diffusion mechanism. The increment in PVA content from 0.05 to 0.20 g did not found to exert very strong effect on the swelling and the structural integrity of these hydrogel; but after 0.20 g of PVA content, a sudden increment in swelling of the hydrogels was observed. This might be due to the fact that PVA as a hydrophilic polymer is able to retain water; hence swelling ratio increases with the increase in weight percent of PVA in the hydrogel matrix. The swelling of these hydrogels prepared with varying amounts of PVA occurred through the non-Fickian type of diffusion mechanism. The increment in sterculia gum amount in the reaction system for the hydrogel synthesis did not find to exert very strong effect; but in general, the increment in gum amount increased the swelling of the resultant hydrogels. This is possibly due to the fact that sterculia gum is hydrophilic gum in nature and higher degree of the gum hydration has occurred, which has increased the number of intimate contacts between the gum particles and water and led to high swelling rate for the radiation-induced sterculia gum-*cl*-poly(VA-*co*-NVP) hydrogel prepared by varying amount of sterculia gum. The increment in swelling with the increase in sterculia gum amounts could also be due to reduction in sterculia-to-NVP ratio and sterculia-to-PVA ratio in the sterculia gum-*cl*-poly(VA-*co*-NVP) hydrogels and formation of loose networks. The swelling of these hydrogels prepared with varying amounts of sterculia gum occurred through the non-Fickian diffusion mechanism.

It was also observed that solvent, water content did not find to exert strong effect on the swelling of these hydrogels. The swelling of the hydrogels occurred through the non-Fickian diffusion mechanism, when these hydrogels were prepared by varying the water content from 10 to 15 ml during synthesis. When the total radiation dose increased from 8.42 to 50.54 kGy during the synthesis of radiation-induced sterculia gum-*cl*-poly(VA-*co*-NVP) hydrogels, the swelling of the resultant hydrogels was found to be increased first and then, it was decreased. This phenomenon denotes that the further increase in total radiation dose has decreased the cross-linking density of the hydrogel networks after the formation of optimum pore size of the hydrogel networks, which has reduced the pore size as well as swelling of the resultant hydrogels. The swelling of these radiation-induced hydrogels prepared with different total radiation doses took place through the case II diffusion mechanism, except for the hydrogels prepared with total radiation dose of 25.17 kGy. The maximum swelling was observed for the optimized radiation-induced sterculia gum-*cl*-poly(VA-*co*-NVP) hydrogel in pH 7.4 buffer than that in other swelling mediums investigated and in case of simulated wound fluid, lesser swelling was observed as compared to that of distilled water, pH 2.2 and pH 7.4 buffers. The decreased swelling of the optimized radiation-induced hydrogel in simulated wound fluid might be due to the very high ionic strength of the simulated wound fluid. An increase in ionic strength might decrease the swelling of the hydrogel network as the difference in concentration of mobile ions between the gel and the solution could be reduced causing a decrease in the osmotic swelling pressure of these mobile ions inside the gel. The swelling of the optimized hydrogel occurred through the non-Fickian diffusion mechanism in different swelling mediums investigated.

The loading of doxycycline hyclate (200 μg/ml) into the optimized radiation-induced sterculia gum-*cl*-poly(VA-*co*-NVP) hydrogel was performed by swelling equilibrium method and the release of drug from the drug-loaded hydrogels was studied in different release mediums (i.e., distilled water, pH 2.2 buffer, pH 7.4 buffer, and simulated wound fluid) at 37 °C. The amounts of doxycycline hyclate released from the doxycycline hyclate-loaded optimized radiation-induced sterculia gum-*cl*-poly(VA-*co*-NVP) hydrogel in pH 7.4 buffer, pH 2.2 buffer, and distilled water were almost comparable. Both swelling and release of drug have been observed less in simulated wound fluid as compared to the other mediums investigated (Fig. 29). The drug release from the drug-loaded optimized radiation-induced sterculia gum-*cl*-poly(VA-*co*-NVP) hydrogel occurred through the non-Fickian diffusion mechanism. Therefore, the optimized radiation-induced sterculia gum-*cl*-poly(VA-*co*-NVP) hydrogel may have potential to act as wound dressings and the drug-loaded sterculia gum-*cl*-poly(VA-*co*-NVP) hydrogel wound dressings can be effective for the double potential action and may show the synergic effect, due to inherent antimicrobial action of sterculia gum and release of antimicrobial agent from the newly synthesized hydrogel.

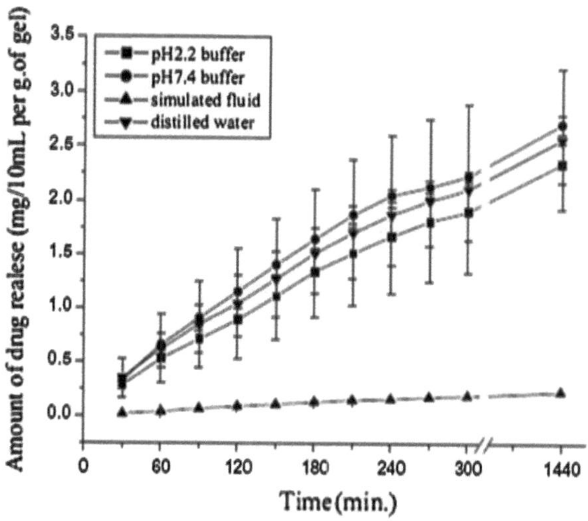

Fig. 29 Release profile of doxycycline hyclate from drug-loaded radiation-induced sterculia gum-*cl*-poly (VA-*co*-NVP) hydrogel in different media at 37 °C [124]. Copyright © 2011 with permission from Elsevier B.V.

15 Conclusion

Recently, various natural polysaccharides are extensively used to develop hydrogels for various biomedical applications to control the hydration rate and swelling, and also to tailor the release profile of various types of drugs. Several attempts to develop sterculia gum-based hydrogels through the modifications of sterculia gum by polymer blending, cross-linking, IPN formation, polymer grafting, etc., also have been undertaken for the use in drug delivery, wound dressing, etc. These sterculia gum-based hydrogels were capable of good swelling in different swelling mediums. The incorporation of various types of drugs (such as antiulcer drugs, antidiarrhoeal drug, antimicrobial drugs, anti-inflammatory drug, antidepressant drugs, etc.), in these sterculia gum-based hydrogels were investigated and found effective for gastroretentive deliveries or wound dressings for slower release of drugs over prolonged period. In view of the medicinal importance of the sterculia gum (such as antiulcer, antidiarrhoeal, bulk laxative, and antimicrobial properties), various types of sterculia gum-based hydrogel systems were designed for the double potential action to attain the synergic effect, due to inherent medicinal action of sterculia gum and release of incorporated drugs from these hydrogels. In some cases like oil-entrapped buoyant beads made of sterculia gum–sodium alginate blends for gastroretentive delivery of aceclofenac, the antiulcer activity of sterculia gum can be helpful to minimize the occurrence of ulceration, which is associated with the prolonged use of the incorporated drug (aceclofenac). This chapter provides comprehensive information of already investigated sterculia gum-based hydrogel systems for various types of drug delivery applications. This information can be supportive for the drug delivery researchers and scientists in applying pertinent strategies to develop newer sterculia gum-based hydrogel systems with improved mechanical properties and capability to control sustained drug release over prolonged period.

References

1. Pal D, Mitra S (2010) A preliminary study on the in vitro antioxidant activity of the stems of *Opuntia vulgaris*. J Adv Pharm Technol Res 1:172–268
2. Pal D, Banerjee S, Ghosh A (2012) Dietary-induced cancer prevention: an expanding research arena of emerging diet related to healthcare system. J Adv Pharm Technol Res 3:16–24
3. Nayak AK, Pal D, Pany DR, Mohanty B (2010) Evaluation of *Spinacia oleracea* L. leaves mucilage as innovative suspending agent. J Adv Pharm Technol Res 1:338–341
4. Nayak AK, Pal D, Pradhan J, Ghorai T (2012) The potential of *Trigonella foenum-graecum* L. seed mucilage as suspending agent. Indian J Pharm Educ Res 46:312–317
5. Hasnain MS, Nayak AK, Singh R, Ahmad F (2010) Emerging trends of natural-based polymeric systems for drug delivery in tissue engineering applications. Sci J UBU 1:1–13
6. Nayak AK, Pal D, Santra K (2015) Screening of polysaccharides from tamarind, fenugreek and jackfruit seeds as pharmaceutical excipients. Int J Biol Macromol 79:756–760
7. Nayak AK, Pal D (2012) Natural polysaccharides for drug delivery in tissue engineering. Everyman's Sci XLVI:347–352
8. Lloyd LL, Kenedy JF, Methacanon P, Peterson M, Knill CJ (1998) Carbohydrate polymers as wound management aids. Carbohydr Polym 37:315–322
9. Barbosa MA, Granja PL, Barrias CC, Amaral IF (2005) Polysaccharides as scaffolds for bone regeneration. ITBM-RBM 26:212–217
10. Nayak AK, Pal D, Santra K (2014) Development of calcium pectinate-tamarind seed polysaccharide mucoadhesive beads containing metformin HCl. Carbohydr Polym 101: 220–230
11. Rana V, Rai P, Tiwari AK, Singh RS, Kenedy JF, Knill CJ (2011) Modified gums: approaches and application in drug delivery. Carbohydr Polym 83:1031–1047
12. Prajapati VD, Jani GK, Moradiya NG, Randeria NP (2013) Pharmaceutical applications of various natural gums, mucilages and their modified forms. Carbohydr Polym 92:1685–1699
13. Waliszewski KN, Aparicio MA, Bello LA, Monroy JA (2003) Changes of banana starch by chemical and physical modification. Carbohydr Polym 52:237–242
14. Manchanda R, Arora SC, Manchanda R (2014) Tamarind seed polysaccharide and its modification-versatile pharmaceutical excipients-a review. Int J PharmTech Res 6:412–420
15. Coviello T, Matricardi P, Marianecci C, Alhaique F (2007) Polysaccharide hydrogels for modified release formulations. J Control Release 119:5–24
16. Wang Q, Xie X, Zhang X, Zhang J, Wang A (2010) Preparation and swelling properties of pH-sensitive composite beads based on chitosan-g-poly (acrylic acid)/vermiculite and sodium alginate for diclofenac controlled release. Int J Biol Macromol 46:356–362
17. Hwang MR, Kim JO, Lee JH, Kim YI, Kim JH, Chang SW, Jin SG, Kim JA, Lyoo WS, Han SS, Ku SK, Yong CS, Choi HG (2010) Gentamicin-loaded wound dressing with polyvinyl alcohol/dextran hydrogel: gel characterization and in vivo healing evaluation. AAPS PharmSciTech 11:1092–1103
18. Pal K, Banthia AK, Majumder DK (2007) Preparation and characterization of polyvinyl alcohol-gelation hydrogel membranes for biomedical applications. AAPS PharmSciTech 8(21)
19. Kuo CK, Ma PX (2001) Ionically crosslinked alginate hydrogels as scaffolds for tissue engineering. Part I: structure, gelation rate and medical properties. Biomater 22:511–521
20. Nayak AK, Pal D (2011) Development of pH-sensitive tamarind seed polysaccharide-alginate composite beads for controlled diclofenac sodium delivery using response surface methodology. Int J Biol Macromol 49:784–793
21. Pal D, Nayak AK (2012) Novel tamarind seed polysaccharide-alginate mucoadhesive microspheres for oral gliclazide delivery. Drug Deliv 19:123–131
22. Nayak AK, Pal D, Santra K (2014) Development of pectinate-ispagula mucilage mucoadhesive beads of metformin HCl by central composite design. Int J Biol Macromol 66:203–221

23. Malakar J, Nayak AK, Jana P, Pal D (2013) Potato starch-blended alginate beads for prolonged release of tolbutamide: development by statistical optimization and in vitro characterization. Asian J Pharm 7:43–51
24. Nayak AK, Pal D, Santra K (2014) Ispaghula mucilage-gellan mucoadhesive beads of metformin HCl: development by response surface methodology. Carbohydr Polym 107:40–41
25. Jana S, Lakshman D, Sen KK, Basu SK (2010) Development and evaluation of epichlorohydrin cross-linked mucoadhesive patches of tamarind seed polysaccharide for buccal application. Int J Pharm Sci Drug Res 2:193–198
26. Maiti S, Ranjit S, Mondol R, Ray S, Sa B (2011) Al^{+3} ion cross-linked and acetalated gellan hydrogel network beads for prolonged release of glipizide. Carbohydr Polym 85:164–172
27. Vieira AP, Ferreira P, Coelho JFJ, Gil MH (2008) Photocrosslinkable starch-based polymers for ophthalmologic drug delivery. Int J Biol Macromol 43:325–332
28. Das A, Wadhwa S, srivastava AK (2006) Cross-linked guar gum hydrogel discs for colon-specific delivery of ibuprofen: formulation and in vitro evaluation. Drug Deliv 13: 139–142
29. Chourasia MK, Jain NK, Jain A, Soni V, Gupta Y, Jain SK (2006) Cross-linked guar gum microspheres: a viable approach for improved delivery of anticancer drugs for the treatment of colorectal cancer. AAPS PharmSciTech 7:E1–E9
30. Malik S, Kumar A, Ahuja M (2012) Synthesis of gum kondagogu- g-poly (N-vinyl-2-pyrrolidone) and its evaluation as a mucoadhesive polymer. Int J Biol Macromol 51:756–762
31. Pandey VS, Verma SK, Yadav M, Behari K (2014) Guar gum-g-N, N′-dimethylacrylamide: synthesis, characterization and applications. Carbohydr Polym 99:284–290
32. Mishra A, Yadav A, Pal S, Singh A (2006) Biodegradable graft copolymers of fenugreek mucilage and polyacrylamide: a renewable reservoir to biomaterials. Carbohydr Polym 65:58–63
33. Kumar A, Singh K, Ahuja M (2009) Xanthan-g-poly(acrylamide): microwave-assisted synthesis, characterization and in vitro release behavior. Carbohydr Polym 76:261–267
34. Vijan V, Kaity S, Biswas S, Isaac J, Ghosh A (2012) Microwave assisted synthesis and characterization of acrylamide grafted gellan, application in drug delivery. Carbohydr Polym 90:496–506
35. Kaur H, Ahuja M, Kumar S, Dilbaghi N (2012) Carboxymethyl tamarind kernel polysaccharide nanoparticles for ophthalmic drug delivery. Int J Biol Macromol 50:833–839
36. Maiti S, Dey P, Banik A, Sa B, Ray S, Kaity S (2010) Tailoring of locust bean gum and development of hydrogel beads for controlled oral delivery of glipizide. Drug Deliv 17: 288–300
37. Kumar A, Ahuja M (2012) Carboxymethyl gum kondagogu: synthesis, characterization and evaluation as mucoadhesive polymer. Carbohydr Polym 90:637–643
38. Das B, Dutta S, Nayak AK, Nanda U (2014) Zinc alginate-carboxymethyl cashew gum microbeads for prolonged drug release: development and optimization. Int J Biol Macromol 70:505–515
39. Gupta S, Sharma P, Soni PL (2005) Chemical modification of *Cassia occidentalis* seed gum: carbamoylethylation. Carbohydr Polym 59:501–506
40. Sharma BR, Kumar V, Soni PL (2004) Carbamoylethylation of guar gum. Carbohydr Polym 58:449–451
41. Kaur H, Yadav S, Ahuja M, Dilbaghi N (2012) Synthesis, characterization and evaluation of thiolated tamarind seed polysaccharide as a mucoadhesive polymer. Carbohydr Polym 90:1543–1549
42. Sharma R, Ahuja M (2011) Thiolated pectin—synthesis, characterization and evaluation as a mucoadhesive polymer. Carbohydr Polym 85:658–663
43. Yadav S, Ahuja M, Kumar A, Kaur H (2014) Gellan–thioglycolic acid conjugate: synthesis, characterization and evaluation as mucoadhesive polymer. Carbohydr Polym 99:601–607
44. Hamcerencu M, Desbrieres J, Khoukh A, Popa M, Riess G (2008) Synthesis and characterization of new unsaturated esters of gellan gum. Carbohydr Polym 71:92–100

45. Tay SH, Pang SC, Chin SF (2012) Facile synthesis of starch maleate monoesters from native sago starch. Carbohydr Polym 88:1195–1200
46. Jana S, Das A, Nayak AK, Sen KK, Basu SK (2013) Aceclofenac-loaded unsaturated esterified alginate/gellan gum microspheres: in vitro and in vivo assessment. Int J Biol Macromol 57:129–137
47. Jana S, Saha A, Nayak AK, Sen KK, Basu SK (2013) Aceclofenac-loaded chitosan-tamarind seed polysaccharide interpenetrating polymeric network microparticles. Colloids Surf B: Biointerf 105:303–309
48. Banerjee S, Chaurasia G, Pal D, Ghosh AK, Ghosh A, Kaity S (2010) Investigation on crosslinking density for development of novel interpenetrating polymer network (IPN) based formulation. J Sci Ind Res 69:777–784
49. Kulkarni RV, Nagathan VV, Biradar PR, Naikawadi AA (2013) Simvastatin loaded composite polyspheres of gellan gum and carrageenan: in vitro and in vivo evaluation. Int J Biol Macromol 57:238–244
50. Kulkarni RV, Mangod BS, Mutalik S, Sa B (2011) Interpenetrating polymer network microcapsules of gellan gum and egg albumin entrapped with diltiazem-resin complex for controlled release application. Carbohydr Polym 83:1001–1007
51. Luo Y, Wang Q (2014) Recent development of chitosan-based polyelectrolyte complexes with natural polysaccharides for drug delivery. Int J Biol Macromol 64:353–367
52. Naidu VGM, Madhusudhana K, Sashidhar RB, Ramakrishna S, Khar RK, Ahmed FJ, Diwan PV (2009) Polyelectrolyte complexes of gum kondagogu and chitosan, as diclofenac carriers. Carbohydr Polym 76:464–471
53. Jana S, Maji N, Nayak AK, Sen KK, Basu SK (2013) Development of chitosan-based nanoparticles through inter-polymeric complexation for oral drug delivery. Carbohydr Polym 98:870–876
54. Peng SC, Chin SF, Tay SH, Tchong FM (2011) Starch-maleate-polyvinyl alcohol hydrogels with controllable swelling behaviours. Carbohydr Polym 84:424–429
55. Rokhade AP, Patil SA, Aminabhavi TM (2007) Synthesis and characterization of semi-interpenetrating polymer network microspheres of acrylamide grafted dextran and chitosan for controlled release of acyclovir. Carbohydr Polym 67:605–613
56. Hoffman AS (2002) Hydrogels for biomedical applications. Adv Drug Deliv Rev 43:3–12
57. Peppas NA (1986) Hydrogels in medicine. CRC Press, Boca Raton
58. Hua S, Ma H, Li X, Yang H, Wang A (2010) pH-sensitive sodium alginate/poly(vinyl alcohol) hydrogel beads prepared by combined Ca^{2+} crosslinking and freeze-thawing cycles for controlled release of diclofenac sodium. Int J Biol Macromol 46:517–523
59. Leung AY (1980) Encyclopedia of common natural ingredients used in foods, drugs and cosmetics. Wiley, New York
60. Cerf DL, Irinei F, Muller G (1990) Solution properties of gum exudates from *Sterculia urens* (karaya gum). Carbohydr Polym 13:375
61. Bera H, Kandukuri SG, Nayak AK, Boddupalli S (2015) Alginate-sterculia gum gel-coated oil-entrapped alginate beads for gastroretentive risperidone delivery. Carbohydr Polym 120:74–84
62. Kulkarni RV, Patel FS, Nanjappaiah HM, Naikawadi AA (2014) In vitro and in vivo evaluation of novel interpenetrating polymer network microparticles containing repaglinide. Int J Biol Macromol 69:514–522
63. Singh B, Sharma N (2011) Design of sterculia gum based double potential antidiarrheal drug delivery system. Colloids Surf B: Biointerf 82:325–332
64. Gauthami S, Bhat VR (1992) A monograph on gum karaya. National Institution of Nutrition, Indian Council of Medical Research, Hyderabad
65. Anderson DMW (1989) Evidence of safety of gum karaya (Sterculia spp.) as a food additive. Food Additive Contamin: Part A 6:189–199
66. Singh S, Sharma V, Pal L (2011) Formation of sterculia polysaccharide networks by gamma rays induced graft copolymerization for biomedical applications. Carbohydr Polym 86: 1371–1380

67. Behall KM, Schofield DJ, Lee K, Powell AS, Mores PB (1987) Mineral balance in adult men: effect of four fibers. Am J Clin Nutri 46:307–314
68. Zide BM, Bevin AG (1980) Treatment of shallow soft tissue ulcers with an infrequent dressing technique. Anal Plastic Surg 4:79–83
69. Huttel E (1983) Treatment of acute diarrhoea in general practice. Therapeutic experiences with karaya bismuth. Die Medizinishe Welt 34:1383–1384
70. Guerre J, Neuman M (1979) Treatment of cronic colonic diseases with a new topical digestive agent, mucilage (karaya gum) combined with polyvinyl polypyrrolidone (PVPP). Med Chirurgie Digest 8:679–682
71. Capron JP, Zeitoun P, Julien DA (1981) A multicenter controlled trial of a combination of kaolin, sterculia gum, meprobamate and magnesium salts, in the irritable bowel syndrome (Authors Transl.). Gastroenterol Clin Biol 5:67–72
72. Meier P, Seiler WO, Stahelin HB (1990) Bulk-forming agents as laxatives in geriatric patients. Schweizerische Medizinische Wochenschrift 120:314–317
73. Anderson DMW, Wang WP (1994) The tree exudate gums permitted in foodstuffs as emulsifiers, stabilisers and thickeners. Chem Ind Forest Products 14:73–83
74. Weiping W (2000) Tragacanth and karaya. In: Williams PA, Philips GO (eds) Handbook of hydrocolloids. Woodhea, Cambridge, pp 155–168
75. Singh B, Sharma V, Chouhan D (2010) Gastroretentive floating sterculia-alginate beads for use in antiulcer drug delivery. Chem Eng Res Des 88:997–1012
76. Siva DA, Brito ACF, Paula RCMD, Feitosa JPA, Paula HCB (2003) Effect of mono and divalent salts on gelation of native, Na and deacetylated *Sterculia striate* and *Sterculia urens* polysaccharide gels. Carbohydr Polym 54:229–236
77. Deshmukh VN, Jadhav JK, Sakarkar DM (2009) Formulation and in vitro evaluation of theophylline anhydrous bioadhesive tablets. Asian J Pharm 3:54–58
78. Park CR, Munday PL (2004) Evaluation of selected polysaccharide excipients in buccoadhesive tablets for sustained release of nicotine. Drug Dev Ind Pharm 30:609–617
79. Munday DL, Philip JC (2000) Compressed xanthan gum and karaya gum matrices: hydration, erosion and drug release mechanism. Int J Pharm 203:179–192
80. Sreenivasa B, Prasanna RY, Mary S (2000) Design and studies of gum karaya matrix tablet. Int J Pharm Excip 239–242
81. Guru PR, Nayak AK, Sahu RK (2013) Oil-entrapped sterculia gum–alginate buoyant systems of aceclofenac: development and in vitro evaluation. Colloids Surf B: Biointerf 104:268–275
82. Singh B, Pal L (2012) Sterculia crosslinked PVA and PVA-poly(AAm)hydrogel wound dressings for slow drug delivery: mechanical, mucoadhesive, biocompatible and permeability properties. J Mech Behav Biomed Mater 9:9–21
83. Singh B, Vashishtha M (2008) Development of novel hydrogels by modification of sterculia gum through radiation cross-linking polymerization for use in drug delivery. Nucl Instr Methods Phys Res Sec B Beam Interact Mater Atoms 266:2009–2020
84. Bera H, Boddupalli S, Nayak AK (2015) Mucoadhesive-floating zinc-pectinate-sterculia gum interpenetrating polymer network beads encapsulating ziprasidone HCl. Carbohydr Polym 131:108–118
85. Lankalapalli S, Kolapalli RM (2012) Biopharmaceutical evaluation of diclofenac sodium controlled release tablets prepared from gum karaya-chitosan polyelectrolyte complexes. Drug Dev Ind Pharm 38:815–824
86. Singh B, Sharma N (2008) Modification of sterculia gum with methacrylic acid to prepare a novel drug delivery system. Int J Biol Macromol 43:142–150
87. Singh B, Chauhan D (2011) Barium ions crosslinked alginate and sterculia gum-based gastrortentive floating drug delivery system for use in peptic ulcers. Int J Polymer Mater 60:684–705
88. Goh CH, Heng PWS, Chan LW (2012) Alginates as a useful natural polymer for microencapsulation and therapeutic applications. Carbohydr Polym 88:1–12
89. Nayak AK, Pal D, Hasnain MS (2013) Development and optimization of jackfruit seed starch-alginate beads containing pioglitazone. Curr Drug Deliv 10:608–619

90. Jana S, Gangopadhaya A, Bhowmik BB, Nayak AK, Mukhrjee A (2015) Pharmacokinetic evaluation of testosterone-loaded nanocapsules in rats. Int J Biol Macromol 72:28–30
91. Sriamornsak P, Sungthongjeen S (2007) Modification of theophylline release with alginate gel formed in hard capsules. AAPS PharmSciTech 8:E1–E8
92. Nayak AK, Das B, Maji R (2012) Calcium alginate/gum Arabic beads containing glibenclamide: development and in vitro characterization. Int J Biol Macromol 51: 1070–1078
93. Nayak AK, Khatua S, Hasnain MS, Sen KK (2011) Development of alginate-PVP K 30 microbeads for controlled diclofenac sodium delivery using central composite design. DARU J Pharm Sci 19:356–366
94. Pal D, Nayak AK (2011) Development, optimization and anti-diabetic activity of gliclazide-loaded alginate-methyl cellulose mucoadhesive microcapsules. AAPS PharmSciTech 12:1431–1441
95. Al-Kassas R, Al-Gohary OMN, Al-Fadhel MM (2007) Controlling of systemic absorption of gliclazide through incorporation into alginate beads. Int J Pharm 341:230–237
96. Banerjee S, Singh S, Bhattacharya SS, Chattopadhyay P (2013) Trivalent ion cross-linked pH sensitive alginate-methyl cellulose blend hydrogel beads from aqueous template. Int J Biol Macromol 57:297–307
97. Yoo S-H, Song Y-B, Chang P-S, Lee HG (2006) Microencapsulation of α-tocopherol using sodium alginate and its controlled release properties. Int J Biol Macromol 38:25–30
98. Nayak AK, Hasnain MS, Beg S, Alam MI (2010) Mucoadhesive beads of gliclazide: design, development and evaluation. Sci Asia 36:319–325
99. Nayak AK, Pal D, Santra K (2013) *Plantago ovata* F. mucilage-alginate mucoadhesive beads for controlled release of glibenclamide: development, optimization and in vitro-in vivo evaluation. J Pharm 2013(151035)
100. Sinha P, Ubaidulla U, Nayak AK (2015) Okra (*Hibiscus esculentus*) gum-alginate blend mucoadhesive beads for controlled glibenclamide release. Int J Biol Macromol 72:1069–1075
101. Rees DA (1981) Polysaccharide shapes and their interactions-some recent advances. Pure Appl Chem 53:1–14
102. Jenkins AD, Kratochivil P, Stepto RFT, Suter UW (1996) Glossary of basic terms in polymer science (IUPAC recommendations 1996). Pure Appl Chem 68:2287–2311
103. Pal D, Nayak AK (2015) Interpenetrating polymer networking systems of natural polymeric blends: Drug delivery, In: Mishra M (ed) Encyclopedia of biomedical polymers and polymeric biomaterials. Taylor & Francis Group, New York. (In Press)
104. Nayak AK, Pal D (2015) Chitosan-based interpenetrating polymeric network systems for sustained drug release. In: Tiwari A, Choi J-W (eds) Advanced theranostics materials. Wiley, New York, pp 207–232
105. Das S, Ng K-Y, Ho PC (2010) Formulation and optimization of zinc-pectinate beads for the controlled delivery of resveratrol. AAPS PharmSciTech 11:729–742
106. Sriamornsak P, Sungthongjeen S, Puttipipatkhachorn S (2007) Use of pectin as a carrier for intragastric floating drug delivery: carbonate salt contained beads. Carbohydr Polym 67: 436–445
107. Sriamornsak P, Nunthanid J, Cheewatanakornkool K, Manchun S (2010) Effect of drug loading method on drug content and drug release from calcium pectinate gel beads. AAPS PharmSciTech 11:1315–1319
108. Munarin F, Tanzi MC, Petrini P (2012) Advances in biomedical applications of pectin gels. Int J Biol Macromol 51:681–689
109. Nayak AK, Pal D (2015) Plant derived polymers: ionically gelled sustained drug release systems. In: Mishra M (ed) Encyclopedia of biomedical polymers and polymeric biomaterials. Taylor & Francis Group, New York. (In Press)
110. Nayak AK, Pal D, Das S (2013) Calcium pectinate-fenugreek seed mucilage mucoadhesive beads for controlled delivery of metformin HCl. Carbohydr Polym 96:349–357
111. Nayak AK, Pal D (2013) Blends of jackfruit seed starch-pectin in the development of mucoadhesive beads containing metformin HCl. Int J Biol Macromol 62:137–145

112. Singh B, Sharma N (2008) Development of novel hydrogels by functionalization of sterculia gum for use in anti-ulcer drug delivery. Carbohydr Polym 74:489–497
113. Singh B, Pal L (2008) Development of sterculia gum based wound dressings for use in drug delivery. Eur Polym J 44:3222–3230
114. Huang JC, Deanin RD (2004) Concentration dependency of interaction parameter between PVC and plasticizer using inverse gas chromatography. J Appl Polym Sci 91:146–156
115. Sen M, Avci EN (2005) Radiation synthesis poly (N-vinyl-2-pyrrolidone)-κ-carrageenan hydrogels and this use in wound dressing applications, I. Preliminary laboratory tests. J Biomed Mater Res A 77A:187–196
116. Sen M, Kantoglu O, Guven O (1999) The effect of external stimuli on the equilibrium swelling properties of poly(N-vinyl 2-pyrrolidone/itaconic acid) polyelectrolyte hydrogels. Polymer 40:913–917
117. Zhao C, Liu X, Nomizu M, Nishi N (2003) Blood compatible aspects of DNA-modified polysulfone membrane-protein adsorption and platelet adhesion. Biomater 24:3747–3755
118. International Organization for Standardization, ISO (2002) Biomedical evaluation of medical devices-Part 4: selection of tests for interactions with blood, 10993-3, TC193
119. Shen W, Li Z, Liu Y (2008) Surface chemical functional groups modification of porous carbon. Recent Patents Chem Eng 1:27–40
120. Lanthong P, Nuisin R, Kiatkamjornwong S (2006) Graft-copolymerization, characterization and degradation of cassava starch-g-acrylamide/itaconic acid super absorbents. Carbohydr Polym 66:229–245
121. Zhai M, Yoshii F, Kume T (2002) Hasim K (2002) Synthesis of PVA/starch grafted hydrogels by irradiation. Carbohydr Polym 50:295–303
122. Maziad NA (2004) Radiation polymerization of hydrophilic monomers for producing hydrogel used in waste treatment processes. Polym-Plast Technol Eng 43:1157–1176
123. Nasef MM, Hegazy EA (2011) Preparation and applications of ion exchange membranes by radiation-induced graft copolymerization of polar monomers onto polar films. Prog Polym Sci 29:499–561
124. Singh B, Pal L (2011) Radiation crosslinking polymerization of sterculia polysaccharide–PVA–PVP for making hydrogel wound dressings. Int J Biol Macromol 48:501–510

Antimicrobial Polymeric Hydrogels

Jaydee D. Cabral

Abstract Microbial infections continue to endanger human health and present a great economic problem to society. To solve this predicament, great efforts to develop macromolecules that can inhibit pathogens without incurring pathogen resistance are needed. The development of antimicrobial polymeric hydrogels has grown considerably as an important alternative in the fight against pathogen drug resistance. This chapter summarizes significant and recent progress in the manufacture and application of antimicrobial hydrogels. Advances in macromolecular sciences have made it possible to modify molecular structure and functionality to generate broad-spectrum antimicrobial activity. As a result, the range of biomedical applications has expanded significantly, from wound dressings, tissue engineering, medical device, and surface coatings; to creams for the treatment and deterrence of multi-drug resistant strains. Both natural and synthetic hydrogels possessing either inherent antimicrobial properties or loaded with antibiotics, antimicrobial peptides, or metal nanoparticles are discussed herein.

Keywords Antimicrobial · Polymer · Hydrogel · Peptides · Nanoparticles

1 Introduction

The progressive rise of antibiotic resistance in pathogenic microorganisms has limited our ability to control bacterial infections. Standard antibiotic treatment frequently leads to resistance and can be attributed to the improper use of antibiotics, the survival of resistant bacteria due to genetic mutation, or by acquired resistance from another bacterium via conjugation, transduction, or transformation [1]. Multiple resistance traits can be collected over time resulting in resistance to entire classes of antibiotics. Compounding the resistance problem is the ever

J.D. Cabral (✉)
Department of Chemistry, University of Otago, Dunedin, New Zealand
e-mail: jcabral@chemistry.otago.ac.nz

S. Kalia (ed.), *Polymeric Hydrogels as Smart Biomaterials*,
Springer Series on Polymer and Composite Materials,
DOI 10.1007/978-3-319-25322-0_6

diminishing global discovery and development of new antimicrobials [2]. Microorganisms mainly involved in resistance, referred to as the ESKAPE pathogens (*Enterococcus faecium, Staphylococcus aureus, Klebsiella pneumniae, Acinetobacter baumanii, Pseudomonas aeruginosa, and enterobacteriaceae*) emphasize their ability to "escape" from customary antibacterial treatments [3]. There exists an urgent need to develop new antibiotics and formulations to keep up with the corresponding growth in global resistance.

Macromolecular antimicrobials have received much attention as next-generation biocides with antimicrobial hydrogels as an important class. Hydrogels exhibit many polymer characteristics, for instance, by remaining in place and intact under physiological conditions. Hydrogels, produced from natural or synthetic polymers, display three-dimensional (3-D) systems that are cross-linked by physical or chemical means. Hydrogels, being highly hydrated materials, are typically soft and compliant allowing good biocompatibility with biological tissue. The first biomedical application of hydrogels was reported by Wichterle and Lin back in the 1960s where cross-linked poly(2-hydroxyethyl methacrylate) (pHEMA) was used as a soft contact lens material [4]. Since then, there has been tremendous growth in hydrogel research and its use in many biomedical applications [5–8]. Their large water content plays a key role in their biocompatibility with applications found in several areas such as wound dressings, surface coatings, bioactive delivery, and in tissue engineering. Each application requires a different set of conditions with its own challenges to overcome, i.e., degradation rates, gel strength, injectability, biodistribution, clearance, etc. The high water content also provides a moist environment to wounds which can be helpful in mounting a cellular immunological response as part of the wound healing process. Unfortunately, this damp environment may also expedite microbial infection. Therefore, hydrogels that can be multi-functional, i.e., promote wound healing and deliver drugs, in addition to possessing antimicrobial activity are of particularly interest.

One methodology takes advantage of either a natural or synthetic polymers intrinsic antimicrobial activity to prepare hydrogels. An alternative method involves the incorporation of conventional antibiotics within the hydrogel itself. The hydrogel, may not possess any inherent antimicrobial capacity, but in this instance can offer other benefits such as biocompatibility, site-specific delivery, and controlled degradation [9]. Numerous classes of innovative materials, such as antimicrobial peptides (AMPs) [10], synthetic cationic polymers [11], and metal nanoparticles [12, 13] have emerged as potential substitutes for conventional antibiotics. The main methodologies to achieve antimicrobial polymeric hydrogels are outlined in Table 1.

An array of chemically or physically cross-linked natural or synthetic hydrogels have been developed using either methodologies. Hydrogels can be further subclassified into different categories depending on various parameters, such as material source, preparation, charge, and mechanical characteristics. This chapter provides an overview of recent antimicrobial polymeric hydrogel strategies with specific focus on (1) naturally derived hydrogels and (2) synthetic hydrogels

Table 1 List of antimicrobial hydrogels

Type of antimicrobial hydrogel	Applications
Intrinsically active due to:	
Peptides	Wound dressing
Chitosan	Tissue engineering scaffolds
Oxidized dextran	
Nanopolymeric materials	
Loaded with drugs:	
Silver	Wound dressing
Copper	surface coatings
Zinc oxide	
Cupric oxide	
Gold	
Cerium	
Conventional antibiotics	

possessing inherent antimicrobial activities, to those (3) loaded with antimicrobials, i.e., antibiotics, antimicrobial peptides, synthetic cationic polymers, and metal nanoparticles.

2 Natural Antimicrobial Hydrogels

Natural biopolymers are polymers created by living organisms. These biomaterials and their derivatives offer a wide range of properties and applications. Biopolymers can be subdivided into three subcategories: (1) polysaccharides such as cellulose, alginate, chitosan, starches, and dextrans, (2) polypeptides such as collagen, gelatin, and keratin, and (3) polynucleotides such as DNA and RNA.

2.1 Polysaccharides and Their Derivatives

Many antimicrobial hydrogels are based on natural biopolymers, such as chitosan. Chitosan and its derivatives have generated much interest due to their biocompatibility, biodegradability, hydrophilicity, intrinsic antimicrobial properties, and low cost [14–16]. Chitosan, derived from the alkaline deacetylation of chitin, is a linear polysaccharide consisting of (1, 4)-linked 2-amino-deoxy-β-D-glucan (Fig. 1). Chitosan is the second most abundant polysaccharide in nature after cellulose. Depending on its application, the chemical characteristics of chitosan can be altered by modifying its degree of acetylation and molecular weight [17]. Under weakly acidic conditions, the primary amino groups of chitosan are readily protonated

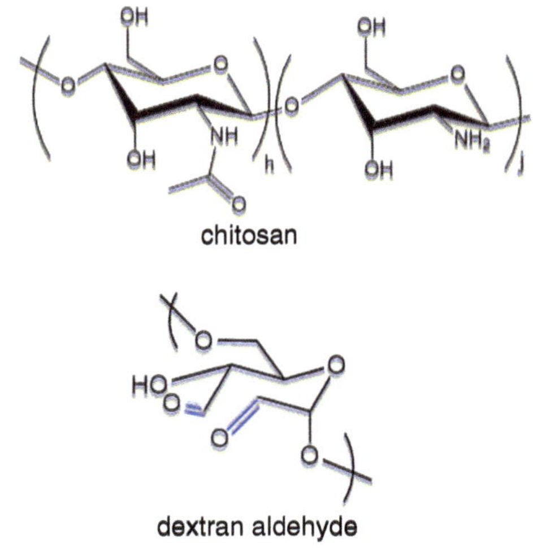

Fig. 1 Structures of commonly used antimicrobial biopolymers

making it polycationic, thereby imparting its intrinsic antimicrobial activity [18]. Chitosan gelation can occur easily via physical or chemical means by mixing or cross-linking with appropriate reagents. Chitosan can be easily modified to fit the needs of a particular biomedical use. For example, to enhance chitosan's antimicrobial properties, functional groups that structurally mimic known antibacterial peptides can be attached. Regions of positively charged functional groups, such as quaternary ammonium or phosponium groups and hydrophobic regions containing alkyl chains have been used. In 2011, a quarternized ammonium chitosan-graft-poly (ethylene glycol) hydrogel was reported to have broad-spectrum antimicrobial activity. The proposed mechanism of the antimicrobial activity of the polycationic hydrogel was thought to be due to the attraction of the anionic microbial membrane into the internal nanopores of the hydrogel. This ionic interaction subsequently led to membrane disruption, and was then followed by microorganism death [19]. Similarly, Jiang et al. developed a quaternary ammonium salt of gelatin using 2,3-epoxypropyl triammonium chloride (EPTAC), and found that an increase in the degree of cationic charges on the polymer correlated to an increase in antimicrobial activity [20]. Another group developed chitosan hydrogels cross-linked with oxalyl-4-(2, 5)-dioxo-2H-pyrrol-1(5H)-yl) benzamide, and were found to possess more antimicrobial efficacy then the parent chitosan molecule. The authors noticed an increase in the degree of cross-linking improved the hydrogel's antimicrobial properties [21]. Venkatesan and coworkers developed novel chitosan–carbon nanotube (CNT) hydrogels which showed increasing antimicrobial activity with increasing CNT concentration. The mechanism of action was thought to be through direct binding of bacterial surface proteins with the CNTs [22]. Aziz et al. reported the development of an antimicrobial chitosan/dextran-based (CD) hydrogel for use in endoscopic sinus surgery [23]. The CD hydrogel was synthesized by reacting

N-succinyl chitosan with dextran aldehyde in a Michael-type addition using a 1:1 mixture [24, 25]. The dextran aldehyde component of the CD hydrogel was found to be the antimicrobial component (Fig. 1), and was thought that addition reactions between the aldehyde moiety of the dextran and bacterial membrane amino groups resulted in membrane disruption.

Another prevalent natural polysaccharide of interest in the development of antimicrobial hydrogels is alginate. Alginate is a biodegradable, linear copolymer composed of two block copolymers, β-D-mannuronic acid (M) and α-L-guluronic acid (G) that are linked by (1–4) glycosidic bonds. Natural polysaccharides, due to their biocompatibility and structural similarity to that of extracellular matrix (ECM) components, i.e., glycosaminoglycans, make them attractive candidates for tissue engineering applications. Alginate has been utilized to synthesize hydrogels by the simple addition of divalent cations, such as Ca^{2+}, which bind to a monomer's carboxylic acid groups providing ionic cross-linking of the polymeric chains, resulting in the gelation of the alginate solution [26]. A novel antimicrobial alginate-based hydrogel containing Ce(III) ions was developed to serve as a synthetic bone substitute. Morias and coworkers showed that Ce(III) significantly improved the antimicrobial activity of the hydrogel without conceding osteoconductivity [27]. Catanzano and coworkers developed an in situ forming ionically cross-linked alginate hydrogel delivery system consisting of a tea tree oil microemulsion to be used as a dressing for infected wounds. These alginate hydrogels were synthesized via a spray-by-spray deposition method and found to exhibit antimicrobial efficacy when tested against *E. coli* [28]. Various alginate solutions have been used for the release of silver nanoparticles [29–31]. Obradovic and coworkers reported the synthesis of nanocomposite Ag/alginate microbeads produced by electrostatic extrusion of alginate colloid solutions containing electrochemically synthesized silver nanoparticles. Both wet and dry microbeads exhibited a growth delay of *S. aureus* and *E. coli* [31]. Alginate solutions, zinc-cross-linked with *N*-succinyl chitosan, were also shown to exhibit antimicrobial efficacy against *S. aureus* and *E. coli* [32].

2.2 Antimicrobial Peptides

In addition to naturally occurring sugars, antimicrobial peptides are considered to be an important weapon in the fight against microbial infection and biofilms. Antimicrobial peptides (AMPs) are endogenous polypeptides made by multicellular organisms to protect against pathogenic microbes [10]. Peptide hydrogels have been developed utilizing the cationic nature of particular amino acids, such as lysine and arginine, in order to disrupt negatively charged bacterial cell membranes through electrostatic interactions. The peptide translocation across the membrane, a membranolytic effect, is the most common mechanism of action [33]. Such a distinct mechanism allows AMPS to elude common resistance mechanisms observed for conventional antibiotics. For example, hydrogels made from epsilon-poly-L-lysine-graft-methacrylamide (EPL-MA) were found to have an impressive broad-spectrum of antimicrobial activity against both bacteria (*E. coli*, *P. aeruginosa*, *S. marcescens,* and *S. aureus*) and fungi (*C. albicans* and *F. solani*) [34]. Protein domains that deliver distinct biological cues or mechanical properties have been copied from nature and synthetically manipulated or attached to other polymers to function as a new material [35]. Cationic antimicrobial peptides have been greatly investigated and considered as a new generation of antibiotics due to their broad-spectrum activities and ability to fight antibiotic-resistant microbes [36]. Most designated antimicrobial peptides are amphipathic. There exists a general recognition that the presence of both a net positive charge and hydrophobicity facilitate peptide contacts with negatively charged membrane lipids [37].

Self-assembled antimicrobial peptides have recently been developed in a "bottom up" approach where known antimicrobial primary sequence peptides are modified with natural and synthetic amino acids to produce inherently antimicrobial hydrogels [38]. A class of β-sheet peptides or β-hairpins formed hydrogels after intramolecular folding and intermolecular assembly, when triggered by changes in an external stimulus, i.e., increases in pH, salt ion concentration, or temperature [35]. Another example of self-assembling hydrogels was the development of the MAX1 or MAX8 peptide (Fig. 2). Self-supporting hydrogels were formed via side-chain hydrophobic associations (via valine residues) among the bilayers, and lateral hydrogen bonding to the peptide backbone (via lysine residues) resulting in the formation of peptide fibrils [39]. MAX1 hydrogel also exhibited antibacterial efficacy and was thought to be due to the high concentration of lysine residues

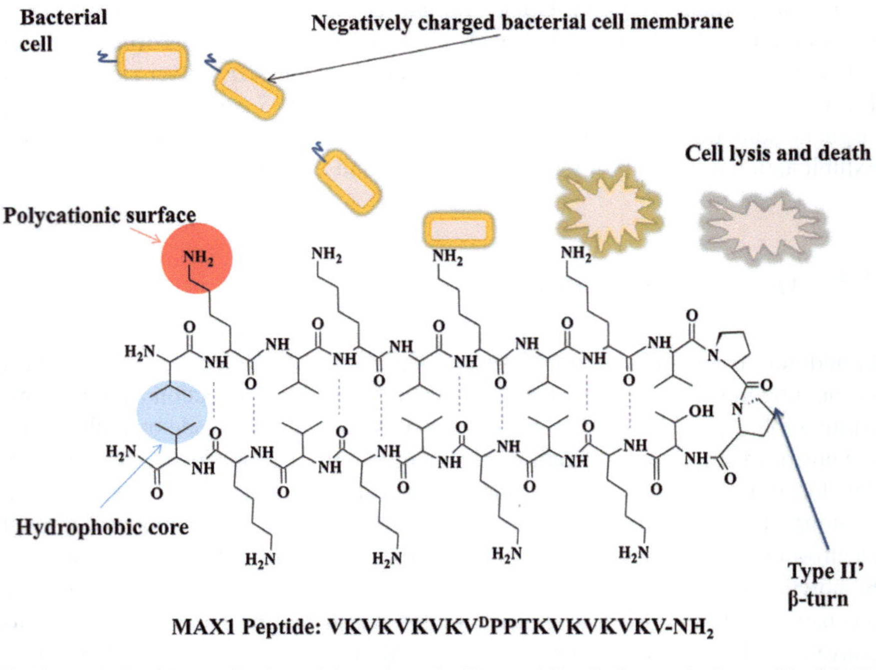

Fig. 2 Antimicrobial mechanism of the action of self-assembling β-sheet cationic peptides MAX1 as shown by the Schneider group [43]

exposed on MAX1 fibril surface disrupting the bacteria's negatively charged membrane surface (Fig. 2) [40]. Another group, Veiga and coworkers, reported on the preparation of arginine (Arg)-rich, self-assembling peptides [41]. They found that the optimal gel contained six Arg residues and functioned by disrupting the bacterial cell wall via dissociation of essential divalent metal ions as the bacteria came into contact with the hydrogel's surface. Another group utilized ultrashort peptides, i.e., composed of two or three amino acids, to produce the self-assembly of ciprofloxin, a sparingly soluble antibiotic, and a hydrophobic tripeptide (^{D}Leu-Phe-Phe) into an antimicrobial nanostructured hydrogel. The ciprofloxin was bound within the hydrogel by noncovalent interactions, and was able to retain its antimicrobial activity over a prolonged release time-frame [42].

In addition to small peptides, polypeptides such as keratin have been developed to make antimicrobial gels. Keratin, a structural protein, has generated much interest due to its intrinsic biocompatibility, biodegradability, mechanical durability, and natural abundance [44]. One group combined chitosan with keratin films that resulted in improved mechanical strength. Furthermore, the chitosan–keratin films also showed antibacterial properties and were found to be good substrates for cell culture. The biological activity of keratin films was also increased by incorporating a cell adhesion peptide, Arg-Gly-Asp-Ser (RGDS), at the free cysteine residues of reduced keratin extracts. These RGDS-carrying keratin films although antimicrobial were proven to be excellent substrates for mammalian cell growth [45].

2.3 Antimicrobial Metal Nanoparticles

Another methodology utilized in the development of antimicrobial hydrogels is the incorporation of metal nanomaterials, such as silver and copper. Hydrogel systems containing silver ions to prevent bacterial growth are of great interest, and has been used in commercial products. For example, SilverMed™ Antimicrobial Hydrogel (MPM Inc.), a hydrolyzed collagen silver impregnated gel, has been approved by the FDA to provide antimicrobial wound care. Collagen was selected as a support polymer due to its hydrophilicity, biocompatibility, non-antigenicity, and mechanical durability, making it useful in many biomedical applications [46]. The antimicrobial activity of metal nanoparticles is attributed to the metal cation causing membrane disruption resulting in subsequent lysis. Metal nanoparticles can be incorporated into polymeric nanocomposites via various approaches. Some of the most common methods include self-assembly processes, nanoprecipitation, single and double emulsification-solvent-evaporation, nanospray, and electrospinning [47]. For polymer hydrogel nanocomposites, the most widely used preparation follows an in situ synthesis where the polymer matrix serves as the reaction medium. A hydrogel's water-rich surrounding has been found to improve metal stabilization and dispersion [48].

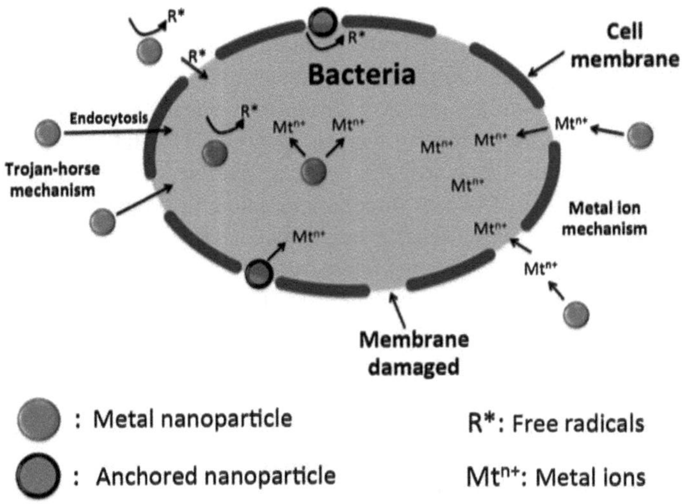

Fig. 3 A summary of the antimicrobial mechanisms of metal nanoparticles as shown by Palza [48]

Copper has been impregnated in polymer matrices, but studied to a lesser extent than silver due to silver's higher stability and efficiency. For example, a copper cellulose nanocomposite was developed and shown to exhibit increased antimicrobial activity with an increase in copper content [49]. Another group reported on the antimicrobial efficacy of covalently attached copper nanoparticles in cotton, cellulose hydrogels [50].

Mechanisms based on reduction potential of metals (Fig. 3) is dependent on the capacity of the metal to participate in redox reactions where electrons are acquired from a donor [51]. These redox-active metals can act as catalytic cofactors generating reactive oxygen species (ROS). An increase in ROS can trigger a pro-inflammatory signaling cascade resulting in programmed cell death [52]. Another mechanism involves ionic mimicry where metal ions can replace original metals present in biomolecules leading to cellular dysfunction [53]. The main mechanism of antimicrobial metal nanoparticle hydrogels is thought to be driven by ion release [54, 53]. Although there are several examples of silver and copper nanoparticles incorporated into polymer matrices, continued research exploring scaling up processes for implant materials at an industrial scale is needed.

Overall, there are several examples of antimicrobial natural hydrogels that exhibit broad-spectrum activity. Yet, there are still challenges that need to be addressed. For example, these hydrogels can be riddled with immunogenicity, batch variance problems, and high cost associated with the development of expensive peptides. More research is needed in order to deliver a consistent and affordable antimicrobial hydrogel product.

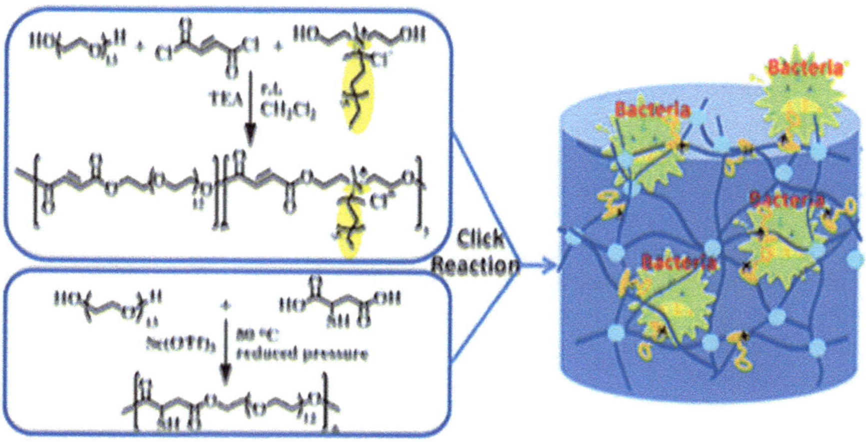

Fig. 4 Syntheses of antimicrobial cationic PEG-type hydrogels as shown by Du [55]

3 Synthetic Antimicrobial Hydrogels

Researchers have turned to synthetic chemistry to address the problems associated with natural biopolymers. Progress in synthetic chemistry has resulted in the development of low-cost and biodegradable antimicrobial hydrogels. These synthetic hydrogels possess broad-spectrum antimicrobial activity, in addition to retaining tuneable rheological and mechanical properties. Examples of hydrogel-forming synthetic polymers include poly(vinyl alcohol) (PVA), poly(ethylene oxide) (PEO), poly(acrylic acid) (PAA), poly(L-lactide) (PLLA), and poly(ethylene glycol) (PEG). Chemically, cross-linked hydrogels can be prepared via various reactions including free radical polymerization, click chemistry, and thiol-ene chemistry [6]. For example, Du and coworkers utilized thiol-ene "click" chemistry to develop a biodegradable, antibacterial, and cationic hydrogel. PEG derivatives with multi-enes and multi-thiols were synthesized by polycondensation of oligo(ethylene glycol) (OEG) with "clickable" monomers. Ammonium groups with long alkyl chains were incorporated into one of the precursors covalently, using dodecyl bis(2-hydroxyethyl) methylammonium chloride as a comonomer (Fig. 4). The authors found that these cationic PEG-type hydrogels were effective against both gram-negative and gram-positive bacteria due to their attached ammonium groups with long alkyl chains [55].

Another group investigated a PEG-based hydrogel as a protective, antibacterial coating for medical equipment. These PEG hydrogels were chemically incorporated with an antimicrobial cationic block copolymer containing quaternary ammonium groups (APC) that were grafted onto silicone rubber; exhibiting effective antifouling and antimicrobial efficacy [56]. Irrespective of the monomer structure, cationic groups, such as APC or protonated amines; and hydrophobic groups, such as alkyl chains, normally present on the polymer chain provide the polymers cationic and hydrophobic properties that are critical for demonstrated antimicrobial efficacy [57].

3.1 Synthetic Cationic Polymers

Due to the demonstrated antimicrobial effectiveness of cation containing polymers, interest in the use of synthetic cationic polymers has grown significantly and is seen as an important substitute to conventional antibiotics. These antimicrobial polymers were designed to imitate the cationic and amphiphilic structures of natural AMPs. Reduced production cost and complexity are key advantages of synthetic polymer materials verses, for example, peptide engineering. Although there are several synthetic cationic polymers reported in the literature (Table 2), there have been limited efforts to integrate these polymers into hydrogels [58]. Several synthesis pathways exist and can be prepared from biodegradable or nonbiodegradable polymer supports.

One example of an antimicrobial, biodegradable, and synthetic hydrogel involved a vitamin E-functionalized polycarbonate synthesis as reported by Lee et al. (Fig. 5). These hydrogels were formed by incorporating positively charged polycarbonates containing propyl and benzyl side-chains with vitamin E moieties into physically cross-linked networks of "ABA"-type polycarbonate and poly (-ethylene glycol) triblock copolymers. The hydrogel was formed though physical cross-links between "flower-like" micellar networks. They were found to be effective against both gram-positive and gram-negative bacteria. In vitro release studies using the antifungal drug fluconazole were found to be effective on *C. albicans* as well [71].

Another example of a synthetic antimicrobial hydrogel preparation was performed using a lactide stereocomplexation upon heating to physiological temperature. The charged hydrogels were developed using noncovalent interactions, and formed from stereocomplexation of biodegradable poly(*l*-lactide)-*b*-poly (ethyleneglycol)-*b*-poly(*l*-lactide) (PLLA-PEG-PLLA) and a charged biodegradable polycarbonate triblock polymer (i.e., PDLA-CPC-PDLA). The moldable aspect of

Table 2 List of synthetic polymers

Synthetic cationic polymers	Reference
Poly(acrylate)	[59]
Poly(norborene)	[60]
Poly(ethyleneimine)	[61]
Poly(arylamide)	[62]
Poly(vinylpyridines)	[63]
Poly(vinylether)	[64]
Poly-β-lactams	[65]
Poly-α-amino acids	[66, 67]
polycarbonates	[68]
Synthetic zwittterionic polymers	Reference
Poly(carboxybetaine)	[69]
Poly(sulfobetaine)	[70]

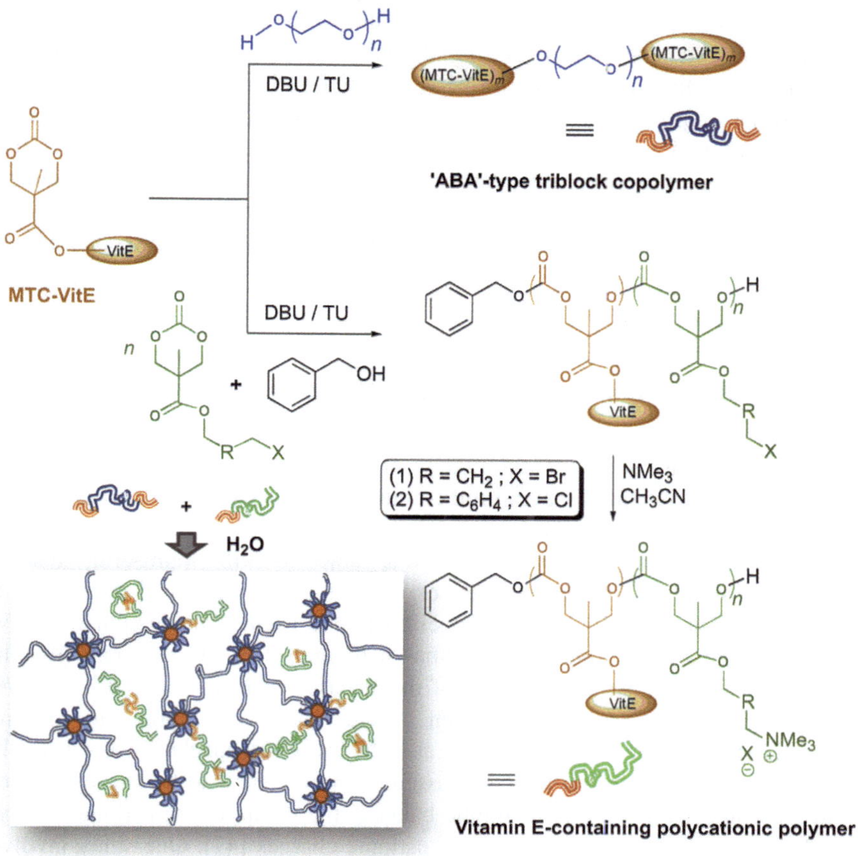

Fig. 5 Syntheses of '(MTC-VE)n-PEG-(MTC-VE)n' and vitamin E-containing polycationic polymers. Schematic illustration of incorporating polycationic polymers into hydrogel system (*inset*) as shown by Lee [72]

these nontoxic, antimicrobial hydrogels were developed with potential applications for injectable, topical, and surface coating needs [73].

3.2 Zwitterionic Antimicrobial Hydrogels

Zwitterionic polymers, such as poly(carboxybetaine) and poly(sulfobetaine), are yet another category of antimicrobial, synthetic, polymeric hydrogels. Jiang and coworkers lead in the development of these novel systems [70]. In 2012, Cao and coworkers reported the synthesis of a smart polymer surface possessing two reversibly switchable equilibrium states, a cationic *N*,*N*-dimethyl-2-morpholinone

(CB-Ring) and a zwitterionic carboxy betaine (CB–OH). The CB-Ring killed bacteria upon contact under dry conditions, while the CB–OH provided an antifouling function by releasing formerly attached bacteria [74]. The Cheng group reported an improvement on the ring-opening/ring-closing strategy by using methacrylate monomers. The hydrogels were formed by UV photopolymerization in the presence of a crosslinker, carboxy betaine dimethacrylate, and an initiator, 2-hydroxy-2-methylpropiophenone, and their respective monomers. Elasticity and stability were shown to increase in addition to maintaining antimicrobial efficacy against *E. coli* [75, 76].

3.3 Synthetic Hydrogels Loaded with Antibiotics

Recent studies have shown that a mixture of synthetic antimicrobial polymers and antibiotics could potentially avoid difficulties of drug resistance. Ng et al. showed that by combining cationic antimicrobial polycarbonates with conventional antibiotics, the developed hydrogel was able to kill multi-drug resistant *P. aeruginosa* [77]. Another strategy explored the distribution of antibiotics by incorporating the drug into the hydrogel backbone itself. Vancomycin was covalently attached by coupling of a PEG-based monomer to create a poly (β-amino ester), and the antimicrobial activity of the released vancomycin confirmed [78]. Nanopolymeric materials have been loaded with antibiotics to produce self-assembled polymers that function as vehicles to encapsulate water insoluble antibiotics. Polycations and their self-assembly were found to be strongly dependent on molecular weight, amphiphilic balance, the nature of the cationic charges, the density, and spatial arrangement of cationic charges [47]. One study involved the development of block copolymers made up of negatively charged polystyrene-*b*-poly(acrylic acid)(PS-*b*-PAA) or positively charged polystyrene-*b*-poly (4-vinylpyridine) (PS-*b*-P4VP) loaded with triclosan and thiocyanomethylthiobenzothiazole. The proposed mechanism of action involved the release of the biocide from the micelle into the bacteria via repeated bacteria/micelle interactions [79].

Another type of modification involves complex formation where metals such as silver (I), copper (II), or zinc (II) are used [80, 81]. The mechanism of action remains to be elucidated, but several important parameters appear to play a role, i.e., slow release of metal ion from the complexes, ligand exchange of metal complexes with bioligands of the microorganism, and formation of hydroxyl radicals. The configuration of functionalized surfaces with antimicrobial-absorbing hydrogel thin films was developed using the layer-by-layer (LbL) technique (Fig. 6). The hydrogel was produced by cross-linking of poly(methacrylic) acid (PMAA) and branched poly(ethyleneimine) (BPEI). These hydrogels were able to absorb and retain antimicrobial substances [82].

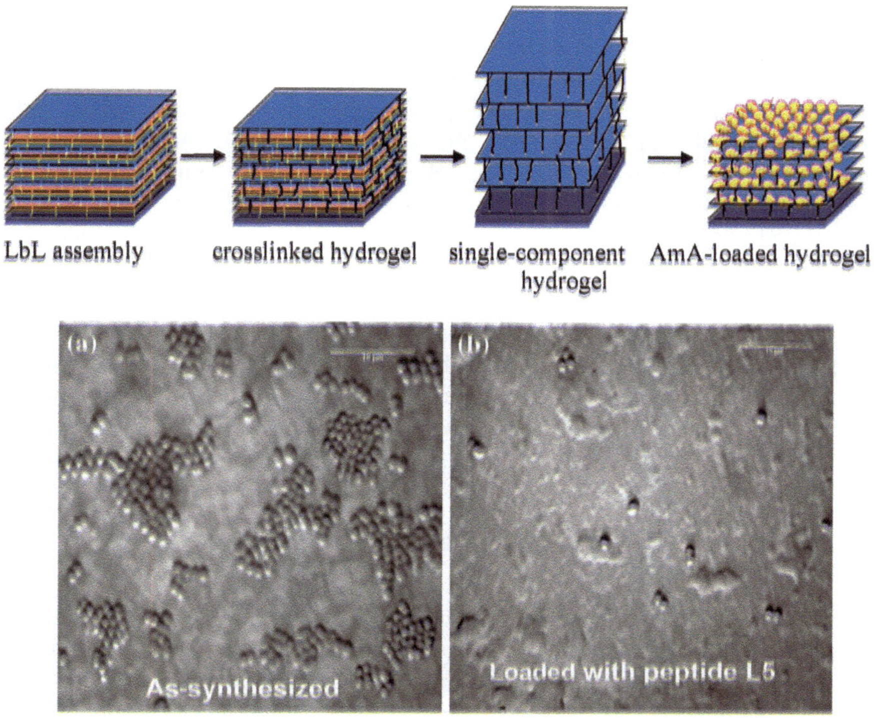

Fig. 6 Schematic formation of loaded cross-linked hydrogel. Representative optical images of **a** cross-linked hydrogel as synthesized and **b** loaded with cationic peptide L5 as shown by Pavlukina [82]

3.4 *Synthetic Hydrogels Containing Antimicrobial Metal Nanoparticles*

Over the last few years, nanotechnology has created a new way to impart antimicrobial behavior by synthesizing highly active metal nanoparticles. Silver nanoparticles are the most widely used fillers in the preparation of polymeric nanocomposites [11]. Generally, these systems use silver salt precursors, such as nitrate, jointly with a reducing agent, in the presence of a polymer gel network [48]. For example, poly(acrylamide-co-acrylic acid) hydrogels were a medium for the formation of silver nanoparticles around 25–30 nm, and demonstrated excellent antibacterial activity which was nanoparticle size-dependent [83]. Ravindra et al. synthesized silver nanoparticles containing poly(*N*-isopropylacrylamide) (PNIPAAm) hydrogels loaded with curcumin. Bacterial inhibition was found to be particle size-dependent as well [84]. Antimicrobial hydrogels based on copper nanoparticles have been investigated to a lesser extent probably due to decreased information available in regard to copper's antimicrobial effects [48]. A hydrogel thin film composed of poly(ethylene glycol diacrylate) modified with copper nanoparticles was used as a coating by means of electrochemical polymerization technique and demonstrated antimicrobial efficacy [85]. Jayaramuda et al. developed a biodegradable, antimicrobial gold nanocomposite hydrogels using acrylamide and wheat protein isolate. The gold nanoparticles were prepared as a gold colloid by reducing HAuCl(4) · XH(2)O with leaf extracts of Azadirachta indica (neem leaf) to form a hydrogel network [86]. Gao et al. synthesized a novel hydrogel formulation containing pH responsive gold nanoparticle-stabilized liposomes embedded into an acrylamide-based hydrogel for topical antimicrobial delivery [87]. The effective release of the nanoparticle-stabilized liposomes was demonstrated in an *S. aureus* model. The addition of metal-based nanoparticles to hydrogel technology embodies a versatile route by combining a nanoparticles strong antimicrobial ability coupled with the mechanical strength and biocompatibility of a hydrogel scaffold. These composite, multifunctional hydrogels could be used in any number of potential biomedical applications.

4 Conclusions and Perspectives

With the decline of antibiotic development and corresponding rise in microbial resistance, health-care associated infections continue to be a constant risk to patient welfare. Adding to the antimicrobial "pipeline" is the emergence of natural and synthetic antimicrobial hydrogels which possess either inherent antimicrobial properties or function as delivery vehicles for antibiotics. This chapter has attempted to present an overview of current trends and ingenuity in the antimicrobial polymeric hydrogel field. The advantage of these types of materials are that they can be readily applied as wound dressings or coatings that prevent bacterial growth as well as offer a comfortable, soft, aqueous, and hygienic environment. The release of different bioactive agents, i.e., such as drugs or proteins that facilitate wound healing, will increase the multi-functionality of hydrogels further and extend their application. More research is needed to test hydrogels against clinically isolated microbes that are multi-drug resistant. As well, a thorough in vitro and in vivo evaluation of a hydrogel's biocompatibility and biodegradability are also required in the development of new antimicrobial hydrogels. As these materials become better understood, they can be controlled and utilized in numerous biomedical applications.

References

1. Salem SAM, Zuel-Fakkar NM (2014) Targeting the molecular basis of resistance. Front Anti-Infect Drug Discov 2:224–268
2. Tomayko JF, Rex JH, Tenero DM, Goldberger M, Eisenstein BI (2014) The challenge of antimicrobial resistance: new regulatory tools to support product development. Clin Pharmacol Ther (N. Y., NY, U. S.) 96:166–168

3. Bassetti M, Merelli M, Temperoni C, Astilean A (2013) New antibiotics for bad bugs: where are we? Ann Clin Microbiol Antimicrob 12, 22/1-22/15
4. Wichterle O, Lim D (1960) Hydrophilic Gels for Biological Use. Nature 185:117–118
5. Gaharwar AK, Peppas NA, Khademhosseini A (2014) Nanocomposite hydrogels for biomedical applications. Biotechnol Bioeng 111:441–453
6. Lau HK, Kiick KL (2015) Opportunities for multicomponent hybrid hydrogels in biomedical applications. Biomacromolecules 16:28–42
7. Patenaude M, Smeets NMB, Hoare T (2014) Designing injectable, covalently cross-linked hydrogels for biomedical applications. Macromol Rapid Commun 35:598–617
8. Whittaker J, Balu R, Choudhury NR, Dutta NK (2014) Biomimetic protein-based elastomeric hydrogels for biomedical applications. Polym Int 63:1545–1557
9. Ng VWL, Chan JMW, Sardon H, Ono RJ, Garcia JM, Yang YY, Hedrick JL (2014) Antimicrobial hydrogels: a new weapon in the arsenal against multidrug-resistant infections. Adv Drug Deliv Rev 78:46–62
10. Seo MD, Won HS, Kim JH, Mishig-Ochir T, Lee BJ (2012) Antimicrobial peptides for therapeutic applications: a review. Molecules 17:12276–12286
11. Muñoz-Bonilla A, Fernández-García M (2012) Polymeric materials with antimicrobial activity. Prog Polym Sci 37:281–339
12. Huh AJ, Kwon YJ (2011) "Nanoantibiotics": a new paradigm for treating infectious diseases using nanomaterials in the antibiotics resistant era. J Control Release 156:128–145
13. Singh R, Smitha MS, Singh SP (2014) The role of nanotechnology in combating multi-drug resistant bacteria. J Nanosci Nanotechnol 14:4745–4756
14. Dash M, Chiellini F, Ottenbrite RM, Chiellini E (2011) Chitosan-A versatile semi-synthetic polymer in biomedical applications. Prog Polym Sci 36:981–1014
15. Kong M, Chen XG, Xing K, Park HJ (2010) Antimicrobial properties of chitosan and mode of action: a state of the art review. Int J Food Microbiol 144:51–63
16. Zhu L, Peng L, Zhang Y-Q (2015) The processing of chitosan and its derivatives and their application for postoperative anti-adhesion. Mini Rev Med Chem 15(4):330–337
17. Rabea EI, Badawy MET, Stevens CV, Smagghe G, Steurbaut W (2003) Chitosan as antimicrobial agent: applications and mode of action. Biomacromolecules 4:1457–1465
18. Lichter JA, Rubner MF (2009) Polyelectrolyte multilayers with intrinsic antimicrobial functionality: the importance of mobile polycations. Langmuir 25:7686–7694
19. Li P, Poon YF, Li W, Zhu H-Y, Yeap SH, Cao Y, Qi X, Zhou C, Lamrani M, Beuerman RW, Kang E-T, Mu Y, Li CM, Chang MW, Leong SSJ, Chan-Park MB (2011) A polycationic antimicrobial and biocompatible hydrogel with microbe membrane suctioning ability. Nat Mater 10:149–156
20. Jiang Q, Xu J, Li T, Qiao C, Li Y (2014) Synthesis and antibacterial activities of quaternary ammonium salt of gelatin. J Macromol Sci Part B: Phys 53:133–141
21. Mohamed NA, Fahmy MM (2012) Synthesis and antimicrobial activity of some novel cross-linked chitosan hydrogels. Int J Mol Sci 13:11194–11209
22. Venkatesan J, Jayakumar R, Mohandas A, Bhatnagar I, Kim S-K (2014) Antimicrobial activity of chitosan-carbon nanotube hydrogels. Materials 7(3946–3955):10
23. Aziz MA, Cabral JD, Brooks HJL, Moratti SC, Hanton LR (2012) Antimicrobial properties of a chitosan dextran-based hydrogel for surgical use. Antimicrob Agents Chemother 56:280–287
24. Cabral JD, Roxburgh M, Shi Z, Liu L, McConnell M, Williams G, Evans N, Hanton LR, Simpson J, Moratti SC, Robinson BH, Wormald PJ, Robinson S (2014) Synthesis, physiochemical characterization, and biocompatibility of a chitosan/dextran-based hydrogel for postsurgical adhesion prevention. J Mater Sci: Mater Med 25:2743–2756
25. Liu G, Shi Z, Kuriger T, Hanton LR, Simpson J, Moratti SC, Robinson BH, Athanasiadis T, Valentine R, Wormald PJ, Robinson S (2009) Synthesis and characterization of chitosan/dextran-based hydrogels for surgical use. Macromol Symp 279:151–157
26. Lee KY, Mooney DJ (2012) Alginate: properties and biomedical applications. Prog Polym Sci 37:106–126

27. Morais DS, Rodrigues MA, Lopes MA, Coelho MJ, Mauricio AC, Gomes R, Amorim I, Ferraz MP, Santos JD, Botelho CM (2013) Biological evaluation of alginate-based hydrogels, with antimicrobial features by Ce(III) incorporation, as vehicles for a bone substitute. J Mater Sci Mater Med 24:2145–2155
28. Catanzano O, Straccia MC, Miro A, Ungaro F, Romano I, Mazzarella G, Santagata G, Quaglia F, Laurienzo P, Malinconico M (2015) Spray-by-spray in situ cross-linking alginate hydrogels delivering a tea tree oil microemulsion. Eur J Pharm Sci 66:20–28
29. Obradovic B, Miskovic-Stankovic V (2013) Silver nanoparticles in alginate solutions and hydrogels aimed for biomedical applications. Nova Science Publishers, Inc., 247–260
30. Ghasemzadeh H, Ghanaat F (2014) Antimicrobial alginate/PVA silver nanocomposite hydrogel, synthesis and characterization. J Polym Res 21:1–14
31. Stojkovska J, Kostic D, Jovanovic Z, Vukasinovic-Sekulic M, Miskovic-Stankovic V, Obradovic B (2014) A comprehensive approach to in vitro functional evaluation of Ag/alginate nanocomposite hydrogels. Carbohydr Polym 111:305–314
32. Straccia MC, Romano I, Oliva A, Santagata G, Laurienzo P (2014) Crosslinker effects on functional properties of alginate/*N*-succinyl chitosan based hydrogels. Carbohydr Polym 108:321–330
33. Teixeira V, Feio MJ, Bastos M (2012) Role of lipids in the interaction of antimicrobial peptides with membranes. Prog Lipid Res 51:149–177
34. Zhou C, Li P, Qi X, Sharif AR, Poon YF, Cao Y, Chang MW, Leong SS, Chan-Park MB (2011) A photopolymerized antimicrobial hydrogel coating derived from epsilon-poly-L-lysine. Biomaterials 32:2704–2712
35. Altunbas A, Pochan DJ (2012) Peptide-based and polypeptide-based hydrogels for drug delivery and tissue engineering. Top Curr Chem 310:135–167
36. Liu SQ, Yang C, Huang Y, Ding X, Li Y, Fan WM, Hedrick JL, Yang Y-Y (2012) Antimicrobial and antifouling hydrogels formed in situ from polycarbonate and poly(ethylene glycol) via Michael addition. Adv Mater 24:6484–6489
37. Fjell CD, Hiss JA, Hancock REW, Schneider G (2012) Designing antimicrobial peptides: form follows function. Nat Rev Drug Discovery 11:37–51
38. Mccloskey AP, Gilmore BF, Laverty G (2014) Evolution of antimicrobial peptides to self-assembled peptides for biomaterial applications. Pathogens 3, 791–821, 31 pp
39. Rajagopal K, Lamm MS, Haines-Butterick LA, Pochan DJ, Schneider JP (2009) Tuning the pH responsiveness of β-Hairpin peptide folding, self-assembly, and hydrogel material formation. Biomacromolecules 10:2619–2625
40. Salick DA, Kretsinger JK, Pochan DJ, Schneider JP (2007) Inherent antibacterial activity of a peptide-based β-Hairpin hydrogel. J Am Chem Soc 129:14793–14799
41. Veiga AS, Sinthuvanich C, Gaspar D, Franquelim HG, Castanho MARB, Schneider JP (2012) Arginine-rich self-assembling peptides as potent antibacterial gels. Biomaterials 33:8907–8916
42. Marchesan S, Qu Y, Waddington LJ, Easton CD, Glattauer V, Lithgow TJ, McLean KM, Forsythe JS, Hartley PG (2013) Self-assembly of ciprofloxacin and a tripeptide into an antimicrobial nanostructured hydrogel. Biomaterials 34:3678–3687
43. Schneider JP, Pochan DJ, Ozbas B, Rajagopal K, Pakstis L, Kretsinger J (2002) Responsive hydrogels from the intramolecular folding and self-assembly of a designed peptide. J Am Chem Soc 124:15030–15037
44. Tanabe T, Okitsu N, Tachibana A, Yamauchi K (2002) Preparation and characterization of keratin-chitosan composite film. Biomaterials 23:817–825
45. Yamauchi K, Hojo H, Yamamoto Y, Tanabe T (2003) Enhanced cell adhesion on RGDS-carrying keratin film. Mater Sci Eng C C23:467–472
46. Sahiner M, Sagbas S, Bitlisli BO (2015) p(AAm/TA)-based IPN hydrogel films with antimicrobial and antioxidant properties for biomedical applications. Journal of Applied Polymer Science 132, n/a-n/a
47. Munoz-Bonilla A, Fernandez-Garcia M (2015) The roadmap of antimicrobial polymeric materials in macromolecular nanotechnology. Eur Polym J 65:46–62

48. Palza H (2015) Antimicrobial polymers with metal nanoparticles. Int J Mol Sci 16:2099–2116
49. Pinto RJ, Daina S, Sadocco P, Pascoal Neto C, Trindade T (2013). Antibacterial activity of nanocomposites of copper and cellulose. Biomed Res Int, 280512
50. Mary G, Bajpai SK, Chand N (2009) Copper (II) ions and copper nanoparticles-loaded chemically modified cotton cellulose fibers with fair antibacterial properties. J Appl Polym Sci 113:757–766
51. Lemire JA, Harrison JJ, Turner RJ (2013) Antimicrobial activity of metals: mechanisms, molecular targets and applications. Nat Rev Microbiol 11:371–384
52. Sintubin L, de Windt W, Dick J, Mast J, van der Ha D, Verstraete W, Boon N (2009) Lactic acid bacteria as reducing and capping agent for the fast and efficient production of silver nanoparticles. Appl Microbiol Biotechnol 84:741–749
53. Ruparelia JP, Chatterjee AK, Duttagupta SP, Mukherji S (2008) Strain specificity in antimicrobial activity of silver and copper nanoparticles. Acta Biomater 4:707–716
54. Ren G, Hu D, Cheng EW, Vargas-Reus MA, Reip P, Allaker RP (2009) Characterisation of copper oxide nanoparticles for antimicrobial applications. Int J Antimicrob Agents 33:587–590
55. Du H, Zha G, Gao L, Wang H, Li X, Shen Z, Zhu W (2014) Fully biodegradable antibacterial hydrogels via thiol-ene "click" chemistry. Polym Chem 5:4002–4008
56. Liu SQ, Yang C, Huang Y, Ding X, Li Y, Fan WM, Hedrick JL, Yang Y-Y (2012b) Antimicrobial and antifouling hydrogels formed in situ from polycarbonate and poly(ethylene glycol) via Michael addition. Adv Mater (Weinheim, Ger.) 24, 6484–6489
57. Li P, Li X, Saravanan R, Li CM, Leong SSJ (2012) Antimicrobial macromolecules: synthesis methods and future applications. RSC Adv 2:4031–4044
58. Wei G, Yang L, Chu L (2010) Progress in researches on synthetic antimicrobial macromolecular polymers. Sheng Wu Yi Xue Gong Cheng Xue Za Zhi 27:953–957
59. Wang Q, Uzunoglu E, Wu Y, Libera M (2012) Self-assembled poly(ethylene glycol)-co-acrylic acid microgels to inhibit bacterial colonization of synthetic surfaces. ACS Appl Mater Interfaces 4:2498–2506
60. Colak S, Tew GN (2012) Dual-functional ROMP-based betaines: effect of hydrophilicity and backbone structure on nonfouling properties. Langmuir 28:666–675
61. Sparks SM, Waite CL, Harmon AM, Nusblat LM, Roth CM, Uhrich KE (2011) Efficient intracellular siRNA delivery by ethyleneimine-modified amphiphilic macromolecules. Macromol Biosci 11:1192–1200
62. Som A, Navasa N, Percher A, Scott RW, Tew GN, Anguita J (2012) Identification of synthetic host defense peptide mimics that exert dual antimicrobial and anti-inflammatory activities. Clin Vaccine Immunol 19:1784–1791
63. Allison BC, Applegate BM, Youngblood JP (2007) Hemocompatibility of hydrophilic antimicrobial copolymers of alkylated 4-vinylpyridine. Biomacromolecules 8:2995–2999
64. Oda Y, Kanaoka S, Sato T, Aoshima S, Kuroda K (2011) Block versus random amphiphilic copolymers as antibacterial agents. Biomacromolecules 12:3581–3591
65. Chakraborty S, Liu R, Lemke JJ, Hayouka Z, Welch RA, Weisblum B, Masters KS, Gellman SH (2013) Effects of cyclic vs. acyclic hydrophobic subunits on the chemical structure and biological properties of nylon-3 co-polymers. ACS Macro Lett 2
66. Engler AC, Shukla A, Puranam S, Buss HG, Jreige N, Hammond PT (2011) Effects of side group functionality and molecular weight on the activity of synthetic antimicrobial polypeptides. Biomacromolecules 12:1666–1674
67. Zhou C, Qi X, Li P, Chen WN, Mouad L, Chang MW, Leong SS, Chan-Park MB (2010) High potency and broad-spectrum antimicrobial peptides synthesized via ring-opening polymerization of alpha-aminoacid-*N*-carboxyanhydrides. Biomacromolecules 11:60–67
68. Engler AC, Tan JP, Ong ZY, Coady DJ, Ng VW, Yang YY, Hedrick JL (2013) Antimicrobial polycarbonates: investigating the impact of balancing charge and hydrophobicity using a same-centered polymer approach. Biomacromolecules 14:4331–4339

69. Zhang L, Cao Z, Bai T, Carr L, Ella-Menye JR, Irvin C, Ratner BD, Jiang S (2013) Zwitterionic hydrogels implanted in mice resist the foreign-body reaction. Nat Biotechnol 31:553–556
70. Jiang S, Cao Z (2010) Ultralow-fouling, functionalizable, and hydrolyzable zwitterionic materials and their derivatives for biological applications. Adv Mater 22:920–932
71. Lee ALZ, Ng VWL, Wang W, Hedrick JL, Yang YY (2013) Block copolymer mixtures as antimicrobial hydrogels for biofilm eradication. Biomaterials 34:10278–10286
72. Lee AL, Ng VW, Wang W, Hedrick JL, Yang YY (2013) Block copolymer mixtures as antimicrobial hydrogels for biofilm eradication. Biomaterials 34:10278–10286
73. Li Y, Fukushima K, Coady DJ, Engler AC, Liu S, Huang Y, Cho JS, Guo Y, Miller LS, Tan JPK, Ee PLR, Fan W, Yang YY, Hedrick JL (2013) Broad-spectrum antimicrobial and biofilm-disrupting hydrogels: stereocomplex-driven supramolecular assemblies. Angew Chem Int Ed 52:674–678
74. Cao Z, Mi L, Mendiola J, Ella-Menye JR, Zhang L, Xue H, Jiang S (2012) Reversibly switching the function of a surface between attacking and defending against bacteria. Angew Chem Int Ed Engl 51:2602–2605
75. Cao B, Li L, Tang Q, Cheng G (2013) The impact of structure on elasticity, switchability, stability and functionality of an all-in-one carboxybetaine elastomer. Biomaterials 34:7592–7600
76. Cao B, Tang Q, Li L, Humble J, Wu H, Liu L, Cheng G (2013) Switchable antimicrobial and antifouling hydrogels with enhanced mechanical properties. Adv Healthc Mater 2:1096–1102
77. Ng VWL, Ke X, Lee ALZ, Hedrick JL, Yang YY (2013) Synergistic co-delivery of membrane-disrupting polymers with commercial antibiotics against highly opportunistic bacteria. Adv Mater (Weinheim, Ger.) 25, 6730–6736
78. Lakes AL, Peyyala R, Ebersole JL, Puleo DA, Hilt JZ, Dziubla TD (2014) Synthesis and characterization of an antibacterial hydrogel containing covalently bound vancomycin. Biomacromolecules 15:3009–3018
79. Vyhnalkova R, Eisenberg A, van de Ven T (2011) Bactericidal block copolymer micelles. Macromol Biosci 11:639–651
80. Ambika S, Arunachalam S, Arun R, Premkumar K (2013) Synthesis, nucleic acid binding, anticancer and antimicrobial activities of polymer-copper(II) complexes containing intercalative phenanthroline ligand(DPQ). RSC Adv 3:16456–16468
81. Nishat N, Rasool R, Parveen S, Khan SA (2011) New antimicrobial agents: The synthesis of schiff base polymers containing transition metals and their characterization and applications. J Appl Polym Sci 122:2756–2764
82. Pavlukhina S, Lu Y, Patimetha A, Libera M, Sukhishvili S (2010) Polymer multilayers with pH-triggered release of antibacterial agents. Biomacromolecules 11:3448–3456
83. Thomas V, Yallapu MM, Sreedhar B, Bajpai SK (2007) A versatile strategy to fabricate hydrogel-silver nanocomposites and investigation of their antimicrobial activity. J Colloid Interface Sci 315:389–395
84. Ravindra S, Varaprasad K, Rajinikanth V, Mulaba-Bafubiandi AF, Venkata Surya Ramam K (2013) Studieson curcumin loaded poly(n-isopropylacrylamide) silver nanocomposite hydrogels for antibacterial and drug releasing applications. J Macromol Sci, Part A: Pure Appl Chem 50:1230–1240
85. Cometa S, Iatta R, Ricci MA, Ferretti C, de Giglio E (2013) Analytical characterization and antimicrobial properties of novel copper nanoparticle–loaded electrosynthesized hydrogel coatings. J Bioactive Compatible Polym 28:508–522
86. Jayaramudu T, Raghavendra GM, Varaprasad K, Sadiku R, Raju KM (2013) Development of novel biodegradable Au nanocomposite hydrogels based on wheat: for inactivation of bacteria. Carbohydr Polym 92:2193–2200
87. Gao W, Vecchio D, Li J, Zhu J, Zhang Q, Fu V, Li J, Thamphiwatana S, Lu D, Zhang L (2014) Hydrogel containing nanoparticle-stabilized liposomes for topical antimicrobial delivery. ACS Nano 8:2900–2907

Biopolymer-Based Hydrogels for Decontamination for Organic Waste

Ajay Kumar Mishra and Shivani Bhardwaj Mishra

Abstract Organic waste material in water is accumulated from domestic and industrial discharge. The nature and type of organic contaminants in wastewater differ based on the industrial products and domestic usage. Some of the organic pollutants occur from soap and detergents, petroleum products, dye and textile industries, drugs and pharmaceuticals, herbicide and pesticide and many more. There are numerous technologies available for the treatment of organics that may include the physical, chemical, photocatalytic and microbial processes. The hydrogels offers eco-friendly substrates that have been investigated for decontamination of organics. There are varieties of hydrogels available for different application in pristine forms as well as synthetic forms. Among these, biopolymers that have been categorized as hydrogels and were used for treating organic pollutants have been discussed in this chapter.

Keywords Biopolymer · Hydrogel · Organic · Contaminants · Treatment

1 Introduction

The ever-growing industrialization for the increasing demands of materialistic world, push forth the challenge of pollution to a new scale. The environmental pollution in all forms such as air, water and soil are rich with organic contaminates and discharged by various sources.

Air pollution is primarily comprised of different types of organics that directly and indirectly enter into air reducing the air quality. The biggest anthropogenic source of the highly toxic organic pollutants is fuel emissions by motor vehicle

A.K. Mishra (✉) · S.B. Mishra
Nanotechnology and Water Sustainability Research Unit, College of Engineering, Science and Technology, University of South Africa, Florida Science Campus, Johannesburg, South Africa
e-mail: ajaykmishra1@gmail.com; mishrak@unisa.ac.za

S. Kalia (ed.), *Polymeric Hydrogels as Smart Biomaterials*, Springer Series on Polymer and Composite Materials, DOI 10.1007/978-3-319-25322-0_7

industry releasing carbon monoxide and carbon dioxide in air. These oxides of carbon directly affects the central nervous system, heart and asthmatic conditions leading to reduced supply of oxygen for the metabolic activity. Volatile organic compounds (VOC) are another set of carcinogenic pollutants that evaporates easily and rapidly into air from solvent use in various industries such as paints, adhesive and aerosols, etc., distribution of fossil fuel, production process like alcoholic drinks and arable farming. Organic pollutants that resist to degrade or biodegrade are persistent organic pollutants (POP) that lead to neurobehavioral disorders, malfunctioning of immune, endocrine and reproductive systems.

The organics pollutants also penetrate the soil and sediments through a number of sources such as mine waste dumping, oil and fuel dumping, the use of herbicide and pesticide, deforestation, acid rain, dye industry and genetically modified plants, etc. The soil pollutants enter into biocycle entering into living bodies affecting the genetic makeup, congenital illness and chronic health problems. These organic beyond the threshold limit effect the growth of plants, soil structure and soil fertility leading to crop damage.

The quality of fresh and ground water is equally affected by the presences of organic contaminants released by domestic sewage (raw and treated), urban run-off, farm and industrial wastewater. These organic lowers the available oxygen and enhances the turbidity thereby reducing the penetration of sunlight that affects underwater photocatalysis affecting the for aquatic life growth.

Hydrogels are three-dimensional polymer networks that have capacity to absorb large quantity of water. The natural hydrogels are usually composed of single component. However, synthetic hydrogels based on biopolymers or polymers can be one, two or multi components yielding amorphous, semi-crystalline, and crystalline type of hydrogels. The hydrogels of both types have found its utility in many industrial applications and most commonly are hygienic products, pharmaceuticals, diagnostics, healthcare and agriculture. Recently, much research is dealing with the use of hydrogels for environmental protection and also many emerging researchers are investigating the hydrogels for wastewater treatment.

This chapter discusses about organic pollutants, those that are responsible for water pollution, biopolymer-based hydrogels. The author further describes about the utility of such biopolymer-based hydrogels for decontamination of organics from waste water.

2 Organic Water Pollutants

Organic water pollutants are the carbon-based chemicals that are released into the environment from various industrial wastewater. Some of the wide range of organic contaminants originates from the industries producing commercial products such as soap and detergent, insecticide and herbicide, dyes, pharmaceutical drugs and metabolite, industrial solvents, perchlorate, food processing waste, petroleum and hygiene and cosmetic products.

2.1 Soap and Detergent Organic Contaminants

Soap and detergents form a raw material for many commercial products that are categorized as bathing soap, hand and body liquid soap, face wash, baby soap, shampoo and conditioners, shaving soap and creams, washing powder and liquids, etc. In a typical soap making process generally known as saponification reaction, the raw material used is a triglyceride that is hydrolyzed by a base to produce salt of a carboxylate/fatty acid (soap). However detergent are the charged surfactants that are used for laundry cleaning. A common cationic surfactants is a quaternary ammonium salt of alkyl benzene where as an anionic is usually alkyl benzene sulphonate. Non-ionic or zwitter ionic may be based on a polyoxyethylene or a glycoside [1]. The final product of soap or a detergent is basically a complex mixture, which is composed of many other ingredients such as additives, perfumes dyes, enzymes, bleaching agents and brightners [2]. These harmful chemicals are xenobiotic and the use of these chemicals leads to allergies of skin or dermatitis [3]. It has been reported that the bioaccumulation of soap and detergents in natural waters induces eutrophication of aquatic life [4]. The water discharged from soap and detergent industry undergoes treatment before the discharge, even then the concentration of these surfactants may range from 1.2 to 9.2 mg/L [5].

2.2 Herbicide

Herbicide is commonly known as weed killer, a chemical compound used to destroy the unwanted weeds that might affect the crops. These chemicals can be natural organic or synthetic organic compounds and its presence in environment may prove to be harmful affecting the living bodies. The natural organics that are frequently used as herbicide is vinegar and corn gluten meal. The synthetic organic compounds that has been used as herbicide are mainly different types of carboxylic acids such as dichlorophenoxyacetic acid, 2-(2,4-dichlorophenoxy) propionic acid, methylchlorophenoxypropionic acid and 3,6-dichloro-2-methoxybenzoic, etc. Researchers across the world have used different methods to treat herbicides as pollutants. In some recent research studies, the herbicide was degraded using chlorine dioxide [6], photodegradation [7] and membrane retention [8].

2.3 Pesticide

Pesticide is another general term used for the chemical or microbes that have ability to prevents or destroy pest that may be insects, mollusks, nematodes, birds and fishes, etc. Pesticides have been identified as endocrine disrupting chemicals posing harmful effects on aquatic life and human beings at the trace levels [9, 10].

Continuous exposure of pesticide can be detrimental leading to cancer and loss of life [11]. Many techniques have been used to remove pesticides which include adsorption by activated carbon [12], bioremediation [13] and biodegradation [14].

2.4 *Pharmaceutical Drugs*

Drugs and pharmaceuticals are those chemical compounds that have the ability to detect, diagnose, prevent or treat the disordered metabolic conditions that cause a disease. Enormous amount of chemicals are used to develop suitable drugs to combat the abnormal functions of the body. The presence of these pharmaceutical drugs in water has increased from nanograms to micrograms that directly affect the normal functions of a living body leading to serious health impacts [15, 16].

2.5 *Dyes*

Dyes are the organics that have ability to impart colour by absorbing certain wavelength from the visible spectrum. These organics are used by various industries such as fabric, textile, food, leather, laser and solvent dyes, etc. The dyes used are may be a natural dye which is extracted from a plant source. Few examples of natural dyes are indigo, saffron, henna and woad. The natural dyes are fairly safe for the environment and life. However, the synthetic dyes which are more popularly used for commercial products have been widely classified based on the type of chromophore in its chemical structure. Some of the categories are phthalocyanine, nitroso, azo, nitro, thiazole and diazonium, etc. The dyes discharged into the wastewater affects the aesthetic value of the water as it resists the penetration of sunlight to the aquatic environment thereby affecting the water bodies. It has been reported that fabric or textile dyes are the most recalcitrant in nature [17–19].

2.6 *Petroleum Products*

The products obtained from petroleum refineries are rich in hydrocarbons. Different types of petroleum products are commonly known as fuel, heavy oil, jet oil, diesel oil, petrol, paraffin wax, asphalt and petroleum coke, etc. The wastewater discharged from the petroleum refineries is heavily loaded with polycyclic aromatic hydrocarbons, phenols and heterocyclic compounds [20]. These organics are rendered as highly toxic for the ecosystems and therefore categorized as organics that are priority substance to be controlled [21, 22].

3 Biopolymer-Based Hydrogels for Decontamination of Organic Waste

3.1 Biopolymer-Based Hydrogels

Natural polymers both plant and animal based such as cellulose, chitosan, collagen, gelatin, alginate, dextrose, hyaluronic acid, silk, fibrin and gums, etc. are well-known hydrophilic gels/hydrogels that have been well documented by researchers. The unique property of these hydrogels is ability of water retention without dissolving in water due to three-dimensional network of cross-linked chains enriched with hydrophilic functional groups [23]. The cross linking may be physical indicating a week interaction via hydrogen bonding/hydrophobic interaction or electrostatic attractions and is reversible in nature. If the nature of crosslinking involve covalent bonding then hydrogel is anticipated as more stronger than the physical ones. The water retention capacity of the hydrogels influences the volume transitions that may be stimulated by external factors such as temperature, pressure, pH and most important is the type of solvent which is in direct contact of these hydrogels [24]. Biopolymer-based hydrogels have received a special recognition in the field of biomedical sciences as these resemble the natural macromolecules. Recognition by cell surface receptors provides an advantage to influence adhesion and proliferation preventing the chronic inflammation, toxicity and immunological bioreactions [25–27]. These are biodegradable and biocompatible for which these have been investigated for various biomedical and environmental applications such as drug delivery systems, targeted drug delivery and tissue engineering and wastewater treatment. Although, these stimulus responsive materials have many benefits and biomedical applications, there are few challenges such as non-adherent, poor mechanical properties that make these hydrogels difficult to handle [28, 29]. The chemical structures of some of the biopolymer hydrogels are shown in Scheme 1.

3.2 Removal of Organic Pollutants Via BioPolymer Hydrogels

Biopolymer-based hydrophilic gels have been used for removing water contaminants specifically by physicochemical techniques where by hydrogels and its composites have been investigated as an adsorbent. After an extensive research for the reports available or published in various scientific journals, it was found that biopolymer hydrogels have been applied to remove the two prominent pollutants that are dyes and drugs. The other organics such as pesticides, herbicides and petroleum products have not been investigated to such an extent. However, a large amount of studies were done for heavy metal removal from the waste water using these natural hydrophilic polymers. Since these have an advantage of being

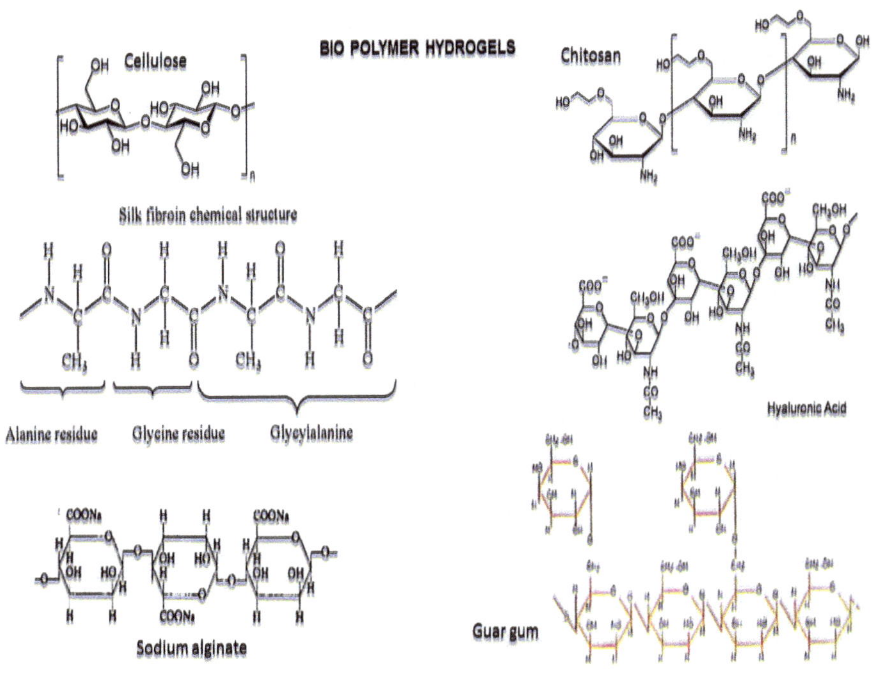

Scheme1 Chemical structures of biopolymer hydrogels

biodegradable and biocompatible, the main focus on application of hydrogels is in the field of biomedical and pharmaceutical sciences. The following sections therefore describes about the recent work done in the field of biopolymers hydrogels applied to remove the dyes, drugs and other organics.

3.2.1 Dyes

Dye industry is continuously in growing demand as per the web reports where it has been disclosed that this industry will have an annual increase of 6 % and is expected to reach $30 billion by 2020 [30]. Dyes are either natural or synthetic but a unique factor in any case is that it is basically made up of carbon skeleton-based chemical structure. There are many industries which make use of different types of dyes for the commercial products. The prominent industries that make use of dyes and pigments are textile and fabric, leather, ink, food, pharmaceutical and nutraceutical. The ever-growing demand of the products from these industries therefore forces the dye industry to target increased amount of dye production. The increased commercialization and market growth of this industry therefore put a pressure on environment pollution due to limited scope and methods of treatment of water before discharge by these industries.

In some recent studies on treatment of dye waste water, chitosan hydrogels were treated with ammonium sulphate solution to be used as an adsorbent for acid orange 7 and acid red 8 [31]. It was reported the adsorption capacities of modified chitosan hydrogels for these dyes were 7.4 and 14.4 times more than the pristine polymer. Desorption of the dyes and regeneration of this hydrogel adsorbent up to 20 times was successfully done without the loss of adsorption capacities. In a different study, the authors prepared lithocholate salts which was studied for its gelatin behaviour and were formed as negatively charged tubular structure [32]. The prepared hydrogel was applied to uptake cationic dyes, such as methylene blue and rhodamine B, and was found that these hydrogels were able to adsorb maximum dye from a solution as shown in Fig. 1.

Semi-interpenetrating networks of sodium alginate-acrylic acid and acrylamide copolymer were developed to remove basic fuchsin and methyl violet from the water. The increase in the crosslinker improved the swelling capacity of these hydrogels and the adsorptions of dyes were found to increase with an increase in feed dye concentration [33]. The researchers described the adsorption mechanism as initially stage to be mass transfer controlled which was followed by diffusion controlled. Table 1 was presented by them to show a comparison with some other similar type of work.

A novel poly(acrylic acid-aniline)-grafted gumghatti-based conducting hydrogel was prepared and studied for adsorption of dyes such as stilbene, malachite green (MG), methyl blue (MB), rhodamine (Rh) and fluorescent sodium (FS), and it was shown that the prepared hydrogels had fascinating stimuli responsive towards electrical conductivity [41]. The additional feature lied in the fact of better adsorption efficiency when compared to semi-interpenetrating networks. This was shown in Fig. 2.

In another study, partially hydrolyzed polyacrylamide/cellulose nanocrystal nanocomposite hydrogels were used to investigate the adsorption kinetics and equilibrium parameter for the methylene blue dye removal [42]. It was reported that these hydrogels had 90 % adsorption efficiency along with qe >40 mg/g and

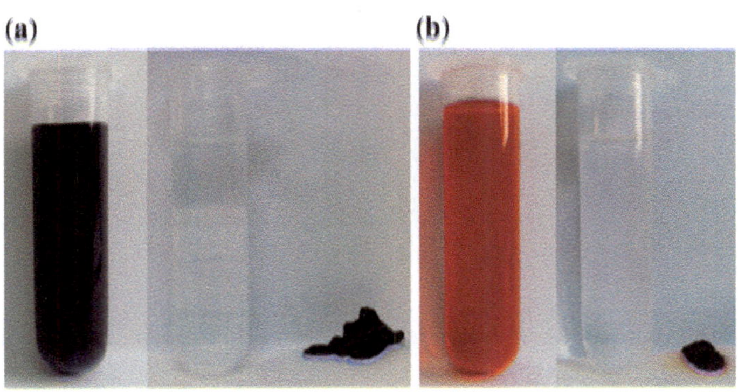

Fig. 1 The methylene blue (**a**) and rhodamine 6G (**b**) solutions before (*left*) and after (*right*) adsorption. Reprinted with permission from [32]. Copyright © 2011, American Chemical Society

Table 1 Comparison of dying adsorption capacity of semi interpenetrating networks of acrylic acid and acrylamide and sodium alginate hydrogels with the reported data

Name of the hydrogel	Dye used in water/temp./conc./pH	Adsorption performance mg/g resin	References
IPN2	2.5 mg/L in water at pH 7 and 25 °C	2.249 for BF, Q_{max} 5.96 1.723 for MV, Q_{max} 3.93	[34]
IPN2	500 mg/L in water at pH 7 and 25 °C	368.70 for BF, Q_{max} 920, Q_{max} 283.76	[34]
Poly (HEMA-g-GA)	700 mg/L in water at pH 5	121.5 Q_{max} 0.189	[35]
Poly (AA-co-AM) attapulgite	200–100 mg/L in water at pH 7	917 for 100 mg/L for MV	[36]
Soya ash	25.9 mg/L at pH 9	4.209 for 5.76 for MV	[37]
Supramolecular and composite gel of agarose	1000 mg/L at pH 7	Removal % 95.1 and 95.7 for MV	[38]
Poly (AA-co-AM)	50 mg/L at pH 7	6.38 for MV	[39]
Poly (VP-co-MA)	500 mg/L at pH 7	4.22 for MV	[40]
CPSA4	2.5 mg/L at pH 7	2.09 for MV with 84 % removal and 2.37 for BF with 95 % removal	[33]
CPSA4	200 mg/L at pH 7	174 for MV with 87 % removal and 188 for BF with 94 % removal	[33]

Reprinted with permission from [33]. Copyright © 2014, Elsevier

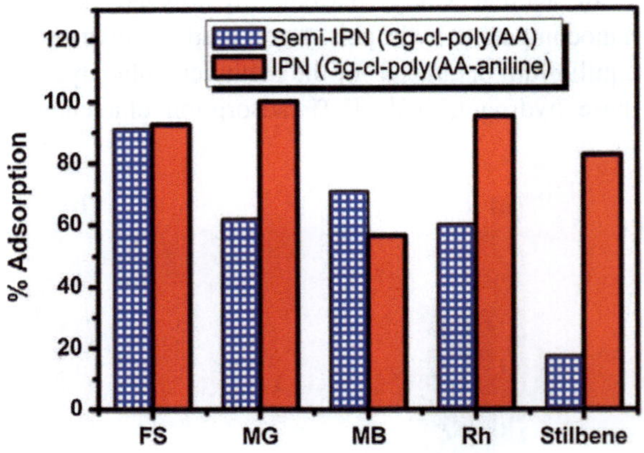

Fig. 2 Comparison of IPN and semi IPN for dye adsorption. Reprinted with permission from [41]. Copyright © 2015, Elsevier

followed pseudo second order model indicating chemisorption. The adsorption mechanism of methylene blue due was shown in Fig. 3.

Carrageenan, a linear sulphated polysaccharide obtained from edible red seaweeds is poly-anionic hydrogel which has been exploited as an adsorbent for

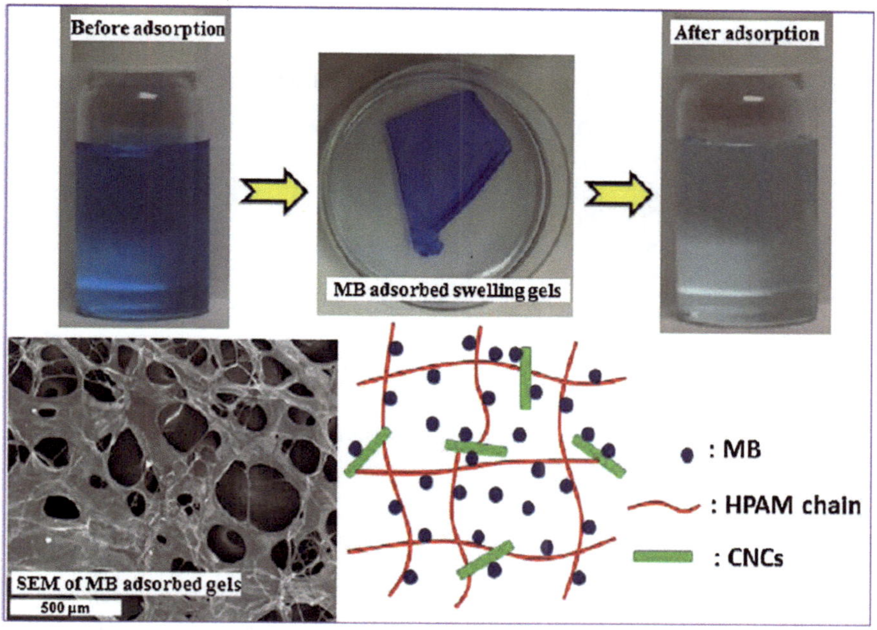

Fig. 3 Adsorption mechanism of partially hydrolyzed polyacrylamide/cellulose nanocrystal nanocomposite hydrogels using methylene blue as an adsorbate. Reprinted with permission from [42]. Copyright © 2014, Elsevier

up-taking cationic dyes [43]. This natural gum was used to form a composite material using polyvinyl alcohol and iron oxide nanoparticles via in situ precipitation method [44]. Crystal violet was the candidate dye as an adsorbate from the aqueous solution. It was concluded that the screening effect of sodium ions and the sulphate ions present on the carrageenan affect the dye adsorption capacity by the hydrogel nanocomposites. The ΔG −20 to 0 kJ/mol indicated primarily physisorption whereas ΔG −80 to −400 kJ/mol confirmed the chemisorption. The scheme representing the overall research work done in Fig. 4.

Chitosan-polyacrylamide-based composite hydrogels were prepared to remove direct blue and methylene blue from the aqueous solution as per the following Fig. 5. Semi-interpenetrating network (sIPN) and full interpenetrating networks (dIPN) were for a comparative study. It was observed that dIPN was able to adsorb the dyes at better fashion when compared to the former due to structural changes occurred when the transformation of amide group into carboxylate group. The dIPN obtained was a monolith with a high porosity with an increase in crosslinker ratio.

Potato starch or modified potato starch was entrapped into polyacrylamide matrix to developed semi-interpenetrating networks which was applied to study the physisorption of methylene blue confirmed from pseudo first-order kinetic model [46]. The kinetic studies indicate that the sorption of methylene blue was considerably affected by the nature of entrapped polymer into polyacrylamide. The studies

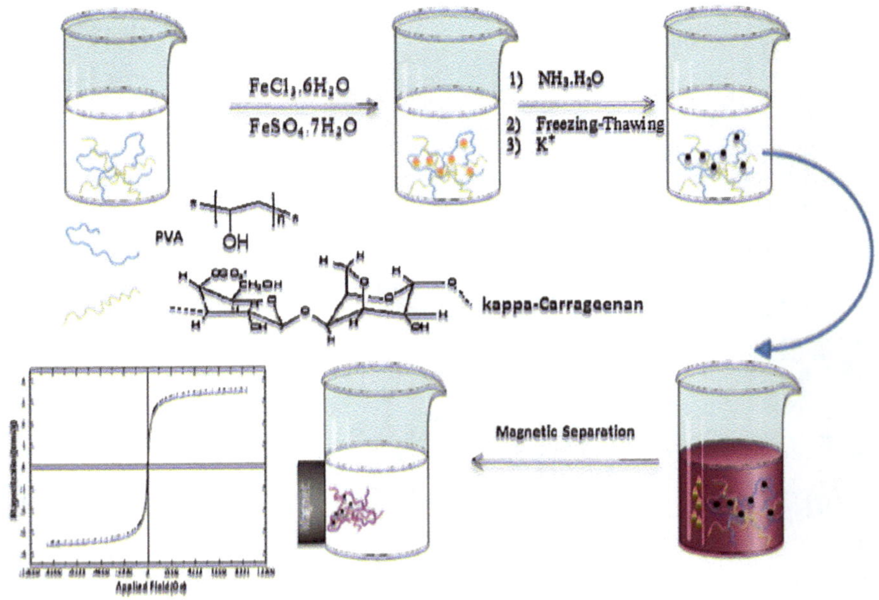

Fig. 4 Scheme showing the preparation of carrageenan-polyvinyl alcohol-iron oxide nanocomposites for the removal of crystal violet. Reprinted with permission from [44]. Copyright © 2014, Elsevier

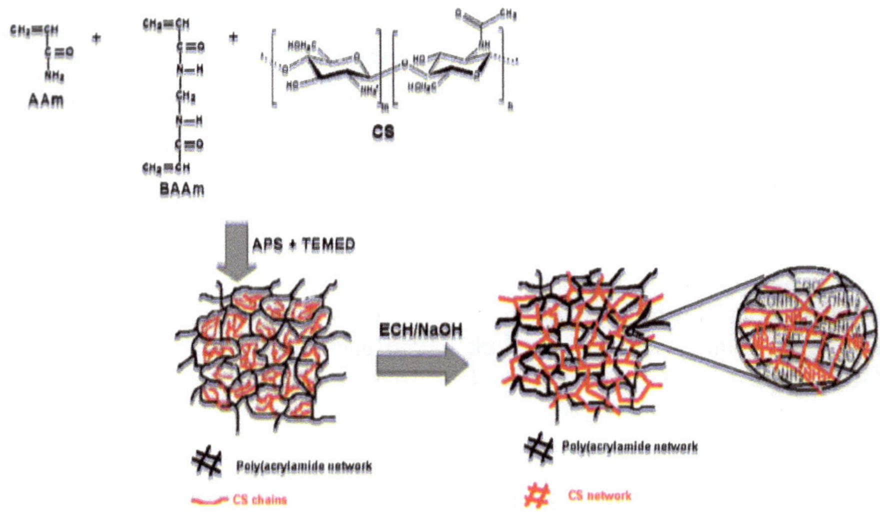

Fig. 5 Preparation of dIPNs from chitosan and polyacrylamide. Reprinted with permission from [45]. Copyright © 2011, Elsevier

were proceeded with desorption cycles that enhanced the possibility of reusability of these composite hydrogels up to four cycles without losing the adsorption efficiency.

Alginate and alginate/polyaspartate hydrogel beads were prepared for methylene blue, malachite green and methyl orange removal. A weak solute-solid interactions were confirmed from Type S isotherm with an adsorption efficiency of 300–700 mg/g of gel [47]. A comparative study of pristine cellulose with cellulose modified with glycidyl methacrylate and sulfosalicylic acid was carried for crystal violet uptake [48]. The modified hydrogel adsorbent showed better conductivity with an adsorption capacity of 218 mg/g which was found to 70 % higher than that of cellulose to remove the dye up to eight cycles.

Recently, a new biocomposite hydrogel was prepared using calcium alginate as a matrix incorporated with vineyard pruning waste. The authors recommended that it would a promising eco-friendly alternative foe the removal of organic wastes [49]. The biocomposite hydrogels were found to be stable and kept elongation with constant roundness and compactness values. These hydrogels successfully adsorb 75 % of the dyes from winery waste water. Besides this study, similar work have been done where immobilization of grape marc peat [50], saw dust [51] and sugarcane molasses [52] onto calcium alginate matrix were used to develop composite hydrogels for the pigments removal from wastewater discharged by agro industries.

Chitosan-*g*-poly (acrylic acid)/vermiculite hydrogel composites were able to adsorb methylene blue from the aqueous water. The adsorption was enhanced with pH, initial time of contact and concentration whereas surfactant concentration, ionic strength and temperature influenced negatively [53]. The electrostatic attraction between the carboxylate group present on the hydrogels composite and the dye molecules were identified as the main cause of adsorption and have the adsorption capacity of reach 1573.87 mg/g.

3.2.2 Drugs

Very few literatures have been found where hydrogel and hydrogel-based nanocomposites have been used to remove pharmaceutical drugs from waste water. In a recent work done, the authors reported the study of measured and predicted equilibrium partition coefficient of pharmaceutical drugs such as acetazolamide, caffeine, hydrocortisone, Oregon Green 488, sodium fluorescein and theophylline in hydroxyethyl methacrylate [HEMA]/methacrylic acid [MAA] [54]. The exclusion factor was found to be $E > 1$ for pure HEMA indicating a strong adsorption and the solute partitioning was influenced by properties like size exclusion and non-specific electrostatic interaction.

Xanthan gum/lignin composite hydrogels were investigated for vanillin carriers [55]. The two natural polymers were cross-linked using epichlorohydrin and were found to have high swelling degree and rate in aqueous solutions. This composite hydrogel had strong intra and intermolecular interactions with vanillin as it allowed slow desorption of about 15–18 % within 100 min attributed to the fact that content of lignin affects the release.

Chitosan-magnesium aluminium silicate nanocomposites were studied for permeability of drugs such as propranolol HCl (PPN), diclofenac sodium (DCF) and acetaminophen (ACT) [56]. It was found that the permeation of negatively charged drugs through this hydrogel composite was primarily a diffusion phenomenon whereas the positively charged drugs was able to concurrently undergo the diffusion and adsorption process.

Modified chitosan was prepared by crosslinking glutaraldehyde and sulfonate (CsSLF) or *N*-(2-carboxybenzyl) (CsNCB), groups to remove pramipexole dihydrochloride which is non-ergoline dopamine agonist from the aqueous water [57]. The adsorption capacity of (CsSLF) was found to be 367 mg/g than that of (CsNCB) 357 mg/g capacity. The interaction of the drug with the modified chitosan is shown in Fig. 6a, b. The authors also provided the comparison of their work with some earlier research shown in Table 2.

3.2.3 Herbicide, Pesticide and Petroleum Products

Rare research work is found in the field of hydrogel as an adsorbent for the removal of herbicide, pesticide or petroleum products. Recently, the authors have developed synthetic hydrogel poly(*N*-vinyl-2-pyrrolidone)/(acrylic acid-co-styrene) [PVP/(AAc-co-Sty)] hydrogels via γ-irradiation as an adsorptive systems. They used this adsorbent to decontaminate the pesticides such as fluometuron (FH), thiophanate methyl (TF) and trifluralin (TI). They reported that the nature of adsorption was influenced by hydrophobic interactions and hydrogen bonds between the hydrogel and the pesticide molecules. The maximum adsorption was observed at a low pH with rise in temperature [61].

Gum tragacanth-acrylic acid-based hydrogel was used to uptake crude oil via adsorption process that had high adsorption rate at pH 3 with 40 % acrylic acid content [62].

3.2.4 Miscellaneous

Hydrogel, functionalized hydrogels and hydrogel composites have been extensively exploited for the uptake of inorganics especially the heavy metal ions from the aqueous solutions. This section will describe some recent research done in this area.

Graft copolymerization was carried out using acrylic acid and xylan rich hemicellulose to develop an adsorbent that could behave as potential candidate for heavy metal removal [63]. Lead, cadmium and zinc ions were the target cations for this modified hydrogel that showed an adsorption capacity of 859, 495 and 274 mg/g, for each of these cations, respectively. Table 3 represents the adsorption/desorption capacity of this hydrogel for heavy metal ions.

Potato starch graft polyamidooxime embedded in chitosan beads was prepared to remove copper ions from aqueous solution [64]. The authors used two different strategies to develop the candidate hydrogel. The first approach was mixing of

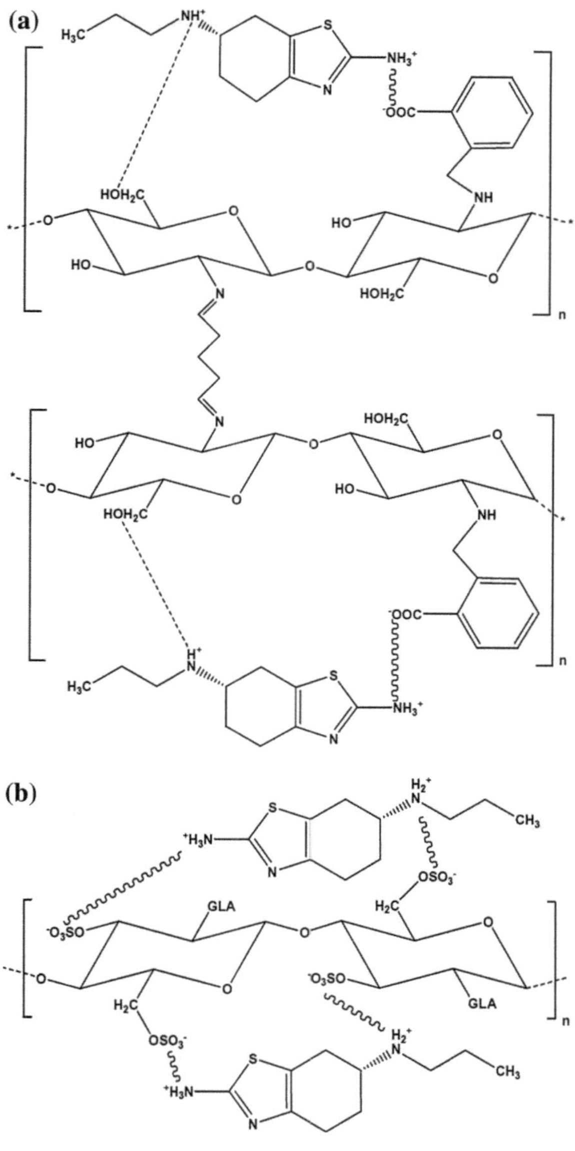

Fig. 6 **a** Interaction of PRM drug with (CsNCB). Reprinted with permission from [57]. Copyright © 2013, Elsevier. **b**. Interaction of PRM drug with (CsSLF). Reprinted with permission from [57]. Copyright © 2013, Elsevier

previously prepared poly(amidoxime) grafted on potato starch in the CS solution which was proceeded with the bead formation and the next approach thorough via mixing the potato starch-g poly(acrylonitrile) (PS-g-PAN) copolymer in the initial CS solution, followed by bead formation. The maximum equilibrium sorption capacity of this hydrogel was found 133.15 mg Cu^{2+}/g for the hydrogel prepared by the first route while the second route yielded hydrogel that showed adsorption efficiency 238.14 mg Cu^{2+}/g.

Table 2 Comparison of adsorption capacities hydrogel composite with other Adsorbents

Adsorbent compound	Drug/adsorption capacity Qm (mg/g)	References
Mesoporous silica SBA-15	Carbamazepine (0.16)	[58]
Mesoporous silica SBA-15	Clofibric acid (0.07)	[58]
Mesoporous silica SBA-15	Diclofenac (0.34)	[58]
Mesoporous silica SBA-15	Ibuprofen (0.41)	[58]
Mesoporous silica SBA-15	Ketoprofen (0.28)	[58]
High-silica zeolite HSZ-385	Sulfamethoxazole (237)	[59]
High-silica zeolite HSZ-385	Sulfathiazole (402)	[59]
High-silica zeolite HSZ-385	Sulfamerazine (302)	[59]
High-silica zeolite HSZ-385	Sulfamethizole (267)	[59]
High-silica zeolite HSZ-385	Sulfadimidine (278)	[59]
Activated carbon	Atenolol (130)	[60]
Activated carbon	Diclofenac (280)	[60]
Non-grafted cross-linked chitosan (Cs)	Pramipexole (181)	[57]
N-(2-carboxybenzyl) grafted chitosan (CsNCB)	Pramipexole (307)	[57]
Sulfonate grafted chitosan (CsSLF)	Pramipexole (337)	[57]

Reprinted with permission from [57]. Copyright © 2013, Elsevier

Table 3 Metal ion sorption and desorption study by xylan type hemicellulose grafted acrylic acid hydrogel

Metal ion	Cycle 1		Cycle 2		Cycle 3		Cycle 4		Cycle 5	
	A[a] (mg/g)	R[b] (%)	A (mg/g)	R (%)	A (mg/g)	R (%)	A (mg/g)	R (%)	A (mg/g)	R (%)
Pb^{2+}	876	98.9	850	99.0	831	98.6	810	97.7	774	96.7
Cd^{2+}	510	99.1	496	98.7	484	98.1	467	97.2	441	96.4
Zn^{2+}	261	99.0	260	98.8	254	98.5	241	98.2	230	97.6

Reprinted with permission from [63]. Copyright © 2012, American Chemical Society
[a]Adsorption capacity of the metal ion, [b]Recovery rate (desorption) of the metal ion

Response surface methodology was used to prepare noval magnetic hydrogel beads with an average diameter of 3.4 mm based on calcium alginate and maghemite with an adsorption capacity of 159.24 mg/g for 500 mg/L initial Cu(II) ion concentration for a period of 6 h [65].

In another fascinating work, calcium alginate/polyacrylamide hydrogel was used for nano-filtration membrane via photo-induced copolymerization. Membranes show the prominent antifouling properties that efficiently resisted the adsorption of proteins [66].

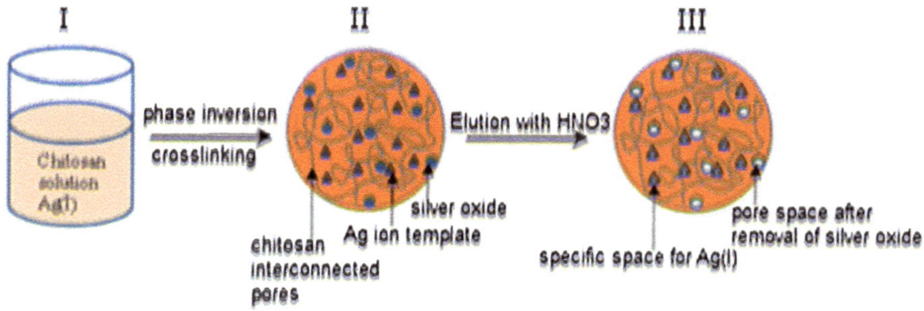

Fig. 7 Scheme representing the formation of molecular ion imprinted chitosan. Reprinted with permission from [67]. Copyright © 2014, American Chemical Society

In another study silver ions were adsorbed from aqueous solution by molecular ion imprinted hydrogels of chitosan that had high adsorption capacity [67]. The scheme shown in Fig. 7 represents the mechanistic path way of the methodology used for the preparation.

4 Summary

The chapter delivers a brief overview in the field of biopolymer-based hydrogels, modified or functionalized biopolymer hydrogels and composite material that have been used for the decontamination of organic wastes. There are various industries that discharge gallons of wastewater enriched with organic pollutants and with growing demand of market with such products, it is estimated there will a quantum leap in the rise of discharged wastes. The organic wastes are predominantly generated by paper making, dye, textile, pharmaceutical and nutraceutical, agro-based and petroleum industries. The important aspect lies in the lack of treatment setup for the wastewater discharge by most of these industries.

Researchers across the world have investigated all the possible treatment methods based on physicochemical, chemical, photocatalytic and biological process to combat the challenge. This chapter dealt with the research work done in the area of biopolymer-based hydrogels and its application for decontamination of organic waste. After extensive literature research done on the recent work in this area suggested that scientists have carried out maximum work in the field of organic dyes followed by heavy metal removal by hydrogels. The ultimate objective in most of the studies was adsorption efficiency of hydrogels. Comparison of the natural polymer-based hydrogels with other potential adsorbents suggested that choice biopolymer hydrogels has the ability to lead the eco-friendly steps towards environment protection.

However, the natural hydrogels has been in demand by the biomedical science especially for developing drug formulation for targeted and slow release, tissue regeneration and molecular engineering.

5 Future Scope

Natural polymers have been the choice of scientists across the world where these have been extensively investigated for possible application in various fields. The biocompatibility, biodegradability, water retention property and supra adsorbent behaviour prioritize its utility for lab scale, pilot scale and industrial scale research. However, the researchers have not explored its capability for decontaminating organic waste discharged from many industries.

The future scope of this work shall focus on the investigation of natural polymers for removal of organics released from as drugs, petroleum products, pesticide and herbicides, etc. The other major possibility which may play a huge role for these biopolymer-based hydrogels is its use to develop catalysts and photocatalysts so that dual benefit of adsorption and photodegradation enhances the removal capacity.

It is important to make a note that any research with commercial application plays a pivotal role for the upliftment of social causes that are directly related to environmental protection and sustainability. To address this, there is a need to utilize these hydrogels that can participate as potential commercial method of water treatment.

Acknowledgements The authors acknowledge University of South Africa for providing research facilities and platform.

References

1. Smulders E, Rybinski W, Sung E, Rähse W, Steber J, Wiebel F, Nordskog A (2002) Laundry detergents. In: Ullmann's encyclopedia of industrial chemistry. Wiley-VCH, Weinheim
2. Yangxin Y, Zhao J, Bayly AE (2008) Chin J Chem Eng 16:517–527
3. Dooms-Goossens A, Blockkeel I (1996) Clinics Dermat 14:67–76
4. Pattusamy V, Nandini N, Bheemappa K (2013) Int J Adv Res 1:129–133
5. Camacho-Muñoz D, Martín J, Santos JL, Aparicio I, Alonso E (2014) Sci Total Environ 468–469:977–984
6. Tian F-X, Xu B, Zhang T-Y, Gao N-Y (2014) Chem Eng J 258:210–217
7. Orellana-Garcia F, Alvarez MA, Lopez-Ramon MV, Rivera-Utrilla J, Sancehz-Polo M (2015) Chem Eng J 267:182–190
8. Plakas KV, Karabelas AJ (2008) J Memb Sci 320:325–334
9. Bayen S, Zhang H, Desai MM, Ooi SK, Kelly BC (2013) Environ Pollut 182:1–8
10. Hirai N, Nanba A, Koshio M, Kondo T, Morita M, Tatarazako N (2006) Aquat Toxicol 79:288–295

11. Alvanaja MCR, Ross MK, Bonner MR, Cancer CA (2013) J Clin 63:120–142
12. Benner J, Helbling DE, Kohler H-PE, Wittebol J, Kaiser E, Prasse C, Ternes TA, Albers CN, Aamand J, Horemans B (2013) Water Res 47:5955–5976
13. Helbling DE (2015) Curr Opin Biotechnol 33:142–148
14. Deng S, Chen Y, Wang D, Shi T, Wu X, Ma X, Li X, Hua R, Tang X, Qing LX (2015) J Hazard Mater [In Press]
15. Mertxell G, Mira P, Damia B (2009) Anal Chem 81:898
16. Togunde OP, Oakes KD, Servos MR, Pawliszyn J (2012) Environ Sci Technol 46:5302
17. Khan Z, Jain K, Soni A, Madamwar D (2014) Int Biodeterior Biodegrad 94:167–175
18. Balapure K, Nikhil B, Madamwar D (2015) Bioresour Technol 175:1–7
19. Cui D, Guo YQ, Lee HS, Wue WM, Liang B, Wanga AJ, Cheng HY (2014) Bioresour Technol 163:254–261
20. Chen C, Wei L, Guo X, Guo S, Yan G (2014) Fuel Process Technol 124:165–173
21. Levchuk I, Bhatnagar A, Sillanpää M (2014) Sci Total Environ 476–477:415–433
22. Rubio-Clemente A, Torres-Palma RA, Peñuela GA (2014) Sci Total Environ 458:201–225
23. Shibayama M, Tanaka T (1993) Adv Polym Sci 109:1–62
24. Khokhlov A, Starodubtzev S, Vasilevskaya VV (1993) Adv Polym Sci 109:123–172
25. Marsich E, Mozetic P, Ortolani F, Contin M, Marchini M, Vetere A, Pacor S, Semeraro S, Vittur F, Paoletti S (2008) Matrix Biol 27:513–525
26. Marsich E, Borgogna M, Donati I, Mozetic P, Strand BL, Salvador SG, Vittur F, Paoletti S (2008) J Biomed Mater Res Part A 84:364–376
27. Tømmeraas K, Köping-Höggård M, Vårum KM, Christensen BE, Artursson P, Smidsrød O (2002) Carbohydr Res 337:2455–2462
28. Chang C, Zhang L (2011) Carbohydr Polym 84:40–53
29. Sannino A, Demitri C, Madaghiele M (2009) Materials 2:353–373
30. Copyright © (2015) The Freedonia Group, Inc.,767 Beta Drive Cleveland Ohio 44143
31. Xu D, Hein S, Loo LS, Wang K (2011) Ind Eng Chem Res 50:6343–6346
32. Wang H, Xu W, Song S, Feng L, Song A, Hao J (2014) J Phys Chem B 118:4693–4701
33. Bhattacharyya R, Ray SK (2015) J Ind Eng Chem 22:92–102
34. Bhattacharyya R, Ray SK (2013) Poly Eng Sci 53:2439–2453
35. Banat I, Nigam P, Singh D, Marchant R (1996) Bioresource Technol 58:217–227
36. Wang Y, Zeng L, Ren X, Song H, Wang A (2010) J Environ Sci 221:7–14
37. Gupta VK, Mittal A, Gajbe J, Mittal J (2008) Colloid Inter Sci 319:30–39
38. Wang J, Wang H, Song Z, Kong D, Chen X, Yang Z (2010) Colloid Surface B 80:155–160
39. Solpan D, Duran S, Saraydin DO (2003) Guven. Radiat Phys Chem 66:117–127
40. Solpan D, Kolge Z (2006) Radiat Phys Chem 75:120–128
41. Sharma K, Kaith BS, Kumar V, Kalia S, Kumar V, Swart HC (2014) Geoderma 232–234:45–55
42. Zhou C, Wua Q, Lei T, Negulescu II (2014) Chem Eng J 251:17–24
43. Mahdavinia GR, Massoudi A, Baghban A, Massoumi B (2012) Iran Polym J 21:609–619
44. Mahdavinia GR, Massoudi A, Baghban A, Shokri E (2014) J Environ Chem Eng 2:1578–1587
45. Dragan ES, Perju MM, Dinu MV (2012) Carbo Polym 88:270–281
46. Dragan ES, Apopei DF (2011) Chem Eng J 178:252–263
47. Jeon YS, Lei J, Kim J-H (2008) J Ind Eng Chem 14:726–731
48. Zhou Y, Zhang M, Wang X, Huang Q, Min Y, Ma T, Niu J (2014) Ind Eng Chem Res 53:5498–5506
49. Vecino X, Devesa-Reya R, Cruz JM, Moldes AB (2015) Carbo Polym 115:129–138
50. Perez-Ameneiro M, Vecino X, Barbosa-Pereira L, Cruz JM, Moldes AB (2014) Carbo Polym 101:954–960
51. Vecino X, Devesa-Rey R, Cruz JM, Moldes AB (2013) Water Air Soil Pollut 224:1–9
52. Gonte RR, Shelar G, Balasubramanian K (2014) Desalin Water Treat 52:7797–7811
53. Liu Y, Zheng Y, Wang A (2010) J Environ Sci 22:486–493
54. Dursch TJ, Taylor NO, Liu DE, Wu RY, Prausnitz JM, Radke CJ (2014) Biomaterials 35:620–629

55. Raschip IE, Hitruc EG, Oprea AM, Popescu MC, Vasile C (1003) J Mol Str 2011:67–74
56. Khunawattanakul W, Puttipipatkhachorn S, Rades T, Pongjanyakul T (2010) Int J Pharma 393:219–229
57. Kyzas GZ, Kostoglou M, Lazaridis NK, Lambropoulou DA, Bikiaris DN (2013) Chem Eng J 222:248–258
58. Bui TX, Choi H (2009) J Hazard Mater 168:602–608
59. Fukahori S, Fujiwara T, Ito R, Funamizu N (2011) Desalination 275:237–242
60. Sotelo JL, Rodríguez AR, Mateos MM, Hernández SD, Torrellas SA, Rodríguez JG (2012) J Environ Sci Health Part B Pesticides Food Contam Agric Wastes 47:640–652
61. Abd El-Mohdy HL, Hegazy EA, El-Nesr EM, El-Wahab MA (2012) J Macro Sci Part A Pure Appl Chem 49:814–827
62. Saruchi BS, Jindal KR, Kumar V (2015) Petro Sci Tech 33:278–286
63. Peng X-W, Zhong L-X, Ren J-L, Sun R-C (2012) J Agric Food Chem 60:3909–3916
64. Dragan ES, Loghin DFA, Cocarta AI (2014) Appl Mater Interfaces 6:16577–16592
65. Zhu H, Fu Y, Jiang R, Yao J, Xiao L, Zeng G (2014) Ind Eng Chem Res 53:4059–4066
66. Zhang X, Lin B, Zhao K, Wei J, Guo J, Cui W, Jiang S, Liu D, Li J (2015) Desalin 365:234–241
67. Song X, Li C, Xu R, Wang K (2012) Ind Eng Chem Res 51:11261–11265

Chitosan and Starch-Based Hydrogels Via Graft Copolymerization

Annamaria Celli, Magdy W. Sabaa, Alummoottil N. Jyothi and Susheel Kalia

Abstract Graft copolymerization is an attractive method for surface functionalization of natural polymers and can be initiated by chemical methods, radiation technique, and other systems. Polymer grafting onto polysaccharides is an effective method for the synthesis of superabsorbents. Depending upon the type of monomers and the conditions employed the properties of graft copolymers vary to a large extent. Chitosan is a nontoxic, biocompatible polysaccharide, and starch is a natural hydrophilic biopolymer. Both these are most abundant natural organic materials which are extensively investigated in the development of biodegradable and environment-friendly materials. Their hydrogels are of utmost importance for wide use in many fields including structural transplants, target drug delivery, tissue engineering, biosensors, adsorbents, etc. In this chapter, the various techniques used for the synthesis of chitosan/starch graft copolymers, their properties and possible applications are discussed in detail.

Keywords Chitosan · Starch · Graft copolymers · Hydrogel · Ion exchangers · Biomedical applications

A. Celli
Department of Civil, Chemical, Environmental and Materials Engineering, University of Bologna, Via Terracini 28, 40131 Bologna, Italy

M.W. Sabaa
Chemistry Department, Faculty of Science, Cairo University, Giza 12613, Egypt

A.N. Jyothi
Division of Crop Utilization, ICAR—Central Tuber Crops Research Institute, Sreekariyam, Kerala 695017, India

S. Kalia (✉)
Department of Chemistry, ACC Wing, Indian Military Academy, Dehradun, Uttarakhand 248007, India
e-mail: susheel.kalia@gmail.com

S. Kalia (ed.), *Polymeric Hydrogels as Smart Biomaterials*, Springer Series on Polymer and Composite Materials, DOI 10.1007/978-3-319-25322-0_8

1 Introduction

Polysaccharides are the basic raw materials for fabrication of micellar carrier because of their excellent properties like biocompatibility, hydrophilic, and biodegradable nature [1]. Lack of these properties can prevent a biomaterial from performing its function and often trigger undesirable side effects. Recently developed biomaterials can actively interact with tissues and organs in comparison to the earlier biomaterials [2, 3]. Polysaccharides-based hydrogels find use in many fields including structural transplants, target drug delivery, tissue engineering, antibacterial agents, biosensors, adsorbents to their excellent properties [4–8].

Chitosan is a cationic polysaccharide and excellent properties of chitosan are being used for controlled releasing of drugs by hydrogels devices [9, 10]. Starch is a natural hydrophilic polysaccharide and a valuable biodegradable material because of its excellent properties. Hydrolysis product of starch and its derivatives are well established in the food and technical industries. Various strategies have been developed for polymer grafting onto starch backbone [11, 12]. Caprolactone was grafted on the active sites of starch using ring-opening polymerization to increase the miscibility of starch and polycaprolactone [13]. Graft copolymers of polysaccharide have been widely used in diversified fields such as in gene delivery, selective water absorption, drug carrier, composites, water purification, etc. Polysaccharides such as starch, cellulose, hydroxyethyl cellulose, sodium aliginate, and partially carboxymethylated guar gum have been grafted copolymerized into water-absorbing polymers [14]. This chapter describes the concept of graft copolymerization and synthesis of chitosan and starch-based hydrogels via graft copolymerization for high-performance applications.

2 Graft Copolymerization

2.1 Concept of Graft Copolymerization

Graft copolymerization is an attractive method to combine properties of two or more polymers in one entity. In graft copolymers, indeed, branches develop from a main backbone (Fig. 1), and main and side chains, which may be constituted by homopolymeric and copolymeric units, differ by chemical structure or composition [15]. Graft copolymers are characterized by branches randomly distributed along the main chain, with a low density, and can be considered a special case of comb-shaped polymers, which have backbone and side chains of the same chemical nature. There is also a special case of cylindrical polymer brush where the density of the branches is very high.

In graft copolymers the chemical nature of the backbone and branches and their structural characteristics, such as molecular weight, number and spatial distribution of the branches, are fundamental to determine some specific behaviors, i.e., in bulk and in solution. More specifically, topological constrains prevent macromolecular

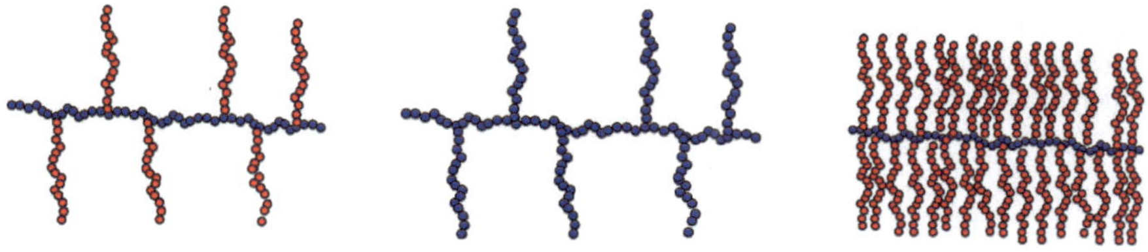

Fig. 1 Scheme of branched polymeric architecture

chain motions [16] and the lack of compatibility between polymers having different chemical characteristics causes repulsions and separation of phases. Some uses of graft copolymers exploit this incompatibility between backbone and side chains: for example, they are used to compatibilize polymer blends, as they tend to settle at the interface between two polymers or other materials, according to their chemical characteristics, improving the interconnections [17]. For example, polypropylene-graft-poly(ε-caprolactone) is a good compatibilizer for blends between PP and many engineering plastics, such as polycarbonate and poly(vinyl chloride) [18].

Moreover, graft copolymers are often amphiphilic materials, which contain both hydrophobic and hydrophilic parts connected each other. In this case, as backbone and grafts are characterized by different solubility, they can be used as emulsifiers, adhesives, moisturizers, or water repellents. In the presence of solvent which exhibits preference for the chemical units present in the grafts, they can assume micellar-type conformations. In particular, in the last few years a notable interest of academic research has been addressed to the micellization behavior of amphiphilic graft copolymers [19, 20], for example, to design drug delivery systems [21–23]. Moreover, the capability of graft copolymers to modify surface properties of materials found numerous applications in biomedical field [24, 25].

Then, the macromolecular architecture must be carefully controlled to obtain successful applications in the different fields. Various copolymerization methods have been developed and now some structural parameters, such as length of main and side chains, can be well adjusted. Otherwise, the control of number and distribution of grafts along the chains is still a challenge.

The synthetic methodologies to prepare graft copolymers have been extensively studied and reviewed [15, 26–30]. These three are the methods that have been optimized with the aim of preparing randomly branched graft copolymers: (a) the "grafting onto," (b) the "grafting from," and (c) the "grafting through" or macromonomer method [31] (Fig. 2). The "grafting onto" method consists of the exploitation of functionalities X that are randomly distributed along the backbone chain. X groups are able to couple with reactive chain ends Y of polymeric branches. However, the incapability between polymers of different chemical nature (backbone and side chains) limits the exact control of the reaction progress, together with the restricted accessibility of the functional groups along the main chain. Nevertheless, this method, first used for the preparation of graft copolymers by ionic living polymerization, has been applied to other copolymerization pathway.

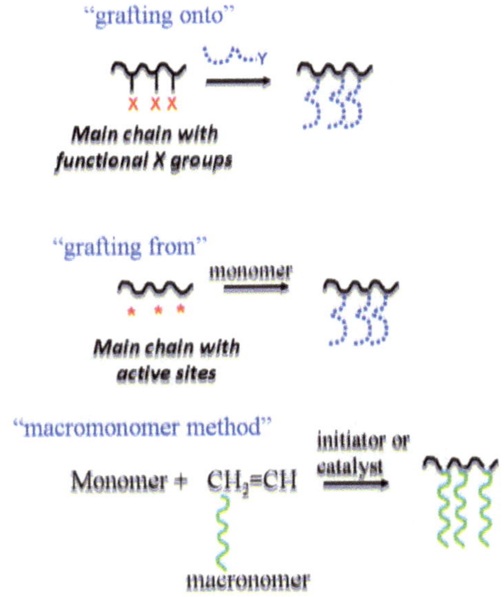

Fig. 2 Scheme of different methods of graft copolymerization

In the "grafting from" method, instead, active sites are generated randomly along the backbone. From these sites the polymerization of a second monomer can initiate creating branches. This method can be applied to a wide range of monomers and polymerization reactions. However, the length and number of grafts are not accessible.

The "grafting through" or macromonomer method is widely used for the preparation of graft copolymers [32]. Macromonomer is an oligomer or a polymer characterized by the presence of a polymerizable chain end, which is used for a polymerization with a low molecular weight monomer. Graft copolymers are then formed. Based on this macromonomer approach it is possible to prepare graft copolymers with different structures, chemical compositions, and properties.

The three approaches described in Fig. 2 are characterized by reactions that can proceed by different mechanisms, such as ionic, radical, etc., as explained in the following paragraph. Moreover, some of these methods, mainly "grafting from," is used for graft copolymerization initiated by photoirradiation or high-energy irradiation techniques. These processes, which are applicable to a large number of polymers, can take place in solution, in bulk or on surfaces.

2.2 *Methods of Graft Copolymerization*

Various polymerization methods are used to produce graft copolymers. In the following a rapid review of the most used methods are reported.

2.2.1 Chemical Methods

Ionic Grafting

Grafting can proceed through an ionic mechanism, involving either cationic or anionic species. In "grafting from method" a large variety of graft copolymers have been obtained over the years via anionic polymerization. The method consists of creating anionic sites along a polymeric backbone. Grafts are built on these sites, thanks to the polymerization of another monomer.

Anionic active sites can be prepared through a metalation of allylic, benzylic, or aromatic C–H bonds, present in the backbone, by using organometallic compounds, such as *s*-BuLi, in the presence of chelating agents that facilitate the reaction.

The metalation of polydienes with *s*-BuLi in the presence of *N*,*N*,*N*′,*N*′-tetramethylethylenediamine (TMEDA) is a significant example. Metalation of polyisoprene (PI) and polybutadiene (PBd) by following this strategy and subsequent polymerization of styrene induce the formation of PI-*g*-PS and PBd-*g*-PS copolymers with well-defined molecular characteristics [28, 29].

In another approach, a polymer, such as PMMA treated with the 18-crown-6 complex of potassium hydroxide to produce carboxylate-active groups, is an anionic macromolecular initiator for ring-opening polymerization (ROP). By reaction with β-butyrolactone in THF at room temperature PMMA-*g*-poly(β-butyrolactone) copolymers were prepared. Grafting efficiency is notably high and the density of grafted chains can be easily controlled [33].

This is an example of graft copolymer synthesis by RPO with anionic mechanism. This approach actually is finding numerous applications. As example, the preparation of graft copolymers characterized by a hydrophobic backbone and water-soluble grafts has been the object of notable research [34]. Graft copolymers with EVOH and polyamide 12 backbone and PEO side chains were prepared by ionization of amide and hydroxyl groups in the main chains by appropriate solvent and catalyst. They served in the following step as initiating sites for the anionic polymerization of ethylene oxide [35].

In the "grafting onto" method a living polymerization mechanism is used to prepare backbone and arms separately. The backbone bears functional groups distributed along the chain that can react with other polymeric chains bearing active chain ends. The functions may already exist along the main chains or can be created by chemical modifications. Backbone and branches are mixed in the desired proportion and under the appropriate experimental conditions, and then the coupling reaction occur and result in the final copolymers.

For example, poly(butadiene-*g*-styrene) graft copolymers are prepared by hydrosilylation of the backbone to introduce chlorosilane groups [36]. Final materials are characterized by a high molecular weight and compositional homogeneity.

Grafting can also proceed through cationic mechanism. For example, tri-alkyl aluminum (R_3Al) and the polymeric backbone in the alide form (P-Cl) react to form carbonium ions along the chain, which can copolymerize [27]:

$$P\text{-}Cl + R_3Al \rightarrow P^+R_3AlCl^-$$
$$P^+ + M \rightarrow PM^+ \rightarrow \text{graft copolymer}$$

Anionic and cationic polymerizations are characterized by the ability to form a "living" polymerization system. There are not specific chain terminator or chain transfer reactions and the growth of the chain is limited only by the amount of the monomer in the system. Therefore, this copolymerization method allows one to control molecular weight and polydispersity index. However, for the classical living polymerization techniques a poor number of monomers can be used as they are sensitive to the most functional groups such as halogens, nucleophilic groups, acidic protons, etc., existing on most vinyl monomers. Moreover, pure solvents and monomers, highly reactive initiators, anhydrous conditions, and low polymerization temperatures are necessary.

Free Radical Grafting

Methods involving classical free radical grafting techniques are very common procedures for the synthesis of graft copolymers [26, 27].

A conventional method to produce free radicals is through redox reactions, which have the advantages of being simple to be carried out, feasible at room temperature and also in aqueous solution, and which produce an extent of grafting controllable by tuning the reaction variables (mixture composition, reaction and temperature, and time).

An example of redox initiator is the Fenton's reagent (Fe^{2+}/H_2O_2), which generates an OH·radical:

$$Fe^{2+} + H_2O_2 \rightarrow Fe^{3+} + OH^- + OH^{\cdot}$$

The hydroxyl radical is able to abstract the H atom from the polymer (P–H), producing the free radical on it (P·). The monomer molecules (M) are radical acceptor and, thus, can be chain initiator for the growth of the graft chains:

$$OH^{\cdot} + PH \rightarrow P^{\cdot} + H_2O$$
$$\text{Initiation: } P^{\cdot} + M \rightarrow P - M^{\cdot}$$
$$\text{Propagation: } P - M^{\cdot} + M \rightarrow P - M_2\cdot$$

The same mechanism, involving the reactions between radicals and polymeric substrate, is found with other redox initiator systems, such as persulfate/Fe^{2+}, persulfate and other reducing agents, such as Ag^+, hydroperoxides/Fe^2. Moreover, also through direct oxidation by transition metal ions with low oxidation potential (e.g., Ce^{4+}, Cr^{6+}, V^{5+}, Co^{3+}) free radical sites are prepared on the polymer backbone.

The use of other redox initiator systems, like metal carbonyls and metal chelates, has also been reported for chemical grafting. Moreover, certain organic compounds (azo, peroxides, hydroperoxides) produce free radicals by heating and transferring them to the main chain. Chemical pretreatments (e.g., ozonation, diazotization, xanthation) and ionizing radiations (gamma rays, X-rays, e-beams) in the presence of air, ozone, UV, and free radical initiators or peroxide groups introduced on the parent polymer have been used.

Radical process has significant advantages, such as to be tolerant of functional groups and impurities can be carried out in wide variety of solvents and to be the leading industrial method to produce polymers. Free radical functionalization of polyolefins is the most studied case. However, conventional methods do not exhibit the basic characteristics of living polymerization such as control over polymer structure and molecular weight due to diffusion-controlled chain terminator and transfer reactions. Cross-linking due to radical recombination, degradation of backbone can also occur.

The introduction of the controlled radical polymerization (CRP) in the 1990s was an important scientific development, which combines the advantages of free radical polymerization with that of living polymerization approaches and opens new important perspectives in design and synthesis of complex branched macromolecular architectures. All the CRP methods are based on finding the conditions for a fast dynamic equilibrium between a small amount of growing free radicals and a large majority of the dormant species. There are currently three main types of living free radical polymerization methods: nitroxide-mediated polymerization (NMP) uses reversible chain termination by means of exchange of a stable radical, a nitroxide group [37–40]. Reversible addition–fragmentation chain transfer (RAFT) polymerization involves reversible chain transfer, rather than chain termination [41–44]. Atom transfer radical polymerization (ATRP) is probably the most efficient method to control the structural parameters of the graft copolymers [24, 30, 45]. In ATRP the radicals, or the active species, are generated through a reversible redox process catalyzed by a transition metal complex (Mt *n*-Y/Ligand, where Y may be another ligand or the counter ion) which undergoes one-electron oxidation with concomitant abstraction of a (pseudo)halogen atom, X, from a dormant species, R–X (Fig. 3).

Polymer chains grow by the addition of the intermediate radicals to monomers in a manner similar to a conventional radical polymerization. Termination reactions also occur in ATRP, mainly through radical coupling and disproportionation; however, in a well-controlled ATRP, a few percent of the polymer chains undergo termination.

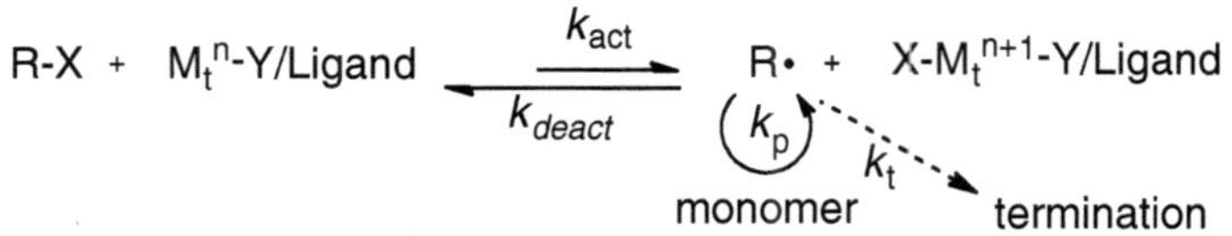

Fig. 3 Mechanism of ATRP. Reprinted from [30], copyright 2001, with permission from American Chemical Society

A successful ATRP will have not only a small contribution of terminated chains, but also a uniform growth of all the chains.

The "grafting from" and "grafting through" methods have been used in conjunction with ATRP to prepare graft copolymers and underscore the versatility of this controlled radical polymerization technique to synthesize a variety of copolymers.

An early example of graft copolymers utilizes the ATRP of vinyl monomers from pendant-functionalized poly(vinyl chloride) (PVC) macroinitiators. The aim was to insert another monomer into the PVC matrix to reduce the brittle characteristics of that polymer. The problem is not to overcome with plasticizers that suffer from some problems such as leaching and phase separation [46, 47]. More recently, Tizzotti et al. [48] in review article covered the literature regarding ATRP graft copolymerization of polysaccharides along with other living polymerization techniques like RAFT and NMP.

Other Chemical Reactions

Graft copolymerization by using ring-opening polymerization (ROP) has gained significant interest in recent years, mainly to prepare amphiphilic copolymers, which are often biodegradable. They may be used as nanocarriers in biomedical applications, such as drug delivery. ROP is a well-established technique to polymerize cyclic monomers such as lactones and lactides. ROP operates through different mechanisms depending on which monomer, initiator, and catalytic system are used. The graft copolymerization by ROP can proceed in homogeneous and heterogeneous conditions by following "grafting to" or "grafting from" mechanisms (Fig. 4) [17].

An alcohol (or hydroxyl group) is generally used as the initiator for ROP: for this reason, for example, ROP is especially interesting for the graft copolymerization of cellulose, which contains numerous hydroxyl groups along the chain. This is an interesting example of "grafting from" procedure in heterogeneous conditions. Indeed, no chemical treatment of the cellulose is necessary prior to the grafting

Fig. 4 Grafting of cellulose with PCL. Reprinted from [17], copyright 2012, with permission from Elsevier

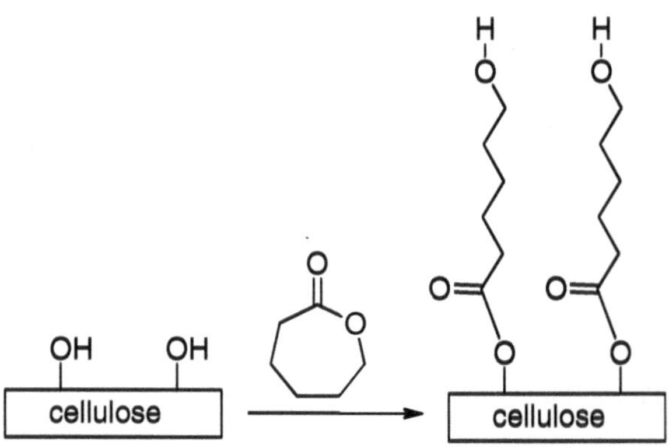

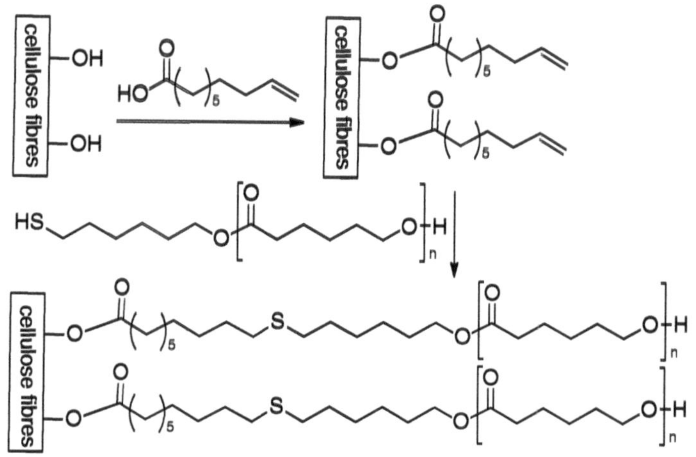

Fig. 5 "Grafting to" via thiol–ene "click" chemistry of PCL onto cellulose. Reprinted from [17], copyright 2012, with permission from Elsevier

reaction [49–51], even if literature reports also approaches where the number of available hydroxyl groups is increased by reaction with specific molecules [52].

Other significant example is the production of PP-graft-poly(ε-caprolactone) copolymers, obtained by two steps: preparation of hydroxylate PP, containing primary or secondary OH groups, followed by anionic ROP of ε-caprolactone [53].

Moreover, thiol–ene chemistry can be considered an eco-friendly route due to the absence of metal catalysts or solvents. Example of using "grafting to," modification of native cellulose by thiol–ene, "click" chemistry in combination with organocatalysis has been reported by Zhao et al. [54]. Initially, the cellulose surface was modified by reaction of 9-decenoic acid in the presence of tartaric acid, and subsequently further reacted with several different molecules with thiol end groups, including PCL (Fig. 5) [17].

2.2.2 Initiation by Irradiation

Graft copolymers can be prepared also by irradiation methods [55–57]. The active sites in the polymer backbone can be obtained by several routes, such as high-energy radiation, ultraviolet (UV) light radiation [58], plasma treatment [59].

Using High-Energy Radiation

High-energy radiation-induced graft copolymerization (RIGC) method has been widely investigated to obtain a bulk modification of polymer films unlike plasma-induced and UV-induced graft copolymerization, which produce only surface modification in polymers.

RIGC is able to induce polymerization in a wide range of temperatures including low regions, in various states of monomers such as in bulk, solution, and emulsion and even at solid state. Additives and catalysts are not needed to initiate the grafting mechanism, which can proceed by production of radicals, cations, and anions. Various types of high-energy radiations can be used, classified into (i) electromagnetic radiations (photons), such as γ-rays and X-rays, and in (ii) particulate radiation (charged particles), such as electrons and swift heavy ions. All these radiations are produced by already available commercial sources. The main source of γ-radiation is the radioactive isotopes such as ^{60}Co and ^{137}Cs. Commercial accelerators (electron beam machines) produce particulate radiations such as electrons. Swift heavy ions are produced by ion accelerators (heavy ion beam machines). Many ions are available, ranging from hydrogen and helium up to ions of heavy elements such as ^{197}Au, ^{208}Pb, ^{209}Bi, and ^{238}U. However, grafting initiated by swift heavy ions is different from that one initiated by -rays or electron beam due to the high electronic stopping power of the particles. Moreover, the values of the grafting yield and the molecular mass distribution vary depending on the kind of ion used.

The amount of the grafted moiety can be controlled by varying the irradiation parameters. Indeed, during irradiation polymers can be subjected to molecular changes, such as chain cross-linking, which causes an increase in the polymer viscosity by the formation of a microscopic network structure. Otherwise, chain scission can take place, inducing a decrement in the molecular weight and substantially changing the polymer material properties. Moreover, small molecule can be produced and structural changes in the polymer can occur. The extent of each one of these reactions depends on the chemical nature of the polymer and irradiation conditions, i.e., irradiation dose and dose rate.

These two are the main methods for RIGC: (1) simultaneous irradiation and (2) pre-irradiation, which can be performed either in the presence of air or under vacuum.

In simultaneous irradiation a polymer backbone is irradiated in the presence of a monomer. The formation of active free radicals may occur both in the polymer and in the monomer. Radicals on polymer backbone induce the growth of side chains, according to the following scheme:

$$\text{Irradiation: } \mathrm{P} \xrightarrow{\gamma\text{-rays}} \mathrm{P}^{\cdot}$$

$$\text{Initiation: } \mathrm{P}^{\cdot} + \mathrm{M} \rightarrow \mathrm{PM}^{\cdot}$$

$$\text{Propagation: } \mathrm{PM}^{\cdot} + n\mathrm{M} \rightarrow \mathrm{PM}^{\cdot}_{n+1}$$

$$\text{Termination: } \mathrm{PM}^{\cdot}_{n} + \mathrm{PM}^{\cdot}_{m} \rightarrow \mathrm{PM}_{n+m}\,(\text{graft copolymer})$$

By combination of two polymeric radicals the deactivation of primary radicals takes place. Moreover, the homopolymerization of the radicals of the monomeric units can be another side reaction which limits the graft copolymer formation. Therefore, some conditions can be adopted to overcome these problems, such as homopolymerization inhibitors or the selection of low dose rates to avoid rapid termination of graft growing chains and the addition of the monomer either in vapor or liquid form to the polymer backbone while it is in a solid form.

Pre-irradiation method involves: (1) irradiation of the polymer backbone to form active radicals and (2) contact of the irradiated polymer backbone with monomer. If irradiation step takes place in air, the generated radicals react with oxygen to form peroxides and hydroperoxides. These stable products are treated with the monomer at high temperature, which causes decomposition of peroxides/hydroperoxides in radicals, initiating grafting. This route can be schematized by the following steps:

Formation of diperoxides/hydroperoxides: $PH + O_2 \xrightarrow{\gamma\text{ - rays}} POOP$ or $POOH$

Thermal decomposition: POOP or $POOH \xrightarrow{\Delta} PO^{\cdot} + OH^{\cdot}$

Initiation: $PO^{\cdot} + M \rightarrow POM^{\cdot}$

Propagation: $POM^{\cdot} + nM \rightarrow POM^{\cdot}_{n+1}$

where PH is the polymer backbone, POOP and POOH are peroxides/hydroperoxide, PO is the primary radical and $POM^{\cdot}$ is the initial chain graft and POM_{n+1} is the graft growing chain.

On the other hand, if irradiation takes place under vacuum or inert atmosphere, the grafting follows the mechanism:

$$PH \xrightarrow{\gamma\text{-rays}} P^{\cdot} + H^{\cdot}$$

$$P^{\cdot} + M \rightarrow PM^{\cdot}$$

The pre-irradiation method has the advantages that the homopolymer formation is little and the grafting can be carried out at any time, away from radiation sources.

Limitations of using high-energy radiation are connected to the facts that the technique is expensive and requires long irradiation times. Moreover, the optimization of the experimental conditions is a huge task and modification of chains (chain scission or cross-linking) can occur.

Using Photoradiation

Grafting can be proceeded by the use of low-energy radiations like UV light, which generates, often in the presence of a photoinitiator, free radicals only on the material surface [58]. Therefore, the grafts will be formed only superficially without modifying or damaging the bulk material. In this case, the reaction proceeds quickly, equipment is simple, the upscaling to industrial level can be easily realized with low costs.

However, conventional photografting methods often do not ensure a good control of the final copolymer architecture. Homopolymerization of the monomers, grafting and cross-linking, degradation of the substrate can occur. Then, in order to overcome these obstacles, "living" or controlled graft copolymerization was developed, where chain transfer and chain terminator mechanisms are absent. The ATRP method may be well-suited for surface modifications and the accurate control of molecular weight, molecular weight distributions, and surface density of grafts has been achieved.

Using Plasma Radiation

Recently, the plasma polymerization technique has received increasing interest [59]. The interaction of plasma with polymeric surface causes electron-induced excitation, ionization, and dissociation. Macromolecular radicals are formed and reactions similar to those produced by high-energy irradiation occur, even if the effects of the plasma are milder than those of conventional energy (e-beam, gamma irradiation). As the changes are confined only to the depth of a few nanometers at the surface, the bulk properties of material, such as degree of polymerization and crystallinity, are not very much influenced.

2.2.3 Other Initiating Systems

Enzymatic grafting is a relatively new method, which have several advantages with respect to chemical and physical pathways. In terms of health and safety, enzymes have eliminated the hazards associated with the use of harsh chemicals and the reactions occur in milder reaction conditions. Moreover, enzyme selectivity and specificity make the reactions very simple and offers the potentiality of a good control of the final macromolecular structure with the advantage that protection and de-protection steps are not necessary. Finally, enzymatic reactions can be used together with chemical reactions. Most of the grafting reactions use oxidoreductase enzymes, which are capable of forming free radicals, thanks to electron removal or due to oxygen supply to the molecules. The use of oxidative enzymes in the grafting reactions is extensively studied for natural polymers. As examples, lipase was used to catalyze ROP from cellulose substrates. Li et al. [60] grafted hydroxyethyl cellulose films with poly(caprolactone) by the use of a lipase and Gustavsson et al. [61] used a lipase for the ROP of ε-CL in close proximity to cellulose fibers in a filter paper. First, the enzyme was immobilized on the filter paper which was used as a substrate, and then polymerization was carried out. FTIR spectroscopy and contact angle measurements showed that PCL is not covalently bonded but coats the cellulose surface.

3 Polysaccharide and Their Hydrogels

3.1 Chitosan-Based Hydrogels

Chitin is a naturally occurring polysaccharide and is considered the second most abundant in nature after cellulose [62]. Chitin consists of units of β-(1,4)-linked-2-acetamido-2-deoxy-D-glucose. It is present in large quantities in the shells of crabs, shrimps and other crustaceans, and in the exoskeletons of insects and mollusks. Chitosan (Ch) is the *N*-deacetylated derivative of chitin [63] obtained by the alkaline hydrolysis of chitin, which is almost never complete [64, 65]. Thus Ch is a copolymer composed of glucosamine and *N*-acetylglucosamine units [66], and the degree of deacetylation indicates whether the biopolymer is chitin or chitosan. Ch is a linear hydrophilic polysaccharide that has received much attention in biological fields. Its physical properties depend on a number of parameters including its molecular weight, its degree of deacetylation (DD), and the sequence of the free amino acetamido groups at C2. Ch is poly [β-(1–4)-2-amino-2-deoxy-D-glucopyranose], its chemical structure is illustrated in Fig. 6. It is an attractive biocompatible, biodegradable, and nontoxic natural biopolymer that exhibits excellent film-forming ability [67].

Ch has many properties that classified it as one of the most important modified natural polymer due to its biodegradability, biocompatibility, its nontoxic nature, and its metal uptake capacity [68]. The presence of the active free amino group at C2 in the structure of Ch affords more functionality toward biotechnological needs in many fields of applications, such as in food, in drug carriers, etc. Polymeric materials based on Ch can be formed into fibers, films, gels, sponges, or nanomaterials [69]. Moreover, Ch was proved to have broad-spectrum antifungal activity toward a variety of fungi [70], in addition to its ability to represent the core of a new generation of drug and vaccine delivery systems [71]. Chemical

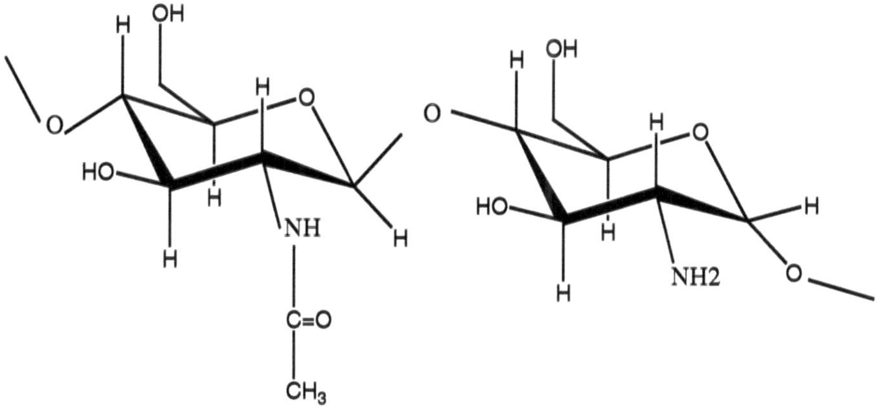

Fig. 6 Chemical structure of chitosan

modification of Ch by grafting technique has found wide field of applications especially as superabsorbent materials, as metal ions adsorption and as ion exchangers, in pharmaceuticals and drug delivery systems, and in the formation of hydrogel materials.

3.1.1 Synthesis of Chitosan-Based Hydrogels Via Graft Copolymerization

Graft copolymerization is a joint polymerization in which the backbone chain is formed from one type of polymer, while the side chains are formed from the other type. The length of the side chains should be less than that of the main chains. Ch has possible applications in many fields such as in biomedicine, in waste water treatment (removal of heavy metal ions and dyes), in polymer membranes, and in flocculation. However, due to its basic nature (resulting from the free $-NH_2$ groups), it is only soluble in dilute acids which limits its applications. For this, there is growing interest nowadays to modify its chemical structure in order to improve its solubility and consequently widen its applications [72]. Among various methods, graft copolymerization of Ch is one of the most attractive technique for modifying its chemical and physical properties. Ch bears two types of reactive groups in its repeated units, the $-NH_2$ group at C2 for the deacetylated units and the $-OH$ group at C3 and C6 for both acetylated and deacetylated units. Grafting onto Ch allows the addition of new functional groups depending on the grafted materials, which can afford more solubility to the grafted copolymer [73]. Free radical initiating method may be considered in terms of irradiation method or chemical method.

Initiation by Irradiation Method

Initiation by irradiation method, gamma rays or an electron beam, is a physical way of initiation, producing the free radical sites at the break points [74]. If this application is done in the presence of a vinyl or an acrylate monomer, copolymerization is initiated and graft copolymer is produced, attached to the Ch at the site of the free radical formation. Irradiation can be applied to a mixture of the Ch and the monomer which will yield some homo polymers, but will also provide for the reaction short-lived free radicals. Low temperature, low moisture content, and the absence of oxygen will favor increased stability of the free radicals. When organic substrates are subjected to high-energy radiation (γ-radiation), free radicals are formed through electron abstraction to form radical cations. The radical formation is concentrated in the vicinity of the incident radiation beam, and it is fairly unselective. Due to the lack of specificity, the formed radical appears to have a number of disadvantages, however, it is considered as a convenient method because there are no synthetic steps to be formed. Initiation by γ-radiation is often used for the initiation of vinyl polymerization (or copolymerization) by both radical and ionic mechanisms. Various vinyl and acrylate monomers are graft copolymerized onto

Ch by means of γ-radiation. Wang et al. [74] graft copolymerized acrylamide (AM) onto Ch by means of γ-radiation. The graft copolymerization was carried out at 20–25 °C. Characterization of the graft copolymer was performed by various techniques, such as FTIR, X-ray diffraction, and TGA. The effect of acetic acid concentration, total irradiation dose, dose rate, and acrylamide concentration on the graft% were also investigated. Moreover, the results of the flocculation experiment showed that the produced graft copolymer was significantly superior to both Ch and PAM.

Pengfei et al. [75] graft copolymerized styrene (St) onto Ch using ^{60}Co γ-radiation. The solvent composition has a marked effect on the degree of grafting. The graft yield was found to increase with the increase in the absorbed dose, while for the same dose, the graft yield of St on Ch was higher than that of chitin.

Initiation by Chemical Methods

Initiation by chemical initiators takes place either by free radical or ionic (cationic or anionic) mechanisms. The most used method of chemical initiation of vinyl graft copolymerization onto Ch is the reaction of Ch with ceric (IV) ions [76], Fenton's reagent [77] or Fe^{2+}/persulfate [78] as redox systems. In these systems, free radicals are produced from the decomposition of the initiators and transferred to the substrate. The latter reacts with the monomer to yield the graft copolymers. Free radicals are produced either by indirect or direct methods.

Ceric Ion-Induced Grafting

Cerium (IV) in a slightly acidic medium is a versatile oxidizing agent used mainly in the graft copolymerization of vinyl and acrylate monomers onto cellulose and starch [79, 80]. It forms a redox pair with the hydroxyl groups at C-2 and C-3 position of the anhydroglucose units of the polysaccharide to yield the macroradicals. The latter is responsible for the attack of the vinyl or acrylate monomer to form the graft copolymer. The use of cerium (IV) was extended to the chemical initiation of chitin and chitosan for the graft copolymerization of various monomers due to the similarity of these polysaccharides with cellulose and starch [81–85]. The mechanism of initiation for Ch is started by complex formation between the Ce(IV) ion with the $-NH_2$ and $-OH$ groups at C-2 and C-3 positions, respectively. The dissociation of the formed complex produced the macroradical which is responsible for the initiation of the grafted copolymer in the presence of the vinyl or acrylate monomers. A typical mechanism for the chemical initiation by cerium ammonium nitrate for the graft copolymerization of vinyl and acrylate monomers onto Ch is represented in Fig. 7.

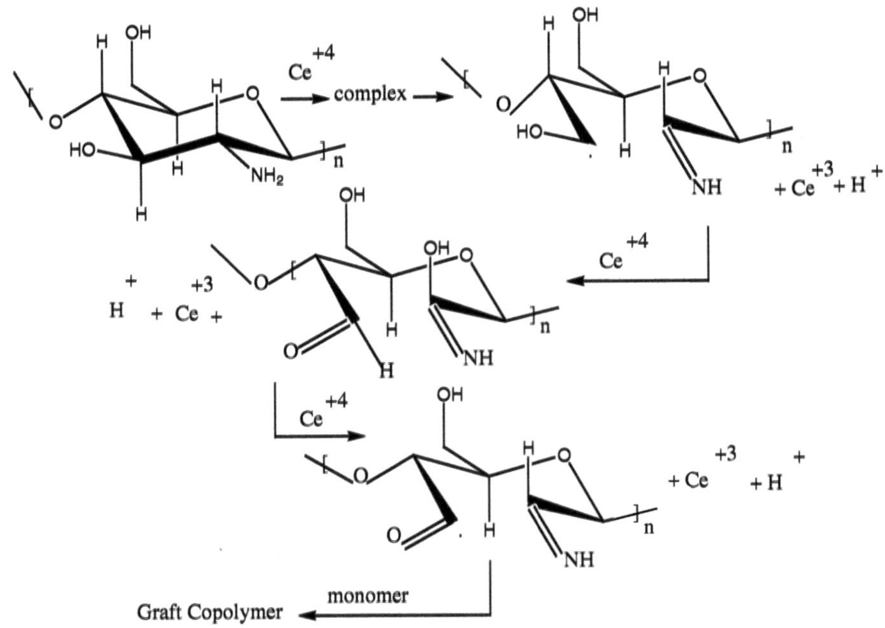

Fig. 7 Graft copolymerization mechanism onto chitosan in the presence of Ce(IV) ions as an initiator

Fenton's Reagent-Induced Grafting

Fenton's reagent (Fe^{2+}/H_2O_2) is a typical redox initiator used frequently for the initiation of graft copolymerization of various monomers onto chitin and chitosan [86]. The reagent involves a redox reaction between the ferrous ions (usually ferrous sulfate) and hydrogen peroxide. This interaction leads to the formation of a single hydroxyl radical which initiates the graft copolymerization through abstraction of a hydrogen atom from the hydroxyl groups of the glycosidic ring of the chitin or Ch (usually from the primary alcohol at C_6). Although hydrogen peroxide can be used alone as an effective initiator for the copolymerization, there are reasons why reducing agent like ferrous ions is preferable to be used. First, a higher yield of radicals is produced at lower temperature in the presence of redox initiator, and second, the chelating properties of Ch with the metal ions promote the formation of OH radicals in the vicinity of the Ch chains, thus lower the opportunity for homopolymer formation. Methyl methacrylate (MMA) was graft copolymerized onto Ch with grafting percentage of 400–500 % in the presence of Fenton's reagent with a homopolymer yield in the range of 20–30 % [77].

A schematic presentation for the graft copolymerization in the presence of Fe(II)/ H_2O_2 redox initiator is illustrated in the following Scheme:

$$H_2O_2 + Fe^{2+} \rightarrow OH + {}^{-}OH + Fe^{3+}$$
$${}^{\cdot}OH + Fe^{2+} \rightarrow {}^{-}OH + Fe^{3+}$$
$${}^{\cdot}OH + H_2O_2 \rightarrow H_2O + {}^{\cdot}OOH$$
$$H_2O_2 + {}^{\cdot}OOH \rightarrow {}^{\cdot}OH + H_2O + O_2$$
$${}^{\cdot}OH + Ch\text{-}H \rightarrow H_2O + Ch^{\cdot}$$
$$Ch^{\cdot} + \text{monomer} \rightarrow Ch\text{-g-copolymer.}$$

Persulfate-Induced Grafting

Ch has been graft copolymerized with vinyl monomers using Fe^{2+} ions/potassium persulfate (Fe^{2+}/KPS) as initiator [87]. The redox initiator in an aqueous medium decomposes by heat to yield sulfate radicals and other free radicals species, as represented in the following scheme:

$$Fe^{2+} + S_2O_8^{2-} \rightarrow Fe^{3+} + SO_4^{2-} + SO_4^{-\cdot}$$
$$SO_4^{-\cdot} + H_2O \rightarrow HSO_4^{-} + {}^{\cdot}OH$$
$${}^{\cdot}OH + Ch\text{-}H \rightarrow H_2O + Ch^{\cdot}$$
$${}^{\cdot}OH + Fe^{2+} \rightarrow {}^{-}OH + Fe^{3+}$$
$$SO_4^{-\cdot} + Fe^{2+} \rightarrow SO_4^{2-} + Fe^{3+}$$

An interesting reaction mechanism with Ch/persulfate system has been proposed by Wang et al. [88], where chitosan's free amino group reacts directly with the persulfate to yield $R{-}{}^{\cdot}NH$, $\cdot OSO_3H$, and $SO_4{}^{2-}$. In the work of Yazdina-Pedram et al. [89, 90] KPS was used to graft Ch by vinyl pyrrolidone and both methyl acrylate (MA) and MMA. They compare the use of KPS alone and in the presence of various reducing agents in the initiation process [90].

Modification of Ch dissolved in 1.5 % acetic acid was carried out in homogeneous phase by graft copolymerization of vinyl butyrate using KPS as a redox initiator. The grafted product is insoluble in common organic solvents and in dilute organic and inorganic acids. The maximum grafting percentage (%G) and grafting efficiency (GE%) values obtained under these conditions were 359 and 94 %, respectively. Moreover, the thermal property measurements (DSC measurements) have proved the higher thermal stability of the grafted Ch as compared with the ungrafted one [91]. Chemical modification of Ch by grafting with PAM was carried out in a homogeneous phase using KPS as redox initiator and in the presence of *N,N*′-methylenebisacrylamide as cross-linking agent. The percentage of grafting was found to depend on the various reaction conditions. At optimized combinations of the reaction variables, a GE of 88 % and a G% of 220 % were attained. When grafting of AM onto Ch was achieved in the absence of the cross-linker, the obtained grafted product was slightly soluble, while in the presence of the cross-linker, it is completely insoluble giving a typical polymeric hydrogel [92].

Graft copolymerization of 2-hydroxyethyl acrylate (HEA) onto Ch using ammonium persulfate (APS) as a redox initiator was carried out in aqueous solution and inert atmosphere. Graft yield up to 300 % was reached depending on the reaction conditions. The polymerization rate was found to be more sensitive to the concentration of the monomer than to the concentration of the initiator. The grafted samples are soluble in water and at alkaline pH, and possess an enhanced hydrophilic character as compared with the parent acetylated Ch [93].

Novel Redox Initiators

A novel redox system, potassium diperiodatocuprate (III)-Ch [94], was used to initiate the graft copolymerization of methyl acrylate (MA) onto Ch in alkaline aqueous solution. Cu (III) was used as an oxidant and Ch as a reductant in this redox system. Cu (III) was an efficient as well as a cheap initiation system, and the initiation mechanism of grafting reaction proceeds according to the following scheme:

$$\text{Chitosan-}\ddot{N}H_2 + \text{Cu (III)} \longrightarrow \text{Chitosan-}NH_2^{\cdot +} + \text{Cu (II)}$$

$$\text{Chitosan-}NH_2^{\cdot +} + {}^{\cdot}OH \longrightarrow \text{Chitosan-}\dot{N}H + H_2O$$

$$\text{Chitosan-}\dot{N}H + \text{MA} \xrightarrow{\text{initiation}} \text{Chitosan-NH-MA}^{\cdot} \xrightarrow{+\text{n MA}} \text{Chitosan-NH-(MA)}_n\text{-MA}$$

Graft Copolymer

3.2 *Starch-Based Graft Copolymers*

Starch is a major storage polysaccharide in higher plants and is found as granules in cereal grains, roots and tubers, pulses, etc. [95]. Starch has been used for centuries as a thickener in food and is a versatile and widely used viscosifier. But more than that, it is a precursor for a very large number of ingredients used in food, textile, paper, pharmaceutical, and adhesive industries. Though starch occurs throughout the plant world, there are only a limited number of plants utilized extensively for the production of commercial starch. Cereals, root and tubers, and legumes are rich sources of starch. Corn (maize) is the major commercial source of starch and other commonly used sources include wheat, potato, tapioca, and rice. The major cereal sources of starch include wheat, rice, maize, sorghum, barley, oats, rye, millets, and the starch content varies in the range 55–79.5 % on dry weight basis [96]. Roots and tubers such as potato (*Solanum tuberosum*), tapioca (cassava, *Manihot esculenta* Crantz), sweet potato (*Ipomoea batatas),* arrowroot (*Maranta arundinacea*), yams

Fig. 8 Structure of amylose and amylopectin molecules

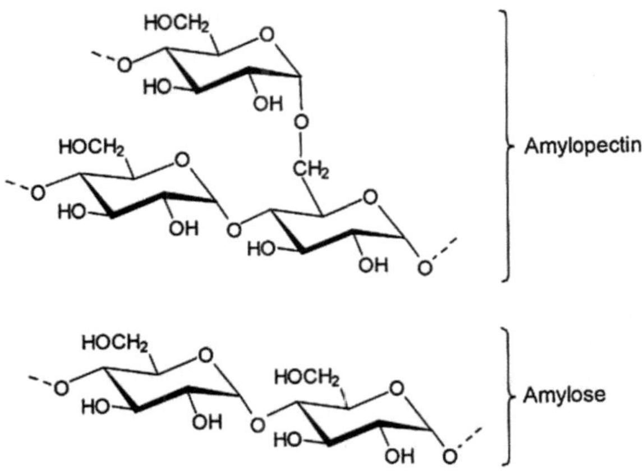

(*Dioscorea* spp.), and taro (*Colocasia esculenta*) are rich sources of starch. In some cereals, the waxy and non-waxy and even sugary genotypes are also available [96]. Starch is composed of two types of α-D-glucose polymers: the linear and helical amylose (20–30 %) and the branched amylopectin (70–80 %). In amylose, the glucose units are linked through α-(1 → 4) linkages, whereas amylopectin consists of linear α-(1 → 4) chains with α-(1 → 6) branch points (Fig. 8). The amylose to amylopectin ratio and the branching in amylopectin depend on the source of starch [97]. Amylose has comparatively lower molecular weight (MW $\sim 10^6$), whereas amylopectin has huge and compact molecules (MW $\sim 10^8$). The α-(1,4) glucan chains in amylopectin are organized as A, B, and C chains. The outer chains (A-chains) are linked to the inner chains (B-chains) through glycosidic bonds and the latter are branched chains. The single C-chain per molecule contains other chains as branches and it has a single terminal reducing group [98].

The starch granular size is unique for a particular plant species. Rice starch granules are relatively small with a granule size of about 2 μm, whereas potato starch granules are very large in size (up to 100 μm). When starch is heated in water, the granules undergo swelling by absorption of water and at some point the swollen granules burst. As a result, the semicrystalline structure of starch will be lost and the smaller amylose molecules leach out into the solution causing an increase in viscosity. This process is called starch gelatinization. When starch paste is cooled or stored for long duration, retrogradation of the amylose takes place resulting in the partial recovery of the semicrystalline structure. This causes thickening of starch paste by expelling water (syneresis). Retrogradation is responsible for the hardening of bread or staling.

3.2.1 Synthesis of Starch-Based Hydrogels Via Graft Copolymerization

In starch graft copolymerization, starch yields free radical sites on the glucan backbone on treatment with initiators in the presence of synthetic monomers resulting in high molecular weight polymer grafts [99]. The starch graft copolymers may be hydrophilic, hydrophobic, or polyelectrolyte in nature depending on the reagents and conditions used for the reaction.

Chemical Methods

The most popular chemical method for the synthesis of starch graft copolymers is the free radical initiated solution polymerization reaction. Other techniques such as bulk polymerization and inverse suspension polymerizations, etc., were also used for the synthesis of graft copolymers. Different chemical initiators were used for the synthesis of grafted starches, which included ceric ammonium nitrate (CAN) [100], ammonium persulfate (APS) [101–103], potassium persulfate (PPS) [104, 105], and benzoyl peroxide [106].

In free radical initiated polymerization reactions, polysaccharide reacts with initiator in two ways. In the first one, the neighboring OH groups of the starch and the initiator reacts to form a redox pair complex, which then dissociates to form carbon free radicals on the starch substrate. The radicals thus formed initiate the graft copolymerization of monomers on the starch [107]. In the second method, the initiator may abstract hydrogen from the OH groups of starch producing initiating radicals on starch backbone, which then reacts with the monomers.

Among the free radical initiators, ceric ion is considered as the most selective initiator, since it reacts directly with the starch backbone, creating radicals at certain points on the backbone where they can effectively start a graft polymer attachment [108]. Here, the homopolymer formation would be suppressed and the reaction takes place effectively even at ambient temperature [109]. The grafting of acrylic acid (AA) onto granular maize starch was carried out by ceric ion initiation and the effect of grafting parameters has been studied [110]. Jyothi et al. [111] synthesized highly water absorbing graft copolymers of cassava starch by polymerization with poly(acrylamide) in the presence of ceric ion initiator. Graft copolymers of starch and acrylonitrile have been synthesized in aqueous solution by ceric ammonium ion initiation at varying concentrations of monomer and initiator [90].

Potassium or ammonium persulfate is an efficient redox initiator for the graft copolymerization of starch in aqueous medium. Heating of an aqueous solution of persulfate causes its decomposition to produce sulfate ion radical and other radical species [107, 112, 113]. The preparation and properties of a number of graft copolymers of starch with vinyl monomers (acrylamide, acrylonitrile, acrylic acid, and methacrylic acid) in the presence of potassium persulfate as the free radical initiator have been reported [113]. Taghizadeh and Mafakhery [105], Bhuniya et al. [104] and Mostafa et al. [114] used potassium persulfate (I) redox system in the graft copolymerization of starch in aqueous medium. Ammonium persulfate (APS), *N,N,N′,N′*-tetramethylethylenediamine (TEMED), and a cross-linker have also been reported to be used for graft copolymerization onto starch.

Starch-graft-poly(methacrylamide) has been synthesized in aqueous medium by using benzoyl peroxide as the free radical initiator [115]. The graft copolymerization of styrene and methyl methacrylate/butyl acrylate onto starch was carried out by using a Fe^{2+}-peroxide redox system [116]. Potassium permanganate–acid system (Mn^{4+}) and manganic pyrophosphate (Mn^{3+}) have been used for the graft copolymerization of vinyl monomers onto starch. Graft polymerization of acrylonitrile onto starch was performed under the initiation of potassium permanganate and the mechanism of reaction has been investigated [117]. The relation between grafting rate and the concentrations of initiator, monomer and starch, as well as the reaction temperature were studied. The initiating ability of potassium permanganate in graft copolymerization reaction of acrylonitrile onto corn starch was studied [118]. Graft copolymerization of *N*-vinyl formamide onto pregelled starch was carried out in the presence of bromate/cyclohexanone redox initiator [119].

Initiation by Irradiation

Irradiation using high-energy γ-radiation is reported to be an efficient method for initiating graft copolymerization reactions on polysaccharides [120]. Even though radiation-based grafting is cleaner and more efficient than chemical initiation methods, there are technical difficulties in handling such reactions. Therefore, reports on earlier studies employing radiation initiation for the synthesis of graft copolymers are scanty. Gamma rays were used as initiator for the graft copolymerization of acrylic monomers onto cassava starch and the effect of various parameters such as monomer–starch ratio, dose of gamma rays (kGy), dose rate (kGy h-1), and the presence of additives were studied [121]. El-Mohdy et al. [122] have also reported the synthesis of starch/AA superabsorbents via γ irradiation. Cassava starch-g-poly (acrylic acid) copolymers were synthesized by γ-ray irradiation from a ^{60}Co source [123]. Graft copolymerization of acrylonitrile onto maize starch was carried out by gamma ray irradiation at different levels of monomer-to-maize starch ratio and total dose (kGy) of gamma rays [124]. Potato starch-graft-poly(acrylonitrile) could be efficiently synthesized using small concentration of ammonium peroxydisulfate (0.0014 M) in aqueous medium under microwave irradiation [125]. Tao and Jiao [126] adopted microwave irradiation technique for the synthesis of multi-grafted starch interpenetrating network structure. Kumar et al. [127] have compared the synthesis of graft copolymers via conventional, microwave-initiated, and microwave-assisted methods.

Other Initiating Systems

Besides the major grafting processes described above, several other initiating systems and methods for synthesizing starch graft copolymers are reported. Pregelled starch has been graft copolymerized with methacrylamide as a reactive monomer using vanadium mercaptosuccinic acid redox pair as an initiation system [128]. Starch-poly(acrylamide) graft copolymers were synthesized in water medium in the presence of horseradish peroxidase (HRP) catalyst/H_2O_2/2,4 pentanedione and acetate buffer at 30 °C and pH 7.0 [129]. It is postulated that peroxide-activated HRP oxidizes PDO to a free radical which then abstracts a proton from starch backbone to give carbonyl radicals, which in turn initiates the grafting reaction and copolymerization. Extrusion technique was also reported for grafting reactions of starch for grafting [130–133].

4 Characterization and Some Properties of Chitosan-Based Hydrogels

4.1 Spectral Analyses

FTIR spectroscopy is a common tool used by many authors [91, 134–138] to characterize the prepared copolymers and acts as a proof for the grafting process. In addition, if it is done quantitatively, it can be considered as an additional proof for the grafting yield. For example, Li et al. [134] characterized the prepared Ch-graft-poly(ethyleneimine) (Ch-g-PEI) by FTIR (Fig. 9). They found that new peaks appeared for the graft at 1468, 1412, and 814 cm^{-1} which are attributed to the absorption of $-CH_2-CH_2-NH-$ moiety. This result gave a strong evidence for the grafting process.

Graft copolymer of Ch with poly[rosin-(2-acryloyloxy)ethyl ester] (Ch-g-PRAEE) has been synthesized and characterized using FTIR spectroscopy by Duan et al. [135], (Fig. 10). Figure 10a represents the characteristic peaks of Ch, while Fig. 10b represents the characteristic peaks of the grafted Ch. The results indicated that in addition to the characteristic peaks of Ch, some new peaks appeared. Thus the peaks at 1728, 1105, and 1248 cm^{-1} are attributed to the C=O, −C−O−, and −O−C− ester functional groups of the grafted PRAEE, respectively. Also, the $-CH_2$ peak of PRAEE appeared at 1546 cm^{-1}. Moreover, peaks of the amide linkage were shifted to lower wave numbers. From these results, it was possible to conclude the successful synthesis of the grafted copolymer.

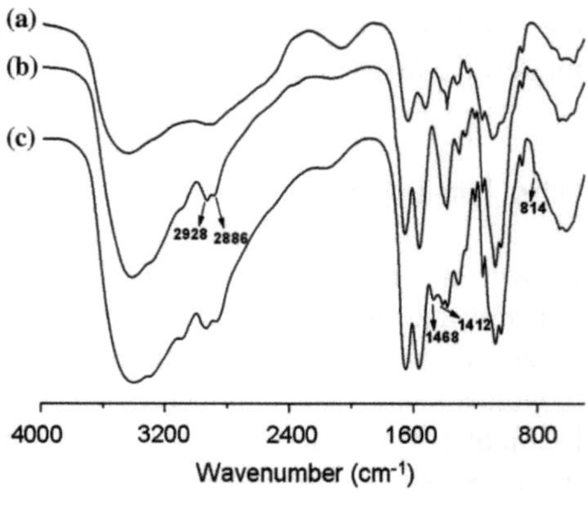

Fig. 9 FTIR spectra of *a* Ch, *b* Ch-SS-COOH, and *c* Ch-g-PEI. Reprinted from [134], copyright 2010, with permission from Elsevier

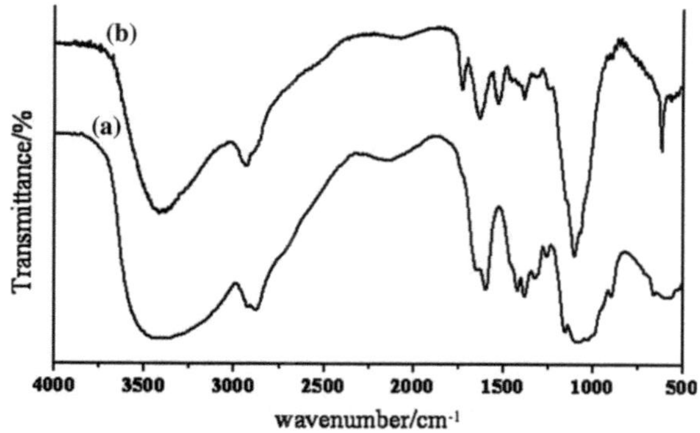

Fig. 10 FTIR spectra of *a* Ch and *b* Ch-g-PRAEE. Reprinted from [135], copyright 2008, with permission from Elsevier

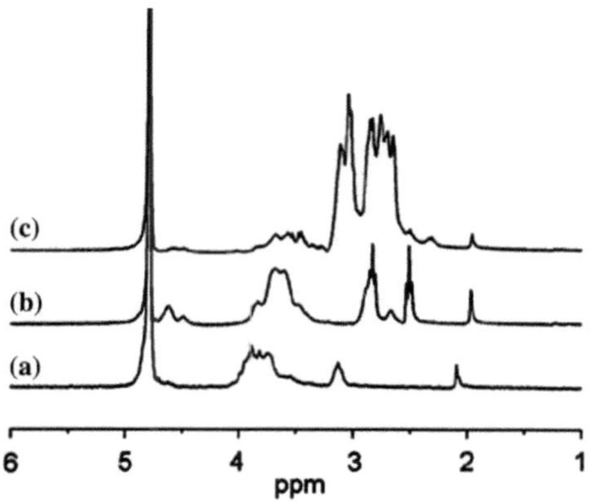

Fig. 11 1H-NMR spectra of *a* Ch, *b* Ch-SS-COOH, and *c* Ch-g-PEI in D_2O. Reprinted from [134], copyright 2010, with permission from Elsevier

Li et al. [134] used ^{1}H-NMR spectroscopy as a fine tool to prove and characterize the grafting of poly(ethylenimine) with a disulfide linkage onto Ch (Ch-g-PEI). The ^{1}H-NMR spectra of the Ch derivatives are represented in (Fig. 11). The proton peaks of $-CH_2CH_2S-$ appeared at 2.8–2.95 and 2.5–2.6 ppm in the spectrum of the intermediate Ch-SSCOOH (Fig. 11b), indicating that 3,3′-dithiodipropionic acid was successfully grafted to the Ch chain. The PEI grafting degree per glucosamine unit was determined by comparing the ^{1}H-NMR signal integrals from $–CH_2$ protons of PEI, which should subtract the absorption areas of $–CH_2CH_2S–$, with integrals of the Ch backbone proton signals.

Liu et al. [139] used the ring-opening polymerization technique for the graft copolymerization of *p*-dioxanone (PDO) onto Ch (Ch-g-PDO) and its chemical structure was determined by ^{1}H-NMR. Comparing the ^{1}H-NMR spectrum of Ch [140, 141] with that of graft copolymer indicated the presence of new proton signals at 4.16, 3.70, and 4.22 ppm, which corresponded to the different methylene groups of PPDO side chains.

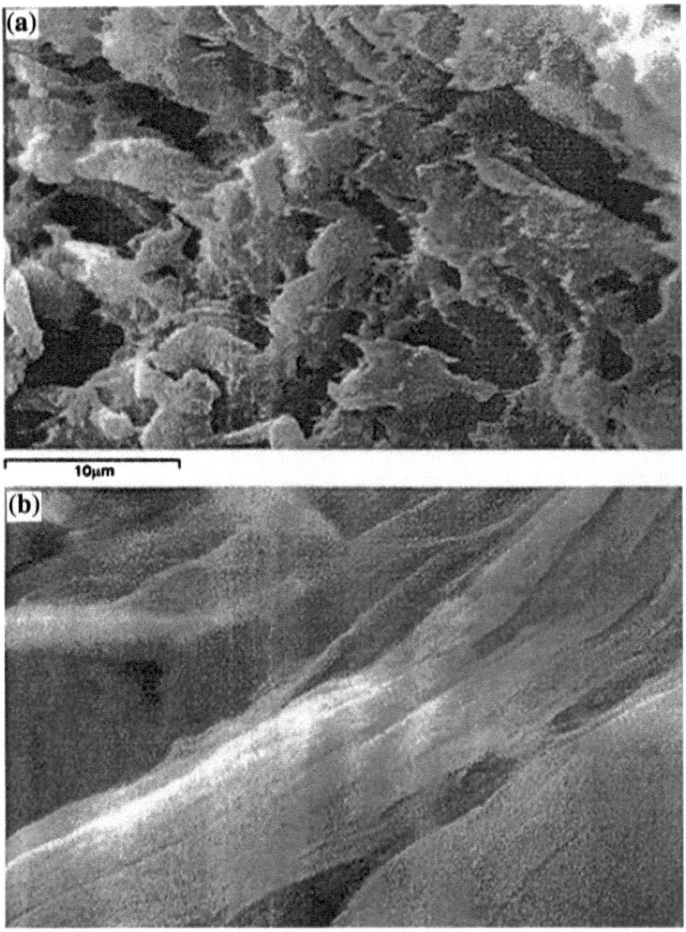

Fig. 12 SEM images of the top view of **a** Ch and **b** Ch-g-PHEA at 3000 magnification. Reprinted from [142], copyright 2008, with permission from Elsevier

4.2 Surface Morphology

Several authors have considered the SEM as a useful tool for proving the grafting process. Thus any change in the morphological structure of the pure Ch is the direct evidence that grafting had occurred [91, 135–137, 142]. For instance, Mum et al. [142] used SEM to characterize the graft copolymerization of 2-hydroxyethyl acrylate (HEA) onto Ch. The results are represented in (Fig. 12). The Ch particles typically present a porous morphology (Fig. 12a), while a nonporous fibrous structure was observed for the Ch-g-PHEA copolymer with percent graft near 100 % (Fig. 12b).

4.3 Thermal Behavior

Thermal analyses (TGA and DSC) are considered by many investigators [135–138, 143] as powerful tools to prove the grafting copolymerization process from one side, and to characterize the thermal stability (or instability) of the resulting graft copolymers as compared with the native matrix. Thus, for example, in the work of Duan et al. [135], the degradation process and thermal stability of Ch and Ch-g-poly [rosin-(2-acryloyloxy)ethyl ester] (Ch-g-PRAEE) copolymer were evaluated by thermogravimetric analysis (TGA) experiments, and the results are illustrated in (Fig. 13). Figure clearly showed three consecutive wight loss stages for both Ch (curve a) and Ch-g-PRAEE (curve b). For the TGA curve of Ch, the first stage indicated a 6.3 wt% loss at temperature range 55–191 °C due to the loss of absorbed and bound water, as a result of the hygroscopic nature of Ch; the second stage occurred in the range 230–327 °C resulted from the scission of the ether linkages in the Ch backbone; while the third stage showed a loss between 327 and 703 °C which is due to the thermal decomposition of glucosamine residue. However, the grafted copolymer (curve b) had different course of the thermal degradation compared to the parent Ch. The first stage at temperature range 30–153 °C is due to the loss of absorbed water; the second stage from 196 to 328 °C which resulted from the scission of the ether linkages in the Ch backbone, while the third stage was from 400 to 470 °C and this is attributed to the thermal decomposition of PRAEE side chains.

This is due to the high crystallinity of Ch, which is responsible for its relatively high thermal stability, as compared with the graft copolymer.

Akgün et al. [91] graft copolymerized poly(vinyl butyrate) onto Ch. Characterization of the graft copolymers was done by differential scanning calorimetry (DSC). The results revealed a difference in thermal stability between Ch and its grafted product. The thermal property of Ch-g-poly(vinyl butyrate) was more stable than that of the ungrafted Ch. El-Sherbiny and Smith [144] graft copolymerized polyethylene glycol (PEG) onto *N*-phthaloylCh (NPHCh). The prepared copolymer was confirmed by studying its thermal characteristics in comparison with PEG-COOH and NHPHCh starting materials. The results indicated the presence of an endothermic peak at around 65 °C for which corresponds

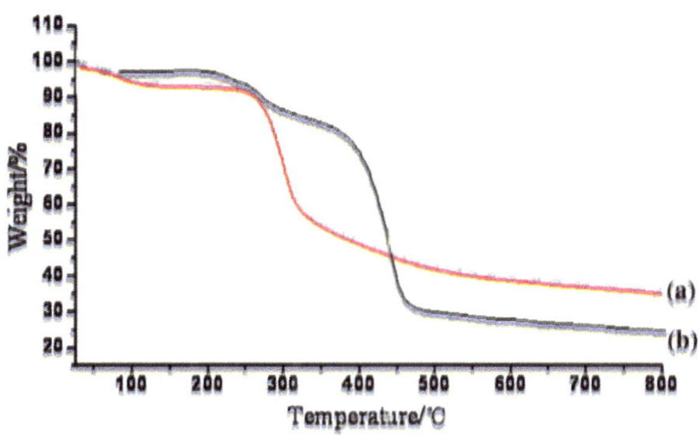

Fig. 13 TGA thermograms of *a* Ch and *b* Ch-g-PRAEE copolymer. Reprinted from [135], copyright 2008, with permission from Elsevier

to its melting process. The same endothermic peak appeared also in the thermogram of the copolymer (PEG-g-NPHCh) at around 58 °C due to melting of the grafted PEG side chains which gave a real proof for the grafting process. Moreover, the copolymer showed an exotherm at 231 °C which was due to the decomposition of the graft copolymer.

4.4 X-Ray Diffraction Analysis

X-Ray diffraction analysis (XRD) is another useful technique used by several researchers [145–147] to elucidate the structure of graft copolymers in the solid state. Liu et al. [145] prepared grafted Ch using binary grafts of hydrophobic poly-caprolactone (PCL) and hydrophilic PEG and characterized the graft copolymers by XRD. The results indicated the disappearance of the signal at $2\theta = 12°$ belonging to the polysaccharide giving evidence for the formation of the graft which suppresses the crystallinity of the parent Ch. Similar results were obtained in the work of Elkhouly et al. [147] when Ch was grafted with poly(acryloylcyanoacetohydrazide).

5 Characterization and Some Properties of Starch-Based Hydrogels

5.1 Spectral Analysis

Fourier transform infrared (FTIR) spectroscopy has been used for the confirmation of graft copolymer formation. The appearance of absorption peaks characteristic to starch and the grafted polymer in the IR spectrum is an evidence for grafting reaction [110, 148–151]. The FTIR spectrum of the native cassava starch was compared with that of cassava starch-graft-poly(acrylamide) for confirming the grafting reaction [111, 151]. In the spectrum of cassava starch, the O–H stretching absorption appears in the region 3390 cm^{-1}, C–H stretching at 2932 cm^{-1} and absorption corresponding to carbonyl group at 1647 cm^{-1}. A triplet absorption band was observed at 1159, 1084, and 1013 cm^{-1} which is peculiar for C–O–C stretching vibration in starch. For the poly(acrylamide) and poly(methacrylamide) grafted starches, the additional peaks found at 3400, 1650, and 1600 cm^{-1} correspond to the stretching vibrations of N–H and C=O groups and bending vibration of N–H group of amide, respectively. These absorption bands are characteristics of $-CONH_2$ group in acrylamide (31). The peak corresponding to the –C–N stretching was found at 1411 cm^{-1}, whereas the weak band in the region 765–710 cm^{-1} is due to the N–H out-of-plane bending. Gao et al. [119] reported the existence of an absorption peak at 2245 cm^{-1} due to CN stretching in PAN, which confirms the graft copolymer starch-g-PAN. Singh et al. [125] reported that the IR spectrum of microwave-synthesized starch-graft-PAN has absorption

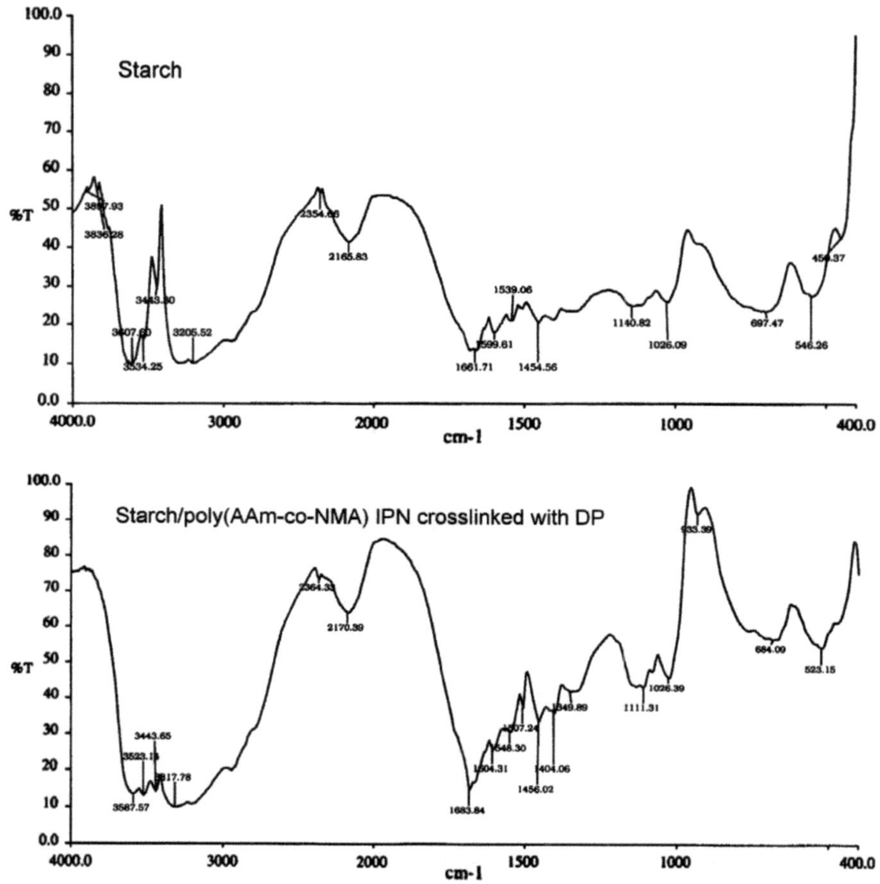

Fig. 14 IR spectra of the starch and starch/poly(AAm-*co*-NMA) semi-IPN hydrogel. Reprinted from [152], copyright 2006, with permission from Elsevier

peaks at 2242–2245 cm^{-1} (–CN stretching) and 1453 cm^{-1} (CH_2 deformation vibration), which could be attributed to grafted PAN chains at starch backbone. The FTIR spectra of cassava starch-graft-poly(acrylonitrile) copolymer showed the existence of a moderate peak at 2240 cm^{-1} which arises from the stretching vibration mode of the nitrile group. FTIR study was used to characterize the semi-IPNs of starch and poly (acrylamide-*co*-sodium methacrylate) [152] (Fig. 14).

The evidence for grafting reaction is also obtained by ^{13}C-NMR spectroscopy (Fig. 15). The esterified products of cassava starch-graft-poly(acrylic acid) with poly(ethylene glycol) 4000 and propylene oxide was characterized by NMR spectroscopy [123]. The presence of hydroxypropyl groups on a modified starch was indicated by the appearance of a peak at a chemical shift of 19.96 ppm and the chemical shift at 52 ppm indicated the C–O– on PEG 4000 chains. A distinct peak at 1.2 ppm in the ^{1}H-NMR spectrum was due to the protons of hydroxypropyl

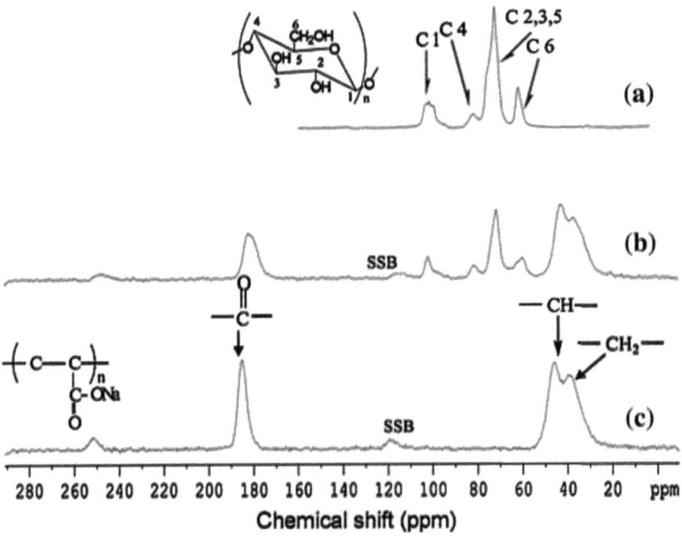

Fig. 15 ^{13}C CP/MAS spectra of **a** pure starch, **b** starch-g-poly(sodium acrylate), and **c** poly (sodium acrylate). Reprinted from [153], copyright 2008, with permission from Springer

groups in the modified starch. High-resolution solid-state ^{13}C NMR spectroscopy was reported to be used for studying the structure of samples such as starch-g-poly (sodium acrylate) superabsorbent, neat starch, poly(sodium acrylate), and blend of starch and poly(sodium acrylate) [153]. The results showed that there was a significant decrease in the crystallinity of starch in grafted starch and starch blends. The ^{1}H spin–lattice relaxation time showed that starch and poly(sodium acrylate) had good compatibility in nanometer scale in grafted starch as well in starch blends. The chemical shift of the carbonyl group of poly(sodium acrylate) in the ^{13}C CP/MAS spectra was found to depend on the composition of grafted starch. This showed that starch and poly(sodium acrylate) have better molecular level compatibility than blended samples.

5.2 Surface Morphology

The surface morphology of starch graft copolymers is investigated using scanning electron microscopy (SEM). The electron micrographs of pure starch exhibited granular structure, whereas grafted poly(methacrylonitrile) exhibited a change in granular structure and the formation of a polymeric coating on the granule surface [154]. Scanning electron micrographs of starch-graft-poly(potassium acrylate-*co*-acrylamide) showed the existence of holes between the fine particles of polymer [150] and it can easily absorb water because of larger surface area. Scanning electron microscopic study of pure starch and starch-graft-poly(butyl acrylamide) with a %G of 33 revealed that the grafting process did not alter the granular nature of starch but

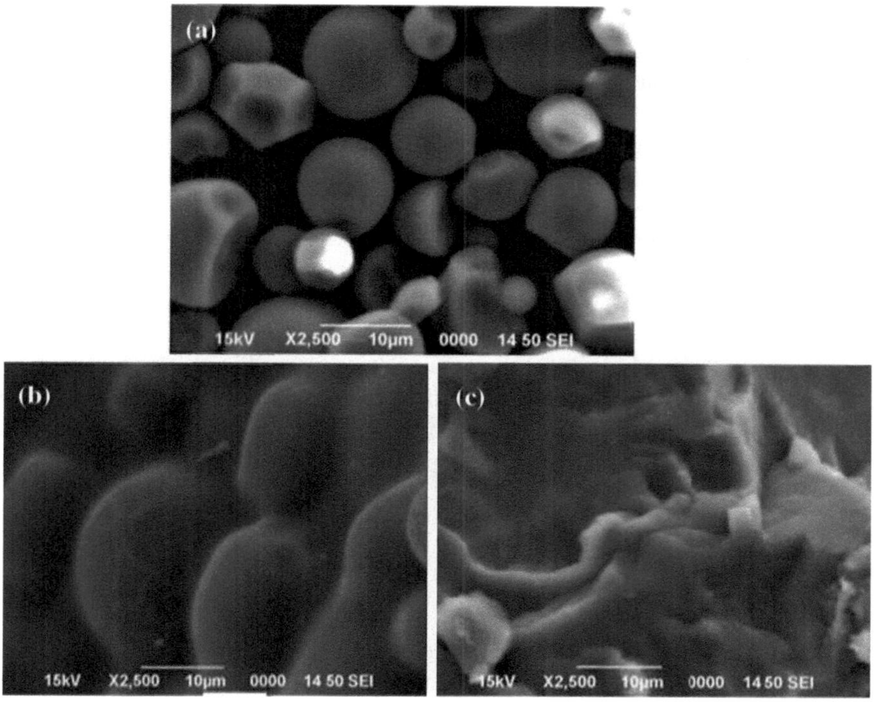

Fig. 16 Scanning electron micrographs of **a** native cassava starch and **b** starch-g-poly(AM) (%G = 31.91), **c** starch-g-poly(AM) (%G = 174.82). Reprinted from [111], copyright 2010, with permission from John Wiley & Sons

it make it more compact and it participates in the orientation of the surface by making some cubic granules [155]. Maharana and Singh [156] reported that starch-g-poly(ethylene) samples showed neither planar nor uniform surface, unlike the virgin polyethylene and increase in grafting percentage resulted in decrease of planarity and uniformity of the surface. The microstructural cross-sectional features of starch/semi-IPN cross-linked with different methylene *bis*acrylamide (MBA) concentrations determined with SEM revealed that hydrogel networks density improved enormously as MBA concentration increased [152]. Jyothi et al. [111] observed that for grafted cassava starch at a high level of grafting (%G = 174.82), there was a complete disappearance of the granular structure (Fig. 16).

5.3 *Melting and Glass Transition Temperature*

In the case of substances like starch which contains amorphous as well as crystalline regions, the midpoint of transition in the differential scanning colorimetric curve is taken as its glass transition temperature. At this point, the majority of the molecules undergo phase transition. Starch graft copolymers usually exhibit lower glass transition temperature than native starch, since the monomer when grafted to starch acts as internal plasticizer resulting in a decrease in T_g. The area and intensity of DSC peaks as well as the heat of fusion decreased significantly with increase in grafting percentage of starch onto PE [156]. Therefore, grafted PE is thermally less stable than the virgin PE. DSC studies showed that the T_g of amylopectin was about 50 °C and it was dropped to −11 °C after grafting [157]. This T_g in grafted samples, which is lower than room temperature is an important factor in the potential application of them as solid electrolytes, since low T_g allows for greater chain mobility which in turn improves solvation and ion conduction. The presence of a second endotherm in the DSC curves around 250 °C for the copolymer synthesized from acrylamide was reported earlier which was due to the fusion of the crystallites [100]. The graft structures of starch-g-poly(1,4-dioxan-2-one) (Starch-g-PPDO) copolymers were reported to affect their thermal and crystallization behavior. The short-grafted chains of PPDO resulted in more defect sites in the crystalline phase of the copolymers and hence its crystal structure was much less perfect than that of PPDO [158].

5.4 Thermogravimetric Analysis

The grafting of vinyl monomers onto starch alters the thermal stability of the latter and in most cases it increases [159]. The thermal stability of graft copolymers can be compared by considering the onset temperature of decomposition and the percentage weight loss at different decomposition stages. Athawale and Lele [154] studied the thermal properties of native starch and graft copolymers of starch with poly(methacrylonitrile) (PMAN). Pure starch showed a characteristic three-step thermogram in thermogravimetric analysis, whereas poly(methacrylonitrile)

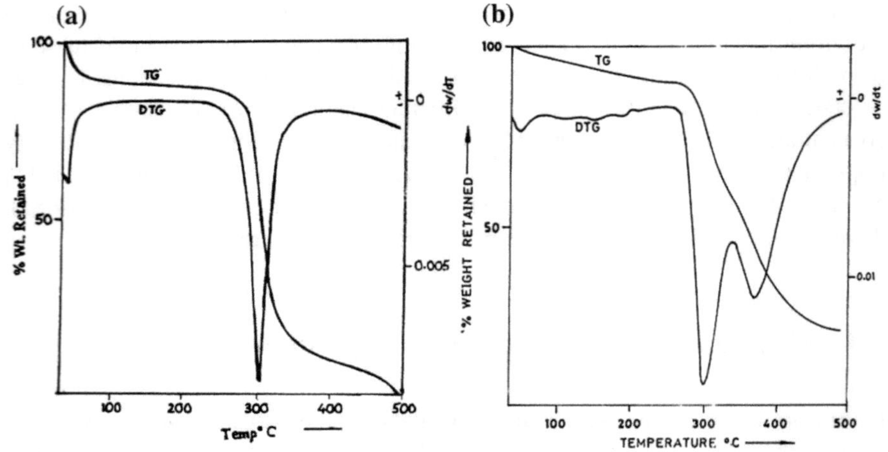

Fig. 17 Primary thermogram and derivatogram for **a** pure granular maize starch and **b** grafted starch with $\%G = 97.3$. Reprinted from [154], copyright 2000, with permission from Elsevier

showed two-stage decomposition pattern (Fig. 17). The thermal stability of starch was not significantly altered in starch-g-methacrylonitrile and the derivatogram exhibits the temperature for maximum decomposition at 309 °C. TGA/DTA thermograms of polyethylene (PE) and PE-g-starch showed a multistep degradation of PE and grafted starch [160]. The maximum decomposition rate of cassava starch appeared at 375 °C [123]. The esterified and etherified cassava starch graft poly (acrylic acid) showed two-stage decomposition. Fares et al. [155] also reported a three-stage decomposition for starch in the thermograms with a major weight loss of 69 %, where the major weight loss occurred at the second stage within the temperature range of 263–336 °C at T_{max} = 316 °C. Poly(butylacrylamide) (BAM) homopolymer showed three decomposition stages with major weight loss within the temperature range of 300–367 °C at T_{max} = 348 °C. Starch-g-BAM copolymer showed two characteristic peaks within the temperature range of 263–332 °C and 332–400 °C with T_{max} at 316 and 343 °C, respectively. Thermal stability of starch studied by TG was reported to increase as a result of grafting [115]. The thermal behavior of starch, poly(ethyl methacrylate) (PEMA), St-g-PEMA, and St-g-PEMA/sodium silicate (SS) was studied [106]. The decomposition temperatures were 210 °C for starch, 230 °C for PEMA and 285 °C for St-g-PEMA/SS, and St-g-PEMA/SS had higher thermal stability.

Thermogravimetric analysis of starch and microwave-synthesized starch-graft-poly(acrylonitrile) showed that grafted starch was more thermally stable than the pure starch [125]. TG and DTA curves of graft copolymers of mixed monomers vinyl acetate and butyl acrylate onto cornstarch showed that starch copolymer has higher thermal stability than pure cornstarch [102]. Thermal stability of cassava starch was reported to increase by grafting poly(acrylamide) onto it [161].

Graft copolymerization significantly alters the solution properties of starch such as viscosity, gelling nature, rheology, etc. Graft copolymers of starch, with hydrophilic grafts generally exhibit higher viscosity, better thermal stability, good film-forming properties, and higher water absorption capacity than native starch.

Hydrophilic, hydrophobic, or polyelectric nature can be imparted to starch by grafting it with suitable monomers. Grafting with acrylic acid and acrylamide polymers causes increased hydrophilicity, whereas acrylonitrile results in more hydrophobic nature to starch. The moisture absorbance was reported to decrease in graft copolymers of starch with AA due to the blockage of active sites and consequent decrease in affinity toward −OH groups of water [162, 163].

6 Applications of Chitosan and Starch-Based Hydrogels

6.1 Graft Copolymers as Absorbents and Ion Exchangers

Environmental contamination is a serious problem that is related to possible serious consequences, including heavy metals contamination of water from various sources. The traditional methods of water purification such as filtration, flocculation, etc., are expensive and unsafe. Adsorption is now recognized as the effective, efficient, and economic method for removal of heavy metal ions from both wastewater and natural water resources.

Recently, Ch has been widely used in water purification due to its excellent properties like chelating capacity for cations, low cost, renewability, and biodegradation [164, 165]. The capacity of Ch to complex metallic ions is one of its most important potentialities. Ch was modified by many chemical treatments to improve its adsorption ability. Adsorption capacity of Ch toward heavy metals was improved by the use of carboxymethylation [166, 167]. Jiang et al. [168] Ch-g-poly (sodium 4-styrenesulfonate) (PSS) ion exchanger was synthesized by nitroxide-mediated polymerization and ion exchanging property of this could be controlled by changing the PSS graft contents.

The results obtained by Sabaa and his group [136] showed that the amount of metal ions uptake (Ni^{2+} and Co^{2+}) by Ch is highly increased by polyacrylonitrile (PAN) grafting and conversion of nitrile groups in grafted copolymers to the more polar amidoxime groups. The results also indicated that carboxymethyl chitosan-grafted polyacrylonitrile (CMCh-g-PAN), on the other hand, adsorb less metal ions (Ni^{2+}, Co^{2+}, Cd^{2+} and Cu^{2+}) from aqueous solutions than the original CMCh, while their amidoxime derivatives showed better results in comparison to both CMCh and their graft copolymers [137].

Zinc ion binding ability of both Ch and Ch-grafted PMMA (Ch-g-PMMA) was studied by Muzzarelli and found to be higher at alkaline pH [169]. In addition, the results obtained by Singh and his group [170] have proved that grafting of PMMA on Ch increased the zinc ions binding capacity by providing the additional binding sites. Amino groups of Ch and ester groups at grafted sites are the best sites for metal ions adsorption. Alkaline medium is best for adsorption because the amino groups remain protonated in the acidic medium and are not available for binding. However, it was shown that at pH >8 for Ch and >10 for graft copolymer, the adsorption decreases.

El-Sherbiny [171] graft copolymerized *N*-acryloylglycine (NAGly) onto CMCh using 2,2-dimethoxy-2-phenyl acetophenone as photoinitiator under nitrogen atmosphere in aqueous solution (Fig. 18). Synthesized graft copolymers were then subjected to the uptake of copper ions from aqueous systems. The obtained results showed that this system may be extended to be used for metal ions uptake and treatment of wastewater.

Hydrogels composed of poly (vinyl alcohol) (PVA) and carboxymethyl chitosan (CMCh) were prepared by Sabaa et al. [172] via ultraviolet (UV) irradiation with the aim to be used as absorbent for heavy metal ions. The results indicated that the prepared hydrogels adsorb more metal ions like Cu^{2+}, Cd^{2+}, and Co^{2+} in comparison to raw CMCh and adsorption capacity was improved with the high PVA contents in the hydrogel.

Traditional chemical and physical methods are very efficient in color removal from wastewater, but these methods are costly [173, 174]. Recent studies have suggested the effective use of graft copolymers as natural adsorbents for color removal [175–177]. Sabaa and his group [136] studied the effect of graft copolymerization of acrylonitrile and its amidoxime derivative onto chitosan on the ability

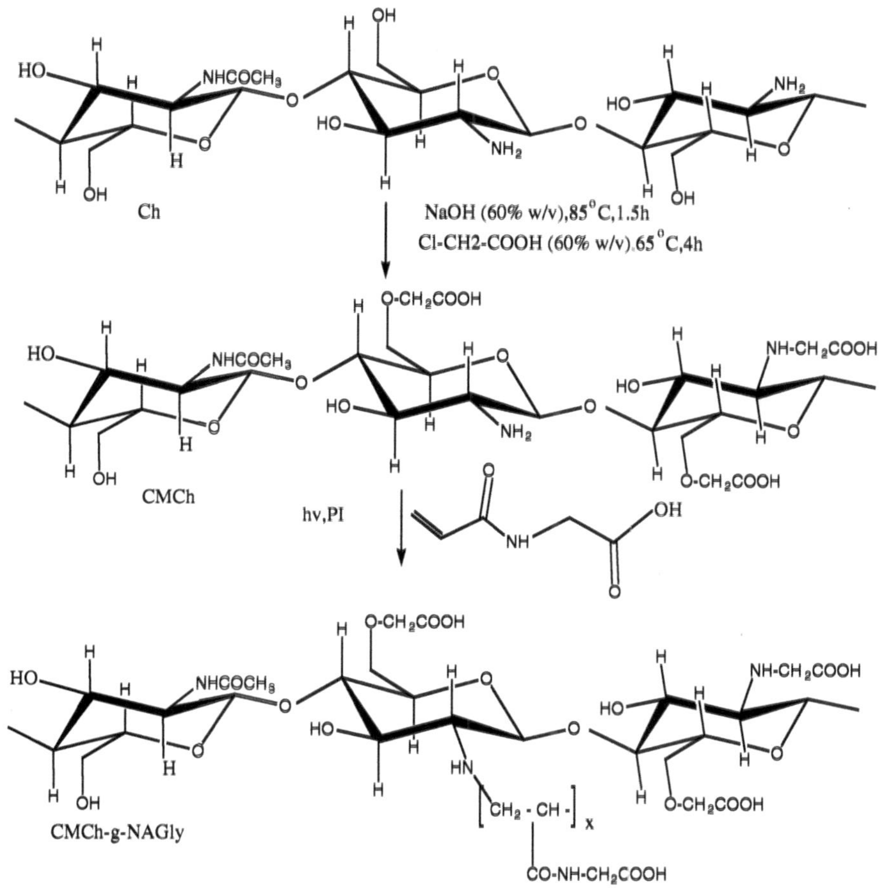

Fig. 18 Preparation of CMCh and CMCh-g-NAGly. Reprinted from [171], copyright 2009, with permission from Elsevier

for removal of Congo red (acidic dye) and Maxilon Blue (cationic dye) from wastewater. The amount of adsorbed dye was calculated as follows [178]:

$$Q = (N_a - N_s)/w$$

where, Q is fixed quantity of dye (mg)/grafted chitosan (g); N_a is the quantity of the original dye (mg); N_s is quantity of the remaining dye after adsorption (mg) and W is the mass of grafted chitosan (g). The amount of dye adsorbed by Ch-g-PAN increases with increase in percentage grafting because of incorporation of more $-CN$ groups into the chitosan backbone in addition to unaltered $-NH_2$ groups. The amidoxime derivative adsorbs more dye than its corresponding graft as the $-C{=}NOH$ group is much more polar than the parent $-CN$ group found in the graft. Also, the chitosan-grafted copolymers and their amidoxime derivatives adsorb more cationic dye in comparison to acidic dye.

Adsorption of dyes by carboxymethyl chitosan (CMCh) was reported in the literature [179]. Sabaa et al. [137] studied the adsorption of acidic and cationic dyes by CMCh, its graft copolymers with acrylonitrile (CMCh-g-PAN) and their corresponding amidoxime derivatives. The reported results indicated that dye adsorption ability of the synthesized samples increase with increase in percentage grafting. Further, the amidoxime samples adsorb more dye (especially acidic dye) due to the introduction of the more polar amidoxime, $-C(NH_2)=NOH$, groups in the polymer backbone, and also due to the basic nature of the $-NH_2$ groups.

Traditional methods used for removal of heavy metals from all kinds of wastewater are relatively expensive. Natural polymers such as starch, gums, glues, alginate, etc., function as bridging flocculants. The dangling-grafted chains of polyacrylamide-grafted polysaccharides have easy accessibility to the pollutants. Starch graft copolymers with longer branches provide better flocculating performance in comparison to the other polysaccharide graft copolymers. Polyacrylamide-based starch graft copolymers are useful flocculants for clays and coal refuse slurries. Enhanced functionality can be imparted to starch by grafting suitable acrylic polymers onto it, which will allow it to be more effective in flocculation, dispersion, and other applications [179].

Starch-g-polyacrylamide was found to be the best in performance among the various polysaccharide graft copolymers [180]. Starch-g-polyacrylamide was found to be better in flocculation and rheological studies in comparison to amylose-g-polyacrylamide [181–183]. Hydrolyzed starch graft poly acrylonitrile (HSPAN) could be potentially used for various applications including adsorption of heavy metal ions such as Cr^{3+} and Co^{2+} [108]. The poly (methacrylamide)-pregelled starch graft copolymers were found to be effective in removing heavy metal ions and followed the order: $Hg^{+2} > Cu^{+2} > Zn^{+2} > Ni^{+2} > Co^{+2} > Cd^{+2} > Pb^{+2}$ [128]. Cross-linked starch graft copolymers with aminoethyl groups were studied for adsorption of copper (II) and lead (II) ions from aqueous solutions. Adsorption capacity was increased with the increase in percentage grafting and metal ions concentration. The adsorption time reaching equilibrium for Cu (II) and Pb (II) was 2 and 1 h, respectively [184]. Poly(*N*-vinyl formamide)-cross-linked pregelled starch graft copolymers with 11.5–55.3 % graft yields were used for removing Cu^{+2}, Pb^{+2}, Cd^{+2}, and Hg^{+2} from their solutions at 200 ppm [114]. The residual removal of heavy metal increased with increase in percentage graft yield irrespective of the nature of metal ions used. It was also observed that synthesized starch graft copolymers were more effective in removing Hg^{+2} than the other metal ions and followed the order: $Hg^{+2} > Cd^{+2} > Pb^{+2} > Cu^{+2}$.

Cornstarch, acrylamide, and sodium xanthate-based graft copolymers were prepared with excellent flocculation capacity using epichlorohydrin as cross-linking agent and ceric ammonium nitrate (CAN) as initiator [185]. An amphoteric starch-graft-polyacrylamide (S-g-PAM) was prepared for treatment of several kinds of industrial wastewater and showed better results in comparison to cationic polyacrylamide (PAM) and hydrolytic PAM [102]. A cationic starch graft copolymer and its composite with alum were found effective in treating wastewater from paper making industry [186].

Flocculating capacity of the graft copolymers and linear polymers was explained by Singh et al. [187]. The graft copolymers have sufficient chain lengths and can easily bind the colloidal particles through bridging and form the flocs. Whereas, polymer segment in linear polymers attached to the surface project into the solution as tails or form part of loops [188].

6.2 *Graft Copolymers as Superabsorbent Polymers*

Superabsorbent polymers (SAPs) can absorb over a hundred times of liquid to their weight [189] and the absorbed liquid can be released slowly. Application of superabsorbent hydrogels is vast which involves use in agriculture, control delivery systems, perfumes, etc. Disposable diapers and sanitary napkins use hydrogel as superabsorbent polymer. Graft copolymerization is best method to prepare Ch- and CMCh-based superabsorbent polymers for various applications [190–192]. Water uptake behavior of Ch was improved by graft copolymerization of acrylonitrile followed by their amidoxime derivatives and swells more water in comparison to raw Ch [136, 137]. Microwave radiation initiated graft copolymerization of partially neutralized acrylic acid was carried out on Ch to synthesize superabsorbent Ch resin. Synthesized resin was found to be an effective superabsorbent material, which absorbs water 704 times of its own dry weight [138].

Alkaline hydrolysis of starch-graft-polyacrylonitrile (SPAN) led to the first commercial SAP. The product, HSPAN, was developed in the 1970s at the Northern Regional Research Laboratory of the US Department of Agriculture and patents were licensed to General Mills Inc., among others [193]. The acrylamide- and acrylic acid-based binary graft copolymers of starch give higher water absorption than the starch graft copolymers of either acrylamide or acrylic acid. Polymer-grafted sago starch was reacted with hydroxylamine in order to have high water absorption capacity [194]. Such products are useful in various products where water absorbency or water retention is important.

Starch graft copolymer-based superabsorbent was found to be useful for agricultural applications. Application of such superabsorbent reduced the water requirement and increases the crop growth and yield, etc. [195]. Starch- and acrylonitrile-based superabsorbents were prepared using a manganic pyrophosphate redox initiation system for various applications [196]. Ethyl methacrylate was grafted on starch to prepare a new superabsorbent composite polymer using sodium acrylate as a cross-linking agent. Synthesized composite was doped with sodium silicate to have higher water absorption capacity [106].

Talaat et al. [197] prepared hydrogel by grafting of starch with acrylonitrile followed by saponification of the synthesized graft copolymer. Afterward, a fertilizing base composed of cornstarch (65–70 %), urea (28–33 %), and other ingredients (2 %) was prepared. Starch-based SAP particle suspension was combined with a liquid fertilizer solution and used as a fertilizer for soil improvements [198].

6.3 Graft Copolymers in Pharmaceutical and Medical Applications

Ch has been widely used in the pharmaceutical and biomedical fields due to its nontoxicity and high biocompatibility [199, 200]. Ch could adhere to the mucosal surface and transiently open the tight junction between epithelial cells. It has been reported that Ch can enhance the penetration of macromolecules across the intestinal and nasal barriers [201]. Ch has been investigated as drug delivery systems for genes and proteins because positively charged Ch can be easily connected with negatively charged DNAs and proteins [202, 203]. Ch is only soluble in acidic medium due to the protonation of the free amino groups in its polymeric chains, which restrict the use of Ch in pharmaceutical and biomedical fields. Several derivatives of Ch have been studied to improve the water solubility of Ch. *N*-Trimethyl chitosan chloride (TMC) is a quaternized derivative of Ch. Various researchers have studied the modifications of Ch by quaternization of the amino groups to enhance the absorption effects [204, 205]. In neutral and physiological environments, the use of TMC could contribute significantly to the delivery of hydrophilic compounds such as protein and gene drugs. Sabaa et al. [206] investigated the antimicrobial activities of a variety of CMCh Schiff bases and their reduced forms as derivatives for CMCh resulting from the reaction of CMCh with a variety of aromatic aldehydes (eighteen derivatives) bearing either electron-donating or electron-withdrawing groups. Both the nature and position of the substituent groups in the aryl ring of the synthesized derivatives affects the antibacterial and the antifungal activities. On the other hand, quaternization of the reduced derivatives using *N*-(-3-chloro-2-hydroxypropyl) trimethyl ammonium bromide (Quat-188) has been synthesized and tested for their antimicrobial activities by Sabaa et al. [207]. The obtained results clearly revealed that all the quaternized derivatives exhibited better antimicrobial properties than that of the original CMCh. These derivatives possess outstanding properties for pharmaceutical and biomedical applications, but limit their use in blood contact problems [208]. Toxicity of many cationic polymers comes from their effect on the plasma membranes [209]. Interaction of cationic polymers with negatively charged cell components and proteins are responsible for other possible toxic mechanisms [210]. High positive charge on quaternized derivative of Ch can be easily contacted with negatively charged blood corpuscles and results in hemolysis and toxicity.

Ch derivatives were modified by biodegradable PEG because of its biocompatibility and capacity to minimize the interaction between cationic polymers and cell membranes [208]. PEG-coated nanoparticles found potential therapeutic application for controlled release of drugs and targeted drug delivery [211–213]. In addition, hydrophilic PEG could form a hydrated shell around the nanoparticles and prevent their quick uptake by the reticuloendothelial system (RES) [214], extend the half-lives of drugs, and change the tissue biodistribution of drugs.

Ch-g-poly(p-dioxanone) (Ch-g-PDO) was synthesized by Liu et al. [139] in bulk by ring-opening polymerization using stannous octanoate ($SnOct_2$) as catalyst.

The degree of substitution (DS) and the degree of polymerization (DP) of the copolymer were influenced by the feed ratio of Ch to PDO. Synthesized copolymer was evaluated as a promising device for controlled drug delivery of Ibuprofen (IBU).

In recent years, nonviral cationic vectors have been applied to protect the DNA from degradation based on the condensation of negatively charged DNA into compact particles essentially by electrostatic interactions [215, 216]. The safety of gene transfer vectors can be improved by choosing the biocompatible and biodegradable materials for cationic polymers [217, 218]. Ch was first applied in the effort of gene delivery [219] and recognized as the best cationic vector [220]. Direct grafting of Ch with poly(ethylenimine) (PEI) was considered to be one of the most prominent methods to improve the transfection efficiency of Ch poly plexes [221–223]. Hydrophilic exterior of the modified Ch polymer decreases the interactions of cationic vector with plasma proteins and erythrocytes [224, 225]. Chemical composition of the synthetic polymer and nanoparticle size of the complex may affect the transfection efficiency [226]. Resulting properties of Ch-g-PEI were due to the conjugation mechanism and grafting ratios of PEI [227]. A novel Ch-g-PEI copolymer was prepared with biocleavage disulfide linkages (Fig. 19) and evaluated as a possible nonviral gene vector due to its excellent properties [134].

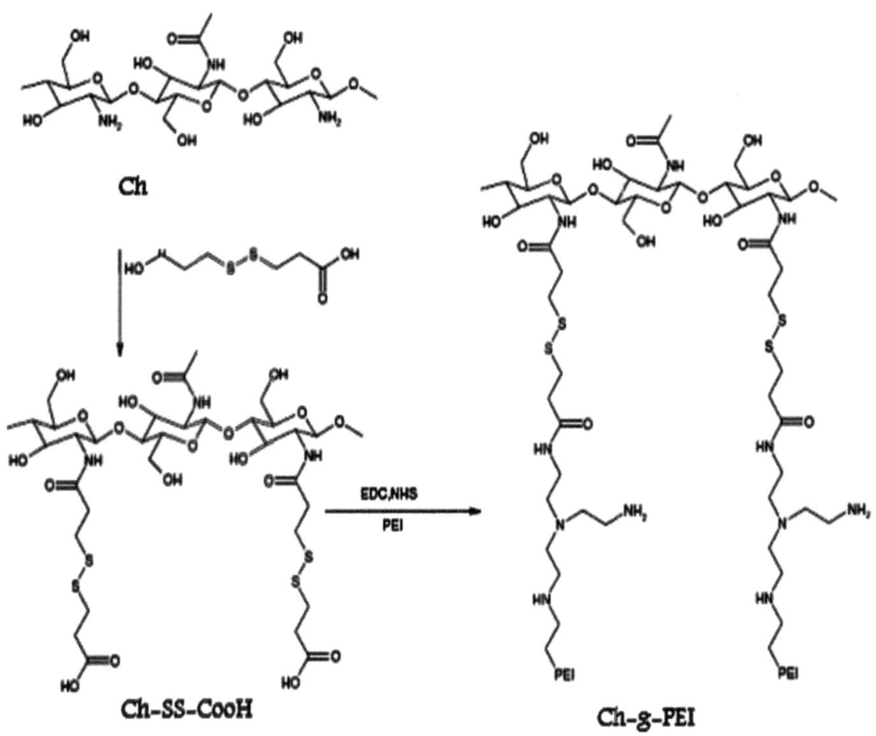

Fig. 19 Synthesis of Ch-g-PEI. Reprinted from [134], copyright 2010, with permission from Elsevier

Rosin-(2-acryloyloxy) ethyl ester (RAEE) was grafted onto Ch by Duan and his coworkers [135] and used as a carrier for fenoprofen calcium and controlled release behavior in artificial intestinal juice. The rate of release of fenoprofen calcium from the carrier becomes very slower than that of Ch in the artificial intestinal juice. A promising approach for sustained pulmonary drug delivery system has been achieved by the work of El-Sherbiny et al. [228]. They encapsulated poly(D,L-lactic-co-glycolic acid) (PLGA) nanoparticles in amphiphilic Ch-g-PEG copolymer and evaluated it as a potential carrier for sustained pulmonary delivery of curcumin.

The Ch graft poly(p-dioxanone) copolymer (Ch-g-PDO) was prepared by Wang and his group [229], starting from *N*-phthaloyl-chitosan (PHCh) [230]. It was then reacted with the already synthesized reactive poly(1,4-dioxane-2-one) tolylene-isocyanate (PPDO-NCO) in DMF homogeneous solution and resulted in grafting of PPDO onto the Ch backbone in the same way as described for the grafting of PPDO onto starch [231]. Finally, free amino group was regenerated by de-protection of the *N*-phthaloyl protecting group in the presence of hydrazine. Pre-polymers of PPDO with different molecular weights can be used to control the PPDO graft chain length. Synthesized copolymer exhibited a significant controlled drug-releasing behavior for sinomenine (7,8-didehydro-4-hydroxy-3,7-dimethoxy-17-methyl-9a,13a,14a-morphinan-6-one) both in artificial gastric juice or in neutral phosphate buffer solution [229].

CMCh was graft copolymerized with poly(ethylene glycol) in the presence of 2,2-dimethoxy-2-phenyl acetophenone (DMPA) as photoinitiator. Synthesized copolymer was cross-linked via methylenebisacrylamide as a cross-linking agent to develop pH-responsive hydrogel (Fig. 20). Synthesized hydrogels were evaluated as good drug delivery systems for in vitro release profiles of 5-fluorouracil (5-FU) as a model drug [232]. 1-Cyanoethanoyl-4-acryloylthiosemicarbazide (CEATS) was graft copolymerized on chitosan using the redox initiators. The graft copolymerization results in improved water swelling capacity and antifungal activity of Ch [233].

Carboxymethyl chitosan (CMCh) and poly(acrylonitrile) (PAN) were cross-linked to prepare the blend hydrogels. Synthesized hydrogels were investigated for their antibacterial behavior toward *E. coli* and hydrogels with high CMCh content exhibited the good antibacterial properties [234].

Hydrolyzed starch-graft-acrylic acid was used as a micro-hydrogel in surgical dressing and maintained relatively a constant temperature on application to skin [235]. Yamamoto et al. [236] have manufactured the wound dressing using a water-absorbable/-swellable starch-graft-acrylic acid copolymer, which acts as barrier for microorganisms. A transdermal therapeutic system based on starch-graft-poly (acrylic acid) was reported with best storage stability and controlled drug release rate [237]. A silicone polymer-grafted starch microparticle system was developed that was efficacious both orally and intranasally [238]. Acrylic acid was graft copolymerized on starch to develop nonirritant bioadhesive drug release systems for buccal application [239]. Saboktakin et al. [240] reported the development of slow release formulations of two anti-inflammatory model drugs, 5-aminosalicylic acid and salicylic acid in the nano gels of poly(methacrylic acid) grafted onto carboxymethyl starch.

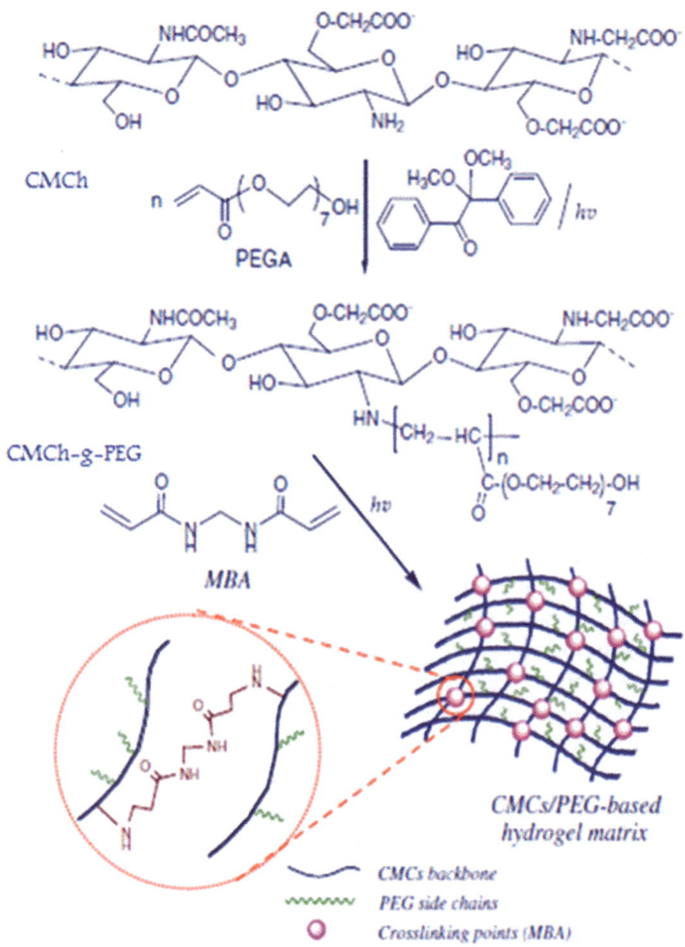

Fig. 20 CMCh-g-PEG and its hydrogel matrix. Reprinted from [232], copyright 2010, with permission from Elsevier

Starch and other polysaccharide-based graft copolymers are the best to develop various stimuli-dependent controlled release systems [120, 241, 242]. Tablets synthesized from graft copolymers of ethyl methacrylate on waxy maize starch and hydroxypropyl starch showed higher crushing strength and disintegration time in comparison to original starch tablets [243].

6.4 Miscellaneous Applications

Graft copolymers can be used in other applications like dehydrating agent for organic solvents and aqueous solutions of polymers such as proteins. It can also be used as an agar substitute in tissue culture media. Graft copolymerization of vinyl acetate and butyl acrylate onto cornstarch was reported to be used as a wood adhesive with superior property and low cost [101]. Starch-based graft copolymers can be used as an effective compatibilizer for starch-based blends [149, 244, 245]. PCL- and PLA-based biodegradable starch graft copolymers can be used as thermoplastics or compatibilizer and exhibit enhanced mechanical performances. Starch graft copolymers are more compatible with the plastic matrix than unmodified starch and plastics filled with graft copolymers exhibit higher tensile strength. Traditional expanded packaging material can be replaced by the extrusion of starch-graft-poly(acrylonitrile) and starch-graft-polymethacrylate [246].

7 Conclusions

Polysaccharide and its derivatives have been widely used in various fields and the properties of polysaccharides can be improved by grafting. Polysaccharide and their graft copolymers are used as drug carrier, gene delivery, composites, selective water absorption from oil–water emulsions, purification of water, etc. Graft copolymers can also be used in other applications like dehydrating agent for organic solvents and aqueous solutions of polymers such as proteins. They can also be used as an agar substitute in tissue culture media. Hydrogels are used as the superabsorbent materials or absorption materials for removal of heavy metal ions or dyes from wastewater. Chitosan has been formulated on films, beads, microspheres, and nanoparticles in pharmaceutical and biomedical fields due to its nontoxic and high biocompatibility. Chitosan and starch-based graft copolymers can be used in different pharmaceutical and biomedical applications.

References

1. Liu ZH, Jiao YP, Wang YF, Zhan CR, Zhang ZY (2008) Adv Drug Deliv Rev 60:1650
2. Amass W, Amass A, Tighe B (1998) Polym Int 47:89–144
3. Langer R, Tirrele DA (2004) Nature 428:487–492
4. Sharma K, Kumar V, Kaith BS, Kumar V, Som S, Pandey A, Kalia S, Swart HC (2015) New J Chem 39:3021–3034
5. Sharma K, Kumar V, Kaith BS, Som S, Kumar V, Pandey A, Kalia S, Swart HC (2015) Ind Eng Chem Res 54:1982–1991
6. Kaith BS, Sharma R, Kalia S (2015) Int J Biolog Macromol 75:266–275
7. Sharma K, Kumar V, Kaith BS, Kumar V, Som S, Kalia S, Swart HC (2015) Polym Degrad Stab 111:20–31
8. Sharma K, Kumar V, Kaith BS, Kumar V, Som S, Kalia S, Swart HC (2014) RSC Adv 4:25637–25649
9. Pontoni L, Fabbricino M (2012) Carbohydr Res 356:86–92
10. Babin M, Ruest, Drowin, Sirosis K, Ouellet S, Gagnon J (2012) Carbohydr Res 351:87–92
11. Martins AF, Piai JF, Schuquel ITA, Rubira AF, Muniz EC (2011) Colloid Polym Sci 289:1133–1144

12. Bagley EB, Fant GF, Burr RC, Doane Wm, Rusell CR (1977) Polym Eng Sci 17: 311–316
13. Beliakove MK, Aly AA, Abdel-Mohdy FA (2004) Starch/Stärke 56:407–412
14. Yadav M, Sand A, Mishra DK, Behari K (2010) J Appl Polym Sci 117:974–981
15. Lutz PJ, Peruch F (2012) Graft copolymers and comb-shaped homopolymers. In: Matyjaszewski K, Moeller M (ed) Polymer Science: a Comprehenvensive Reference, vol 6. Elsevier pp 511–542
16. Yu HY, Qin ZY, Wang LF, Zhou Z (2012) Carbohydr Polym 87:2447–2454
17. Carlmark A, Larsson E, Malmström E (2012) Eur Polym J 48:1646–1659
18. Chung TC, Rhubright D (1994) Macromolecules 27:1313–1319
19. Kreig A, Lefebvre AA, Hahn H, Balsara NP, Qi S, Chakraborty AK, Xenidou M, Hadjichristidis N (2001) J Chem Phys 115:6243–6251
20. Pitsikalis M, Woodward J, Mays JW, Hadjichristidis N (1997) Macromolecules 30:5384–5389
21. Xing J, Deng L, Li J, Dong A (2009) Intern J Nanomed 4:227–232
22. Ren Y, Jiang X, Yin G, Yin J (2010) J Polym Sci Part A: Polym Chem 48:327–335
23. Hruby M, Konak C, Kucka J, Vetrik M, Filippov SK, Vetvicka D, Mackova H, Karlsson G, Edwards K, Rihova B, Ulbrich K (2009) Macromol Biosci 9:1016–1027
24. Siegwart DJ, Oh JK, Matyjaszewski K (2012) Prog Polym Sci 37:18–37
25. Albertsson AC, Varma IK (2003) Biomacromolecules 4:1466–1486
26. Bhattacharya A, Ray P (2009) Basic features and techniques. In: Bhattacharya A, Rawlins J, Ray P (ed) Polymer grafting and crosslinking. Wiley, New Jersey, pp 7–64
27. Bhattacharya A, Misra BN (2004) Prog Polym Sci 29:767–814
28. Hadjichristidis N, Pitsikalis M, Pispas S, Iatrou H (2001) Chem Rev 101:3747–3792
29. Hadjichristidis N, Iatrou H, Pitsikali M, May J (2006) Prog Polym Sci 31:1068–1132
30. Matyjaszewski K, Xia J (2001) Chem Rev 101:2921–2990
31. Hadjichristidis N, Pitsikalis M, Iatrou H (2005) Adv Polym Sci 189:1–124
32. Hadjichristidis N, Pitsikalis M, Iatrou H, Pispas S (2003) Macromol Rapid Commun 24:979–1013
33. Kowalczuk M, Adamus G, Jedlihski Z (1994) Macromolecules 27:572–575
34. Neugebauer D (2007) Polym Int 56:1469–1498
35. Dérand H, Jannasch P, Wesslén B (1998) J Polym Sci Part A: Polym Chem 36:803–811
36. Xenidou M, Hadjichristidis N (1998) Macromolecules 31:5690–5694
37. Solomon DHJ (2005) J Polym Sci, Part A: Polym Chem 43:5748–5764
38. Solomon DH, Rizzardo E, Cacioli P (1985) Polymerization process and polymers produced thereby US Patent 4,581,429
39. Hawker CJ (1994) J Am Chem Soc 116:11185–11186
40. Hawker CJ, Bosman AW, Harth E (2001) Chem Rev 101:3661–3688
41. Chiefari J, Chong YKB, Ercole F, Krstina J, Jeffery J, Le TPT, Mayadunne RTA, Meijs GF, Moad CL, Moad G, Rizzardo E, Thang SH (1998) Macromolecules 31:5559–5562
42. Perrier S, Takolpuckdee P (2005) J Polym Sci Part A: Polym. Chem 43:5347–5393
43. Moad G, Rizzardo E, Thang SH (2006) Aust J Chem 59:669–692
44. Barner L, Davis TP, Stenzel MH, Barner-Kowollik C (2007) Macromol Rapid Commun 28:539–559
45. Coessens V, Pintauer T, Matyjaszewski K (2001) Prog Polym Sci 26:337–377
46. Wang JS, Matyjaszewski K (1995) J Am Chem Soc 117:5614–5615
47. Wang JS, Matyjaszewski K (1995) Macromolecules 28:7901–7910
48. Tizzotti M, Charlot A, Fleury E, Stenzel M, Bernard J (2010) Macromol Rapid Commun 20:1751–1772
49. Hafren J, Cordova A (2005) Macromol Rapid Commun 26:82–86
50. Cordova A, Hafren J (2005) Nord Pulp Pap Res J 20(4):477–480
51. Hafren J, Cordova A (2007) Nord Pulp Pap Res J 22:184–187

52. Lonnberg H, Zhou Q, Brumer H III, Teeri T, Malmstrom E, Hult A (2006) Biomacromolecules 7:2178–2185
53. Chung TC, Rhubright D (1994) Macromolecules 27:1313–1319
54. Zhao GL, Hafren J, Deiana L, Cordova A (2010) Macromol Rapid Commun 31:740–744
55. Nasef MM, Hegazy E-SA (2004) Prog Polym Sci 29:499–561
56. Nasef MM, Güven O (2012) Prog Polym Sci 37:1597–1656
57. Barsbay M, Güven O (2009) Radiat Phys Chem 78:1054–1059
58. Deng J, Wang L, Liu L, Yang W (2009) Prog Polym Sci 34:156–193
59. Wavhal DS, Fisher ER (2002) J Membr Sci 209:255–269
60. Li J, Xie W, Cheng HN, Nickol RG, Wang PG (1999) Macromolecules 32:2789–2792
61. Gustavsson Malin T, Persson Per V, Iversen T, Hult K, Martinelle M (2004) Biomacromolecules 5:106–112
62. Roberts GAF (1992) Chitin Chemistry, 1st edn. Macmillan, London
63. Pillai CKS, Paul W, Sharma CP (2009) Progr Polym Sci 34(7):641–678
64. Khor E (2002) Chitin: a biomaterial in waiting. Curr Opin Solid State Mater Sci 6:313–317
65. Van Luyen D, Huong DM (1996) In: Salamone JC (ed) Polymeric materials encyclopedia, vol 2. CRC Press, Boca Raton, Fl, pp 1208
66. Roja G, Floores JA, Rodriguez A, Ly M, Maldonado H (2005) Sep Purif Technol 44:31–36
67. Peniche C, Argüelles-Monal W, Peniche H, Acosta N (2003) Macromol Biosci 3 (10):511–520
68. Varma AJ, Deshpande SV, Kennedy JF (2004) Carbohydr Polym 55:77–93
69. Dutta PK, Tripathi S, Mehrotra CK, Dutta J (2009) Perspectives for chitosan based antimicrobial films in food applications. Food Chem 114:1173–1182
70. Bautista-Bănos S, Hermàdez-Lauzardo AN, Velàzquez-del Valle MG, Hermàdez-López M, Ait Barka E, Bosquez-Molina E, Wilson CL (2006) Crop Protection 25:108–118
71. Kato Y, Onishi H, Hachida Y (2003) Curr Pharm Biotechnol 4:303–309
72. Sugimoto M, Morimoto M, Sashiwa H (1998) P Carbohydr Polym 36(1):49–59
73. Jayakumar R, Prabaharan M, Reis RL, Mano JF (2005) Carbohydr Polym 62:142–158
74. Wang J, Chen Y, Zhang S, Yu H (2008) Bioresour Technol 99:3397–3402
75. Pengfei L, Maolin Z, Jilan W (2001) Radiat Phys Chem 61(2):149–153
76. Mino G, Kaizerman S (1958) J Polym Sci 31(122):242–243
77. Lagos A, Reyes J (1988) Polym Sci. Polym Chem Edn 26:985–991
78. Kataoka S, Ando T (1981) Kobunshi Ronbunshu 38:797–799
79. Athawale VD, Rathi SC (1999) J Macromol Sci Rev Macromol Chem Phys C 39(3):445–480
80. Berlin Ad A, Kislenko VN (1992) Prog Polym Sci 17:765–825
81. Pourjavada A, Mahdavinia GR, Zohuriaan-Mehr MJ (2003) J Appl Polym Sci 90:3115–3121
82. Poujavadi A, Zohuriaan-Mehr, Mahdavinia GR (2004) Polym Adv Technol 15(4):173–180
83. Mahdavinia GR, Pourjavadi A, Hosseinzadeh H, Zohuriaan MJ (2004) Eur Polym J 40:1399–1407
84. Jenkins DW, Hudson SM (2001) Chem Rev 101(11):3245–3274
85. Pourjavadi A, Mahdavinia GR, Zohuriaan-Mehr MJ, Omidian H (2003) J Appl Polym Sci 88 (8):2048–2054
86. Yazdani-Pedram M, Lagos A, Retuert J, Guerrero R, Riquelme P (1995) J Macromol Sci Pure Appl Chem A32(5):1037–1047
87. Kataoka S, Ando T (1984) Kobunshi Ronbunshu 41:519–524
88. Yi W, Jing-xian Y, Kun-yuan QIU (1994) Acta Polymerica Sinica 1:188–195
89. Yazdani-Pedram M, Retuert J (1997) J Appl Polym Sci 63(10):1321–1326
90. Retuert J, Yazdani-Pedram M (1993) Polym Bull 31(5):559–562
91. Akgün S, Ekici G, Mutlu N, Besirli N, Hazer B (2007) J Polym Res 14:215–221
92. Yazdani-Pedran M, Lagos A, Retuert PJ (2002) Polym Bull 48:93–98
93. Mun GA, Nurkeeva ZS, Dergunov SA, Nam IK, Maimakov TP, Shaikhutdinov EM, Lee SC, Park K (2008) React Funct Polym 68:389–395
94. Zohuriaan-Mehr MJ (2005) Iranian Polym J 14(3):235–265
95. Buléon AP, Planchot CV, Ball S (1998) Int J Biol Macromol 23:85–112

96. Vegh KR (2009) Starch bearing crops as food sources. In: Cultivated plants primarily as food sources, vol 1
97. Singh N, Singh J, Kaur L, Singh Sodhi N, Singh Gill B (2003) Food Chem 81:219–231
98. Bertoft E, Piyachomkwan K, Chatakanonda P, Sriroth K (2008) Carbohydr Polym 74:527–543
99. Athawale VD, Lele V (2000) Starch/Stärke 52:205–213
100. Ikhuoria EU, Folayan AS, Okieimen FE (2010) Int J Biotech Mol Bio Res 1(1):10–14
101. Wu YB, Lv CF, Han MN (2009) Adv Mater Res 79–82:43–46
102. Song H, Wu D, Zhang R-Q, Qiao L-Y, Zhang S-H, Lin S, Ye J (2009) Carbohydr Polym 78 (2):253–257
103. Bernd H, Bernd T, Thomas H, Bernhard J (2004) US Patent 20040170596
104. Bhuniya SP, Rahman MDS, Satyanand AJ, Gharia MM, Dave AM (2003) J Polym Sci Part A: Polym Chem 41:1650–1658
105. Taghizadeh MT, Mafakhery S (2001) J Sci I R Iran 12(4):333–338
106. Sahoo PK, Rana PK (2006) J Mater Sci 41:6470–6475
107. Zohuriaan-Mehr MJ, Kabiri K (2008) Iran Polym J 17:451–477
108. Fanta GF, Doane WM (1986) Grafted starches. In: Wurzburg OB (ed) Modified starches: properties and uses. CRC Press, Boca Raton, pp 149–178
109. Athawale VD, Rathi SC (1999) J Macromol Sci, Rev Macromol Chem Phys 39(3):445–480
110. Athawale VD, Rathi SC, Lele V (1998) Eur Polym J 34(2):159
111. Jyothi AN, Sreekumar J, Moorthy SN (2010a) Starch/Stärke 62:18–27
112. Bhattacharyya SN, Maldas D (1984) Prog Polym Sci 10:171–270
113. Hebeish A, El Alfy E, Bayazeed A (1988) Starch/Stärke 40:191–196
114. Mostafa KM, Samarkandy AR, El-Sanabary AA (2010) J Polym Res 17:789–800
115. Celik M (2006) J Polym Res 13:427–432
116. Meshrama MW, Patila VV, Mhaskeb ST, Thorat BN (2009) Carbohydr Polym 75(1):71–78
117. Zhang L, Gao J, Tian R, Yu J, Wang W (2003) J Appl Polym Sci 88(1):146–152
118. Gao J, Yu J, Wang W, Chang L, Tian R (1998) J Appl Polym Sci 68:1965–1972
119. Mostafa KM, Samarkandy AR, El-Sanabary AA (2010) J Polym Res 17:789–800
120. Maiti S, Ranjit S, Sa B (2010) Int J Pharm Tech Res 2(2):1350–1358
121. Kiatkamjornwong S, Chomsaksakul W, Sonsuk M (2000) Radiat Phy Chem 59(4):413–427
122. Abd El-Mohdy HL, El-Sayed A, HegazyaAbd El-Rehima HA (2006) J Macromol Sci Part A: Pure Appl Chem 43(7):1051–1063
123. Kiatkamjornwong S, Thakeow P, Sonsuk M (2001) Polym Degrad Stab 73:363–375
124. Abdel-Aal SE, Gad YH, Dessouki AM (2006) J Hazard Mater 129(1–3):204–215
125. Singh V, Tiwari A, Pandey S, Singh SK (2007) Peroxydisulfate initiated synthesis of potato starch-graft-poly(acrylonitrile) under microwave irradiation eXPRESS Polym Lett 1:51–58
126. Tao M, Jiao X (2009) J China Agric Univ 14(3):118–122
127. Kumar R, Setia A, Mahadevan N (2012) Int J Recent Adv Pharm Res 2(2):45–53
128. MostafaKh.M, Samerkandy AR, El-Sanabay AA (2007) J Appl Sci Res 3(8): 681–689
129. Shogren RL, Willet JL, Biswas A (2009) Carbohyr Polym 76:189–191
130. Willett JL, Finkenstadt VL (2003) Polym Eng Sci 43:1666
131. Willett JL, Finkenstadt VL (2006) J Polym Environ 14:125–129
132. Finkenstadt VL, Willett JL (2005) Macromol Phy Chem 206:1648–1652
133. Frost K, Kaminski D, Shanks R (2009) Extrusion grafting of starch with reactive dyes to form sheets with reduced retrogradation. In: 33rd annual condensed matter and materials meeting, Wagga, NSW, Australia, 4–6 Feb 2009, http://www.aip.org.au/Wagga
134. Li Z, Guo J, Zhang J, Zhao Y, Lv L, Ding C, Zhang X (2010) Carbohydr Polym 80 (1):254–259
135. Duan W, Chen C, Jiang L, Li GH (2008) Carbohydr Polym 73(4):582–586
136. Mohamed RR, Sabaa MW (2010) J Appl Polym Sci 116:413–421
137. Sabaa MW, Mohamed NA, Ali R, Abd El Latif SM (2010) Polym-Plast Technol 49:1055–1064
138. Huacai G, Wan P, Dengke L (2006) Carbohydr Polym 66:372–378

139. Liu G, Zhai Y, Wang X, Wang W, Pan Y, Dong X, Wang Y (2008) Carbohydr Polym 74 (4):862–867
140. Detchprohm S, Aoi K, Okada M (2001) Macromol Chem Phys 202:3560–3570
141. Zhong Z, Kimura Y, Takahashi M, Yamane H (2000) Polymer 41:899–906
142. Mum GA, Nurkeeva ZS, Dergunov SA, Nam IK, Maimakov TP, Shaikhutdinov EM, Lee SC, Park K (2008) React Funct Polym 68:389–395
143. Sabaa MW, Mohamed NA, Mohamed RR, Khalil NM, Abd El Latif MS (2010) Carbohydr Polym 79:998–1005
144. El-Sherbiny IM, Smyth HDC (2010) Int J Pharm 395:132–141
145. Liu L, Xu X, Guo S, Han W (2009) Carbohydr Polym 75:401–407
146. Kang H, Cai Y, Liu P (2006) Carbohydr Res 341:2851–2857
147. ElKhholy SS, Khalil KD, Elsabee MZ (2011) J Polym Res 18:459–467
148. Trimnell D, Fanta GF, Salch JH (1996) J Appl PolymSci 60(3):285–292
149. Kiatkamjornwong S, Mongkolsawat K, Sonsuk M (2002) Polymer 43:3915–3924
150. Lu S, Duan M, Lin S (2003) J Appl Polym Sci 88:1536–1542
151. Lanthong P, Nuisin R, Kiatkamjornwong S (2006) Carbohydr Polym 66:229–245
152. Keshava Murthy PS, MuraliMohan Y, Sreeramulu J, MohanaRaju K (2006) React Funct Polym 66:1482–1493
153. Zhang Q, Xu K, Wang P (2008) Fiber Polym 9(3):271–275
154. Athawale VD, Lele V (2000) Carbohydr Polym 41:407–416
155. Fares MM, El-faqeeh AS, Osman ME (2003) J Polym Res 10(2):119–125
156. Maharana T, Singh BC (2006) J Appl Polym Sci 100:3229–3239
157. Dragunski DC, Pawlicka A, Carlos S (2001) Mat Res 4(2):77–81, 1516–1439
158. Wang XL, Yang KK, Wang YZ, Chen DQ, Chen SC (2004) Polymer 45:7961–7968
159. Singh V, Tiwari A, Pandey S, Singh SK (2006) Starch/Starke 58:536–543
160. Gupta B, Anjum N (2001) J Appl Polym Sci 82:2629
161. Jyothi AN, Sajeev MS, Moorthy SN, Sreekumar J (2010) J Appl Polym Sci 116(1):337–346
162. Pathania D, Sharma R (2012) Adv Mat Lett 3:136
163. Pathania D, Sharma R, Kalia S (2012) Adv Mat Lett 3(2):259–264
164. Babel S, Kurniawan TA (2003) J Hazard Mater 97(1–3):219–243
165. Guibal E (2004) Sep Purif Technol 38(1):43–74
166. Chen XG, Park HJ (2003) Carbohydr Polym 53(4):355–359
167. Hon DNS, Tang LG (2000) J Appl Polym Sci 77(10):2246–2253
168. Jiang J, Hua D, Jiang J, Tang J, Zhu X (2010) S Carbohydr Polym 81:358–364
169. Muzzarelli RAA (1973) Analytical application of chitin and chitosan. In Belcher R, Freiser H (eds) Natural chelating polymers; alginic acid, chitin and chitosan. Pergamon Press, New York, pp 177– 227
170. Singh V, Tripathi DN, Tiwari A, Sanghi R (2006) Carbohydr Polym 65(1):35–41
171. El-Sherbiny IM (2009) Eur Polym J 45:199–210
172. Sabaa MW, Mohamed RR, Eltaweel SH, Seoudi RS (2012) J Appl Polym Sci 123:3459–9469
173. Crinia G, Gimberta F, Roberta C, Martelb B, Adama O, Morin-Crinia N, De Giorgia F, Badot P (2008) J Hazard Mater 153:96–106
174. Kim T, Park C, Shin E, Kim S (2004) Desalination 161:49–58
175. Daneshvar N, Ashassi-Sorkhabi H, Tizpar A (2003) Purif Technol 31(2):153–162
176. Basibuyuk M, Forster CF (2003) Proc Biochem 38(9):1311–1316
177. Chong Chao A, Sing Shyu Sh, Chuang Lin Y, Long Mi F (2004) Bioresour Technol 91:157–162
178. Kapdan IK, Kargi F (2002) Proc Biochem 37(9):973–981
179. Tripathy T, De RB (2006) J Phys Sci 10:93–127
180. Fanta GF, Burr RC, Doane WM, Russel CR (1972) J Appl Polym Sci 16:2835
181. Karmakar GP, Singh RP (1996) Synthesis and characterisation of starch-g-acrylamide copolymers for improved oil recovery, SPE 37297, Houston
182. Karmakar NC, Sastry BS, Singh RP (2002) Bull Mat Sci 25(6):477–478

183. Karmakar GP (1994) Flocculation and rheological properties of grafted polysaccharides. Ph. D. Thesis, IIT, Kharagpur
184. Kuilin D, Na J, Yaqin Z, Dawei Y, Duanmin H (2006) Chem J Internet 8(11):68
185. Hao X, Chang Q, Duan L, Zhang Y (2007) Starch/Stärke 59:251–257
186. Li H, Xu Q, Zhang D (2011) Adv Mater Res 356–360:1990–1993
187. Singh RP, Tripathy T, Karmakar GP, Rath SK, Karmakar NC, Pandey SR, Kannan K, Jain SK, Lan NT (2000) Curr Sci 78(7):798–803
188. Dickinson E, Eriksson L (1991) Adv Colloid Interface Sci 34:1–29
189. Omidian H, Rocca JC, Park K (2005) J Control Rel 102:3–12
190. Liu JH, Wang Q, Wang AQ (2007) Carbohydr Polym 70:166–173
191. Sun LP, Du YM, Shi XW, Chen X, Yang JH, Xu YM (2006) J Appl Polym Sci 102:1303–1309
192. Pang HT, Cheng XG, Park HJ, Cha DS, Kennedy JF (2007) Carbohydr Polym 69:419–425
193. Buchholz FL, Peppas, NA (1994) Superabsorbent polymers science and technology american chemical society, Washington, DC, p 573
194. Lutfor MR, Sidik,WanYunus WMZ, AbRahman MZ, Mansoor A, Jelas H (2001) Carbohydr Polym 45:95–100
195. Doane SW, Doane WM (2004) Starch graft copolymers and methods of making and using starch graft copolymers for agriculture WO/2004/033536
196. Qunyi T, Ganwei Z (2005) Carbohydr Polym 62:74–79
197. Talaat HA, Sorour MH, Aboulnour AG, Shaalan HF, Ahmed EM, Awad AM, Ahmed MA (2008) J Agric Environ Sci 3(5):764–770
198. Savich MH, Olson GS, Clark EW (2009) WO/2009/014824
199. Calvo P, Remunan-lopez C, Vila-Jato JL, Alonso MJ (1997) J Appl Polym Sci 63:125–132
200. Giunchedi P, Genta B, Muzzarelli RAA (1998) Biomaterials 19:157–161
201. Borchard G, Lueben HL, De Boer GA, Verhoef JC, Lehr CM, Junginger HE (1996) J Control Rel 39(2–3):131–138
202. Richardson SCW, Kolbe HVJ, Duncan R (1999) Int J Pharm 178:231–243
203. Janes KA, Calxo P, Alonso MJ (2001) Adv Drug Delivery Rev 47:83–97
204. Thanou MM, Kotzé AF, Scharringhausen T, Luessen HL, De Boer AG, Verhoef JC et al (2000) J Control Rel 64(1–2):15–25
205. Xu YM, Du YM, Huang RH, Gao LP (2003) Biomaterials 24:5015–5022
206. Mohamed NA, Sabaa MW, El-Gandour AHH, Abdel-Aziz MM, Abdel-Gawad OF (2013) J Am Sci 9(3):247–266
207. Mohamed NA, Sabaa MW, El-Gandour AHH, Abdel-Aziz MM, Abdel-Gawad OF (2013) Int J Biol Macromol 60:156–164
208. Amiji MM (1997) Carbohydr Polym 32(3–4):193–199
209. Choksakulnimitr S, Masuda S, Tokuda H, Takakura Y, Hashida M (1995) J Control Rel 34 (3):233–241
210. Fischer D, Li Y, Ahlemeyer B, Krieglstein J, Kissel T (2003) Biomaterials 24:1121–1131
211. Gerf R, Minamitake Y, Perracchia MT, Trubetskoy V, Torchilin V, Langer R (1994) Science 263:1600–1603
212. Peracchia MT, Gref R, Minamitake Y, Domb A, Lotan N, Langer R (1997) J Control Rel 46 (3):223–231
213. Quellec P, Gref R, Perrin L, Dellacherie E, Sommer F, Verbavatz JM (1998) J Biomed Mater Res 42:45–54
214. Hu Y, Jiang XQ, Ding Y, Zhang LY, Yang CZ, Zhang JF et al (2003) Biomaterials 24:2395–2404
215. Mintzer MA, Simanek EE (2009) Chem Rev 109:259–302
216. Nguyen DN, Green JJ, Chan JM, Langer R, Anderson DC (2009) Adv Mater 21 (16):847–867
217. Hussain SM, Braydich-Stolle LK, Schrand AM, Murdock RC, Yu KO, Mattie DM et al (2009) Adv Mater 21(16):1549–1559

218. Ravi Kumar MNV, Muzzarelli R A A, Muzzarelli C, Sashiwa H, Domb AJ (2004) Chem Rev 104 (12):6017–6084
219. Mumper RJ, Wang J, Claspell JM, Rolland AP (1995) Novel polymeric condensing carriers for gene delivery, In: Proceedings of the international symposium on controlled release of bioactive materials, vol 22, p 178
220. Liu WG, Yao KD (2002) J Control Rel 83(1):1–11
221. Jiang HL, Kim YK, Arote R, Nah JW, Cho MH, Choi YJ (2007) J Control Rel 117 (2):273–280
222. Lu B, Xu XD, Zhuo RX, Cheng SX, Zhuo RX (2008) Biomacromolecules 9(10):2594–2600
223. Wong K, Sun G, Zhang X, Dai H, Liu Y, He C, Leong KW (2006) Bioconjugate Chem 17 (1):152–158
224. Kunath K, Von Harpe A, Fischer D, Petersen H, Bickel U, Voigt K, Kissel T (2003) J Control Rel 89(1):113–125
225. Neu M, Fischer D, Kissel T (2005) J Gene Med 79(8):992–1009
226. Jeong JH, Kim SW, Park TG (2007) Prog Polym Sci 32(11):1239–1274
227. Kircheis R, Wightman L, Wagner E (2001) Adv Drug Delivery Rev 53(3):341–358
228. El-Sherbiny IM, Smyth HDC (2012) Mol Pharm 9:269–280
229. Wang X, Huang Y, Zhu J, Pan Y, He R, Wang Y (2009) Carbohydr Res 344(6):801–807
230. Kurita K, Ikeda H, Yoshida Y, Shimojoh M, Harata M (2002) Biomacromolecules 3(1):1–4
231. He R, Wang X, Wang Y, Yang K, Zeng J, Ding S (2006) Carbohydr Polym 65(1):28–34
232. El-Sherbiny IM, Smyth HDC (2010) Carbohydr Res 345(14):2004–2012
233. Elkholy SS, Khalek HA, Elsabaa MZ (2012) J Macromol Sci Part A 49:720–728
234. Mohamed RR, Seoudi RS, Sabaa MW (2012) Cellulose 19:947–958
235. Kuraray Co Ltd (1982) Jpn Pat 82 47,339. Abstr 97(1982) 44362b
236. Yamamoto K, Watanabe T, Yamamoto T (1994) Jpn Pat 06,00, 201 Abstr 120 173553g
237. Mori M (1997) US Patent 569 577
238. McDermott MR, Heritage PL, Bartzoka V, Brook MA (1998) Immunol Cell Biol 76:256–262
239. Geresh S, Gdalevsky GY, Gilboa I, Voorspoels J, Remon JP, Kost J (2004) J Control Rel 94:391–399
240. Saboktakin MR, Maharramov A, Ramazanov MA (2007) Nat Sci 5(3):67
241. Silva I, Gurruchaga M, Goñi I (2009) Carbohydr Polym 76:593–601
242. Shaikh MM, Lonikar SV (2009) J Appl Polym Sci 114(5):2893–2900
243. Marinich JA, Ferrero C, Jiménez-Castellanos MR (2009) Eur J Pharm Biopharm 72 (1):138–147
244. Chen L, Qiu XY, Xie ZG, Hong ZK, Sun JR, Chen XS, Jing XB (2006) Carbohydr Polym 65:75–80
245. Choi E-J, Kim C-H, Park J-K (1999) J Polym Sci Part B: Polym Phy 37:2430–2438
246. Riaz MN (1999) Cereal Foods World 44:705–709

MIX
Papier aus verantwortungsvollen Quellen
Paper from responsible sources
FSC® C105338

If you have any concerns about our products,
you can contact us on
ProductSafety@springernature.com

In case Publisher is established outside the EU,
the EU authorized representative is:
Springer Nature Customer Service Center GmbH
Europaplatz 3, 69115 Heidelberg, Germany

Printed by Libri Plureos GmbH
in Hamburg, Germany